# Geoscience and Remote Sensing

# Geoscience and Remote Sensing

*Editor*

**Marek Lagunov**

scitus
academics

**Geoscience and Remote Sensing**
Edited by **Marek Lagunov**

ISBN: 978-1-68117-271-2
Library of Congress Control Number: 2016937441


SCITUS Academics LLC,
www.scitusacademics.com
Box No. 4766, 616 Corporate Way,
Suite 2, Valley Cottage,
NY 10989



**Notice**

Reasonable efforts have been made to publish reliable data and views articulated in the chapters are those of the individual contributors, and not necessarily those of the editors or publishers. Editors or publishers are not responsible for the accuracy of the information in the published chapters or consequences of their use. The publisher believes no responsibility for any damage or grievance to the persons or property arising out of the use of any materials, instructions, methods or thoughts in the book. The editors and the publisher have attempted to trace the copyright holders of all material reproduced in this publication and apologize to copyright holders if permission has not been obtained. If any copyright holder has not been acknowledged, please write to us so we may rectify.

Printed in United States of America on Acid Free Paper

# Preface

Remote sensing refers to the activities of recording/ observing/ perceiving (sensing) objects or events at far away (remote) places. In remote sensing, the sensors are not in direct contact with the objects or events being observed. The information needs a physical carrier to travel from the objects/events to the sensors through an intervening medium. The electromagnetic radiation is normally used as an information carrier in remote sensing. The output of a remote sensing system is usually an image representing the scene being observed. A further step of image analysis and interpretation is required in order to extract useful information from the image. The human visual system is an example of a remote sensing system in this general sense. Geoscience addresses critical issues such as energy, meteorology, water and mineral resources, stewardship of the environment, oceanography, reducing natural hazards for society, planetary science and more. It can be considered to be a branch of planetary science, but with a much older history. There are both reductionist and holistic approaches to Earth sciences. Therefore, the geoscience comprehends earth, atmospheric, oceanography, pedology, petrology, mineralogy, hydrology and geology. This book, Geoscience and Remote Sensing, deals with latest and futuristic developments in remote sensing novel theory and applications. This book shall give readers a good insight frontier works in remote sensing information processing. It should be of immense valuable tool for students, professors and researchers.

# Table of Contents

# Chapter 1

# Remote Sensing of Forest Health

**Jyrki Tuominen[1], Tarmo Lipping[1], Viljo Kuosmanen[2] and Reija Haapanen[3]**

[1] *Tampere University of Technology, Finland*

[2] *Geolocical Survey of Finland, Finland*

[3] *HaapanenForestConsulting, Finland*

## 1. INTRODUCTION

Global forest health is declining. The main reasons for this unfortunate development include climate change, air pollution and increased human activities. There is a need to monitor and quantitatively measure the change in forest health. Forest health can be defined in many different ways depending on the perspective one takes. The relationship between the cause and the symptom of forest health deterioration is complex mainly because the same symptom can often be induced by multiple different stressors.

Modern state of the art remote sensing technologies provide the means for wide coverage measurement of forest health with reasonable accuracy. In the assessment of forest health by means of remote sensing, features called Vegetation Indices (VIs) are usually extracted from the data. VIs are combinations of surface reflectances at two or more wavelengths designed to highlight a particular property of vegetation. In addition to specific VIs some attempts to develop a general forest health index combining the assessments of the various properties have been published.

In this chapter we first discuss the term 'forest health' as well as the various causes and symptoms of forest health deterioration. We then argue about the role of remote sensing in forest health monitoring and present the most common VIs used for the assessment of forest health. The VIs' capability to detect different types of forest damage is assessed in case studies; the results for Ni contaminated and pest inflicted forest areas are presented in sections 7 and 8, respectively. Finally, future possibilities in applying the remote sensing technologies to forest health monitoring are discussed.

## 2. FOREST HEALTH

Despite its widespread use, the term "forest health" is vaguely defined in the literature, making its application to forest management difficult (Kolb et al., 1994). The definition of forest health is always a matter of perspective. Social, economic and ecological perspectives are taken most frequently. Looking from different perspectives the definitions of healthy forest can even appear contradictory.

Social perspective emphasises people's needs for healthy living and recreational environment. In many countries people are becoming more and are more concerned about their environment. Since the 1960s, people have become more concerned about their environment, due to intensified management of natural resources, increased availability of information and structural changes in the society, i.e. decreased importance of primary production and increased leisure time. Healthy forests are needed for aesthetical pleasure and variety of outdoor activities. Economical perspective is quite complex. Historically short-term needs have usually exceeded long-term needs in forest industry. Production of commodities and services has outweighed ecological considerations in the past. Nevertheless, long-term economical benefits can only be obtained by practicing sustainable forest management. Ecological perspective is focused on ecosystems instead of human needs. This perspective emphasises the fact, that counterproductive interaction between ecosystems and humans should be minimized. The potential should exist for all biotic and abiotic elements to be present with sufficient redundancy at appropriate spatial and temporal scales across the landscape. Human intervention should not impact ecosystem sustainability by destroying or significantly degrading components that affect ecosystem capabilities.

Utilitarian perspective versus ecosystem perspective is another categorization presented in literature (Kolb et al., 1994). The utilitarian perspective emphasizes forest conditions that directly satisfy human needs. The ecosystem perspective emphasises the maintenance of sustainable ecosystems over the landscape. Often different perspectives do overlap, however, on occasions they may be contradictory. Depending on the perspective, the condition of the same forest can be viewed as healthy or unhealthy. For example, a common component in ponderosa pine forests is dwarf mistletoe. It reduces the growth of ponderosa pine and increases its mortality. The existence of dwarf mistletoe is harmful from economical perspective. However, abundance and species richness of birds is higher when dwarf mistletoe is present. Thus, the existence of dwarf mistletoe is desirable from the ecological perspective (Kolb et al., 1994).

The first definition of forest health usually cited in the scientific literature is by Aldo Leopold (1949): "Health is the capacity of the land for self-renewal. Conservation is our effort to understand and preserve this capacity." Although Leopold's definition is concerned also with other issues than forest health, it has founded the basis for all later definitions. Since Leopold's definition many new ones have been presented in the scientific literature. Some of these definitions deserve more in-depth consideration.

O'Laughlin et al. (1994) defined forest health in the following way: "Forest health is a condition of forest ecosystems that sustains their complexity while providing for human needs". This definition is an effort to take all the different perspectives of forest health, i.e., social, ecological and economical, into account. The question often asked when discussing forest health is: 'Can forest health be measured?': According to O'Laughlin et al. (1994) the answer is: "Objective indicators of forest condition can be specified and measured, but forest health assessments contain subjective value judgements which must be clearly recognized."

Some definitions of forest health like that by Monnig & Byller (1992), for example, emphasize ecological perspective: "A healthy forest is an ecosystem in balance"(Monnig & Byler, 1992). Human social and economical needs are not accounted for and the condition of ecosystems matters for its own sake. Kolb, Wagner & Covington (1994) defined forest health as follows: "The term forest health should be restricted to the examination of the role of biotic and abiotic agents in ecosystem processes." Several characteristics for such a system were mentioned: resistance to dramatic change in populations of important organisms within the ecosystem not accounted for by predicted successional trends; a functional balance between supply and demand of essential resources; and a diversity of seral stages, cover types and stand structures that provide habitat for many native species and all essential processes.

Climate change, in particular increased temperatures and levels of atmospheric carbon dioxide, as well as changes in precipitation and the occurrence of extreme climate events, is having notable impacts on the world's forest health. Increased temperatures may relieve forest stress during colder seasons but increase it during warmer seasons. Impacts of increased temperatures vary widely among different climatic zones. Climate change has both direct and indirect impacts on forest health (Moore &Allard, 2008). For example, climate change has strong influence on forest pests which can be considered as a direct impact. Pests can rapidly react to changing climate because of their short generation times. Drought is a good example of a cause that is influencing forest indirectly. Drought can change tree physiology in a way that it is more vulnerable to certain insect and pest species. Such changes are, e.g., sugar content of foliage, changes in leaf colour, change of leaf thickness and structural changes of foliage.

# 3. CAUSES AND SYMPTOMS OF DETERIORATED FOREST HEALTH

In the scope of this chapter it is only possible to address the complex relationship between causes and symptoms of deteriorated forest health briefly. The relationship is complex mainly because there are often multiple stressors causing a certain symptom (Ferretti, 1997). One of the major causes of deteriorated forest health is air pollution, particularly acid rain and ground-level ozone. There are two major types of pollutant threatening forest health: photochemical oxidants of which ozone is the primary compound, and nitrogen pollutants. Ozone is toxic (plant-killing) to sensitive plant species. Nitrogen is the primary growth-limiting nutrient, yet it is also a pollutant when in excess. The emission levels of these two pollutants are expected to increase significantly globally (Moore & Allard, 2008).

Some causes of deteriorated forest health are controversial. Depending on the perspective they can be seen as beneficial or negative. The role of disease agents (pathogens) and insects, for example, is controversial. They are essential to the function of dynamic ecosystems as they recycle nutrients and create habitats for different species. They can also negatively affect forest health, increase mortality and create growth losses. Diseases and insects influence the health of forests, trees outside forests and other wooded lands. Globally, all ecosystems with tree cover are under increasing threat, as the periods between sequential outbreaks are rapidly decreasing because of a range of factors including climate change and lack of proper forest management (Moore & Allard, 2008).

The impact of forest fires may also be controversial. Although they are usually seen as a threat to forest health, these natural events are key elements in many forest ecosystems as well. Another major cause for deteriorated forest health is posed by droughts. The consequences of a long-lasting drought can be very severe. In addition to direct impacts, they have indirect ones, for example pest outbreaks are associated with droughts. Drought can also increase the risk of forest fires.

Another cause of forest health deterioration is invasive species. Any species non-native to a particular forest ecosystem and whose introduction and spread causes deteriorated forest health can be considered invasive species. A major cause of increasing number of invasive species is increaced human activity. Transport vechiles act as carriers of seeds and plants. Sometimes invasive species can be intentionally introduced to an ecosystem to provide economic or environmental benefits. These species have later spread and caused serious problems in forests ecosystems.

There is an abundance of symptoms of deteriorated forest health. Some symptoms are easy to detect and follow while others might be very difficult to monitor. Discoloration is a good example of a symptom, which is relatively easy to monitor. Discoloration is usually also a very useful index of forest health. Growth losses, in turn, are difficult to measure from large forest areas. Usually sample plots are used, but, as the problem may be sporadic and local, it may not be caught with the sampling design. The new remote sensing technologies offer possibilities to large-scale forest growth monitoring.

Needle or leaf loss is a common symptom, which is also difficult to monitor unless the loss is severe. The needle loss symptom is often observed from needle retention. Needle retention provides an index of the number of years that needles are retained. It is only useful as a measure of needle loss if the loss occurs progressively from the oldest to the youngest needles (Innes, 1993). Defoliation can be estimated by observing the form of the tree crown. Mechanical damage to the crown is usually caused by wind. Butt and stem damage is another mechanical damage usually caused by animals like rabbits and squirrels, for example. A common nominator for most of the causes and symptoms of deteriorated forest health is, that almost all are expected to become more frequent.

# 4. THE ROLE OF REMOTE SENSING IN FOREST HEALTH MONITORING

First attempts to introduce aerial photographs as a remote sensing tool in forestry were made in 1887. An airborne balloon was used as a photographic platform to produce photographs of forests in the vicinity of Berlin (Van Laar & Acka, 2007). The objective was to examine the possibility of preparing forest maps from aerial photographs. The forest was classified and described on the basis of visual examination of the photographs. Airborne photography from aircraft was introduced during World War 2. After the war the techniques developed for military use became available for civil applications. Since then aerial photography has been widely used in forestry applications. Forest inventory and measurement has been the most widely used application in forestry remote sensing. Earlier, stereo imagery consisting of pairs of oriented aerial images was used to measure individual trees. Today LiDAR based 3D-measurements have made stereo images obsolete (Clement, 2004).

Storm damage studies using aerial photographs were probably the first forest health applications using remotely sensed data. After the introduction of false colour photographs and early multispectral satellites it was possible to study red edge related observables (Barret & Curtis, 1997). By the red edge the difference between the reflectance maximum at near-infrared region and corresponding minimum at visible red region typical for all green vegetation due to chlorophyll absorption is meant. Red edge related indices such as normalized difference vegetation index (NDVI) and leaf area index (LAI) were applied. The introduction of hyperspectral sensors and the increased spatial resolution of multispectral data available enabled new forest health applications: detection of root diseases (Leckie et al., 2004), detection of pest inflicted damages (Vogelmann & Rock, 1989) and assessment of photosynthetic efficiency (Gamon et al., 1992).

Recent advances in high spectral resolution hyperspectral technology have enabled a whole new approach to forest health studies - the remote chemistry. The most advanced technology to study the health of forest is foliar chemistry. Estimates of the foliar chemistry of canopies allow a better understanding of the functioning of forest ecosystems since many biochemical processes such as photosynthesis, respiration and litter decomposition, are related to the foliar chemistry of trees (Huber et al., 2008). The use of high spectral resolution data has enabled to study many forest health relevant observables, such as the concentration of nitrogen (Fourty et al., 1996), carbon (Daughtry et al., 2001) and leaf pigments (Gitelson et al., 2002).

In the concept of forest health remote sensing nowdays means airborne or satellite imaging of forest areas. Depending on the number of wavelength channels used images obtained by remote sensing are categorized into aerial photographs, multi- and hyperpectral images. Other methodologies such as radar and Light Detection and Ranging (LIDAR) also used in remote sensing of forests. It has been shown that remote sensing can provide useful and relevant forest information (Solberg, 1999). Historically, the potential of remote sensing for forest health studies remained limited for a variety of reasons. Most of the remote sensing data suffered from insufficient spatial, spectral, or temporal resolution. Modern remote sensing instruments have overcome these problems opening new possibilities for forest health assessment. Especially promising is the modality of hyperspectral imaging.

Traditionally, forest health monitoring is performed by means of field studies. In many cases systematically monitored sample plots are used. There is no conflict between field studies and remote sensing. Both techniques are needed and they have their own important role in forest health monitoring. The advantages of remote sensing over conventional field studies include better spatial coverage, shorter sampling intervals, efficiency of data acquisition as well as access to remote or restricted areas. Climate change and increasing pollution set new requirements to forest health monitoring. State of the art remote sensing technology and methods offer an efficient solution to meet those requirements.

Disadvantages of remote sensing in forest health assessment mainly arise from quality and interpretation issues. The quality of modern remote sensing instruments has improved remarkably, however, unfortunately the advancements in algorithm development and verification are not at the same level. Algorithms used to retrieve end-user products such as chlorophyll content have to be validated using field measurements. Algorithms are used worldwide, but the field measurements used in their validation often cover only some specific area. Therefore it is not unusual that algorithms fail to produce reliable results at certain climate zone or geographic location. Remote sensing data products are sometimes difficult to interpret and that makes otherwise reliable data useless. Reliable verification of airborne remote sensed data can only be done using adequate ground truth measurements during the flight operation.

Practically all remote sensing algorithms require the data to be atmospherically corrected. Atmospheric correction is a procedure where the filtering effects of the atmosphere are compensated for. Optical properties of the atmosphere are time and place variant making correction procedure complex and difficult. Atmospheric models used in the correction procedure are not precise enough to ensure correct results in all cases. Atmospheric correction should be made with utmost care in order to produce quality results. Results should always be verified using known ground targets.

Another remarkable difficulty is the so called “mixed pixel problem”. This means that usually in remote sensing data one pixel represents many different materials. In forest areas one pixel often represents tree canopy, soil and some other materials like rock, for example. Then vegetation indices measuring certain forest property can give false results because the percentage of tree canopy is too low. In dense tropical forest the problem is not so serious as in boreal forest areas with low tree density. Because of the mixed pixel problem remote sensing usually provides general estimate of forest health over larger area rather than precise condition of a single tree.

# 5. VEGETATION INDICES

In the assessment of forest health by means of remote sensing, features called Vegetation Indices are usually extracted from the data. VIs are combinations of surface reflectances at two or more wavelengths designed to highlight a particular property of vegetation. They are derived using the reflectance properties of vegetation. Each of the VIs is designed to accentuate a particular vegetation property. VIs are usually developed by means of empirical laboratory measurements of the property to be studied as well as correlation analysis of remotely sensed data. VIs can be categorized into narrow or broadband ones according to the bandwidth of used wavelength channels. The use of narrowband VIs requires data of high spectral resolution, i.e., hyperspectral data. The use of many VIs is limited because they saturate in dense vegetation areas (Mutanga & Skidmore, 2004). Some narrowband VIs can overcome this problem. Another problem with VIs is nonlinearity (Jiang et al., 2006). The relationship between VI value and measured property is nonlinear which makes the use of VI somewhat difficult. Here we present the most important categories of VIs and some examples of VIs for each category. The ability of the VIs to detect different types of stressed forest areas are tested by means of two case studies presented in sections 7 and 8.

## 5.1. Greennes (Chlorophyll Concentration) Vis

Greennes VIs are designed to measure the general quantity and vigor of green vegetation. They measure many different aspects: chlorophyll concentration, canopy area and canopy structure. VI value is always determined by the combination of these different effects. Greennes VIs are based on the measuring of reflectance peak in near-infrared region (NIR). Red wavelength where the chlorophyll absorption is strongest is used as a reference.

Normalized difference vegetation index (NDVI) is the most frequently used and most well know VI. It simply measures the reflectance peak at NIR region. NDVI is a good overall measure of green vegetation, but it has problems with saturation and non-linearity (Jiang et al., 2006). NDVI is defined according to the equation

$$NDVI = \frac{\rho_{NIR} - \rho_{RED}}{\rho_{NIR} + \rho_{RED}}, \tag{1}$$

where $\rho_x$ represents reflectance at wavelength band

SS

(Tucker, 1979).

Enhanced Vegetation Index (EVI) is designed to be used in dense vegetation areas. NDVI saturates in densely vegetated areas (Huete et al., 1997). In order to overcome this problem blue reflectance is used to compensate the effects of background soil and atmospheric scattering effects. EVI is defined according to the equation

$$EVI = 2.5\left(\frac{\rho_{NIR} - \rho_{RED}}{\rho_{NIR} + 6\rho_{RED} - 7.5\rho_{BLUE}}\right). \tag{2}$$

The Red Edge Normalized Difference Vegetation Index (RENDVI) is a broadband version of the NDVI. While NDVI uses the minimum and maximum reflectances of the red edge region, the RENDVI employs wavelength bands along the red edge (Sims&Gamon, 2002). RENDVI is very sensitive to small changes in canopy chlorophyll content. It can only be calculated based on hyperspectral data. RENDVI is defined according to the equation

$$RENDVI = \frac{\rho_{750} - \rho_{705}}{\rho_{750} + \rho_{705}}. \tag{3}$$

Malenovsky et al. (2006) presented a new approach to canopy chlorophyll content measurement in the form of an index called Area under curve Normalized to Maximal Band depth between 650-725 nm (ANMB). Most of the VIs use simple band ratios to calculate the measured property. In ANMB the surface integral and maximum band depth are calculated. In that way the whole red edge region is estimated instead of a few bands. In the first phase the area under the reflectance curve between 650 and 750 nm is integrated according to the equation

$$AUC_{650-725} = \frac{1}{2}\sum_{j=1}^{n-1}(\lambda_{j+1} - \lambda_j)(\rho_{j+1} + \rho_j), \tag{4}$$

where $\rho_x$ and $\rho_{j+1}$ are reflectances at the $j$ and $j$+1 bands, $\lambda_j$ and $\lambda_{j+1}$ are wavelengths of the $j$ and $j$+1 bands, and *n* is the number of the used spectral bands. The ANMB index is computed according to

$$ANMB_{650-725} = \frac{AUC_{650-725}}{MBD_{650-725}}, \quad (5)$$

where $MBD_{650-725}$ is the maximal band depth of the reflectance, placed at one of the spectrally stable wavelengths of the strongest chlorophyll absorption peaks around 675-680nm.

## 5.2. Water Content Vis

Water content VIs are designed to provide an estimate of canopy water content. Water content is an important vegetation property which correlates with vegetation health. Water content VIs are based on the fact that there are well known water absorption features in the near-infrared and shortwave infrared regions. The use of water content VIs requires high spectral resolution data.

The Water band index (WBI) is a simple reflectance measurement that is sensitive to changes in canopy water content. WBI is utilizing well know water absorption feature at 970nm. The ratio of the reflectance at 970nm to that at 900nm is measured (Penuelas et al., 1995). WBI is defined according to the equation

$$WBI = \frac{\rho_{900}}{\rho_{970}}. \quad (6)$$

Normalized difference water index (NDWI) is sensitive to changes of canopy water content. It uses two different bands, 857 and 1241 nm having similar but slightly different water absorption properties (Gao, 1995). The scattering of light by canopy enhances the weak water absorption at 1241nm. NDWI is defined according to the equation

$$NDWI = \frac{\rho_{857} - \rho_{1241}}{\rho_{857} + \rho_{1241}}. \quad (7)$$

The Moisture Stress Index (MSI) is a simple reflectance measurement that is sensitive to increasing canopy water content. The strength of the absorption at 1599 nm increases when the canopy water content increases. The absorption at 819 nm is used as a reference because it is nearly unaffected by changing water content in the canopy (Ceccato et al., 2001). MSI is defined according to the equation

$$WBI = \frac{\rho_{1599}}{\rho_{819}} . \qquad (8)$$

## 5.3. Leaf Pigment Vis

Leaf pigment VIs are measuring the amount of stress-related pigments in vegetation. Carotenoids and anthocyanins are pigments which are present in higher concentrations in stressed vegetation. Carotenoids are pigments that protect vegetation from high light condition. Anthocyanin pigment content is high in senescence and in new leafs.

Anthocyanin reflectance index 700 (ARI_700) is sensitive to anthocyanin amount in vegetation (Gitelson et al., 2001). ARI_700 is defined according to the equation

$$ARI\_700 = \left(\frac{1}{\rho_{550}}\right) - \left(\frac{1}{\rho_{700}}\right) . \qquad (9)$$

Anthocyanin reflectance index NIR (ARI_NIR) is similar to ARI_700, but it uses one additional NIR band (Gitelson et al., 2001). ARI_NIR is capable in detecting higher anthocyanin concetrations than ARI_700. ARI_NIR is defined according to the equation

$$ARI\_NIR = \rho_{800}\left(\frac{1}{\rho_{550}}\right) - \left(\frac{1}{\rho_{700}}\right) . \qquad (10)$$

Carotenoid Reflectance Index 550 (CRI_550) measures the amount of carotenoids in canopy. CRI_550 calculates the difference of two bands sensitive to carotenoid amount (Gitelson et al., 2002). CRI_550 is defined according to the equation

$$CRI\_550 = \left(\frac{1}{\rho_{510}}\right) - \left(\frac{1}{\rho_{550}}\right). \tag{11}$$

Carotenoid Reflectance Index 700 (CRI_700) is a similar reflectance measurement as CRI_550, but it uses NIR band instead of the green one. CRI_700 is designed to measure higher caroteinid concentrations compared to CRI_550. CRI_770 is defined according to the equation

$$CRI\_550 = \left(\frac{1}{\rho_{550}}\right) - \left(\frac{1}{\rho_{700}}\right). \tag{12}$$

### 5.4. Carbon Vis

Vegetation contains many types of carbon: cellulose, lignin, sugar and starch. Cellulose is used to form cell walls of vegetation tissues. Lignin is used in structurally strong parts in vegetation. There is large amount of carbon in dead or senescent vegetation. These VIs can be used to observe the state of senescence of vegetation.

Normalized Difference Lignin Index (NDLI) is designed to estimate the amount of lignin in vegetation. Reflectance at 1754 nm is primarily determined by lignin concentration of the canopy (Serrano et al., 2002). Reflectance at 1680 nm is used as a reference. NDLI is defined according to the equation

$$NDLI = \frac{\log(1/\rho_{1754}) - \log(1/\rho_{1680})}{\log(1/\rho_{1754}) + \log(1/\rho_{1680})}. \tag{13}$$

The Cellulose Absorption Index (CAI) is a vegetation index indicating surfaces containing dry wood material. Absorptions in the 2000 nm to 2200 nm range are sensitive to cellulose (Daughtry et al., 2004).

CAI is defined according to the equation

$$CAI = 0.5(\rho_{2000} + \rho_{2200}) - \rho_{2100}. \tag{14}$$

## 5.5. Light Use Efficiency Vis

The light use efficiency VIs are providing a measure of the efficiency with which vegetation is able to use incident light for photosynthesis. Light use efficiency correlates with carbon uptake efficiency and growth rate. Light use efficiency VIs can be used in precision forestry to estimate growth rate and production.

The Photochemical Reflectance Index (PRI) is a reflectance measurement that is sensitive to changes in carotenoid pigments in canopy. The amount of carotenoid pigments indicates photosynthetic light use efficiency and the rate of carbon dioxide uptake (Gamon et al., 1992). It can be used to estimate vegetation productivity and stress. PRI is defined according to the equation

$$PRI = \frac{\rho_{531} - \rho_{570}}{\rho_{531} + \rho_{570}}. \qquad (15)$$

The Structure Insensitive Pigment Index (SIPI) is a reflectance measurement designed to maximize the sensitivity to the ratio of bulk carotenoids to chlorophyll while minimizing the affects of variation in canopy structure (Penuelas et al., 1995). SIPI uses three bands: blue, NIR and the maximum chlorophyll peak at 680 nm. SIPI is defined according to the equation

$$SIPI = \frac{\rho_{800} - \rho_{445}}{\rho_{800} - \rho_{680}}. \qquad (16)$$

The Red Green Ratio (RG Ratio) index is a reflectance measurement that indicates the relative expression of leaf redness caused by anthocyanin to that of chlorophyll. The Red Green Ratio has been used to estimate the course of foliage development in canopies. The RG Ratio index is an indicator of stress, leaf production and may also indicate flowering in some cases. The ratio is calculated as the mean of all bands in the red range divided by the mean of all bands in the green range. RGRI is defined according to the equation

$$RGRI = \frac{\rho_{RED}}{\rho_{GREEN}}. \qquad (17)$$

# 6. GENERAL FOREST HEALTH INDICES

Forest health is likely to deteriorate due to climate change, pollution and increased human activities. This unfortunate development sets new demands for forest health assessment. There is a need to obtain a quantitative and reliable assessment of the general condition of forest health. Most of the VIs measure a certain property of forest that doesn't necessarily represent the general condition of the forest. A general forest health index capable of detecting a variety of different damage types would be very desirable. Several methods for obtaining general forest health index have been proposed in the literature.

NDVI index is generally used as a forest health index. Xiao & McPherson (2005) presented a method using NDVI with multispectral data to asses tree health in urban environment. Although NDVI is usually a good VI for general assessment of forest health it leaves opportunities for improvement. NDVI can detect chlorophyll content and defoliation, but there are problems regarding saturation with dense vegetation and non-linearity.

Solberg et al. (2005) presented a method based on combining the LiDAR and hyperspectral data. Variation of canopy chlorophyll mass per area is used to measure forest health. NDVI calculated from hyperspectral data is used as a measure of chlorophyll mass while the Leaf area index (LAI) calculated from the LiDAR data is used as a measure of canopy area. The proposed method produced promising results.

In hyperspectral data analysis the ENVI software package is widely used. ENVI features a forest health tool, which creates a spatial map showing the overall health and vigor of a forested region. The spatial map is showing health index using 9 different classes. Forest health tool uses three different vegetation indices in the classification process. Each VI represents one of the following VI categories: greennes, leaf pigments, canopy water content and light use efficiency. Forest health tool has produced good results in the detection of contaminated forest areas (Tuominen et al., 2008). It constitutes a promising effort to develop a tool to calculate general forest health index using combined VIs.

Forest health can be evaluated using remote sensing by measuring chlorophyll content, defoliation and content of certain pigments (Solberg, 1999). A good general forest health index should measure at least those variables. In addition, water content and photochemical indices could be used. Even advanced foliar chemistry indices yet in experimental phase could be utilized. General forest health index integrating across different damage types is feasible, but a lot of research has to be done. Comprehensive algorithm verification with different forest and damage types would be required.

# 7. CASE STUDY 1: DETECTION OF TALC DUST CONTAMINATED FOREST AREAS

The objective of this study was to evaluate how well talc dust contaminated forest areas can be detected using the various vegetation indices. Contamination is one of the major causes of deteriorated forest health. The level of contamination in forest areas can be influenced byrelated to industrial enterprises. Therefore, it is important to provide reliable forest health information in order to support decision making in all administrative levels.

## 7.1. Test Area

The test area contains Boreal forest around the Lahnaslampi talc mine in North-East Finland. Geographic location as well as the true colour image of the test area are shown in Fig 1. Forest in the test area is dominated by Pine, Spruce and Birch trees. Due to large scale mining of crumbly ore, talc, carbonate etc. dust from the rock piles and tailings is carried by the wind to the environment (Kuosmanen et al., 2004). Magnesium and nickel are components of dust emitted from the Lahnaslampi talc plant and mine. Magnesium is bound to talc and magnesite, whereas sources of Ni are both sulphides and Ni bearing Mg-silicates, e.g. talc. Elevated concentrations of Ni and Mg in both moss and humus samples reflect the amount of precipitated talc dust along the prevailing wind direction (Helminen & Räisänen, 2002). Another source of environmental impact is constituted by the seepage of acid waters through rock piles and tailing ponds. Generally, the level of vegetation stress around the Lahnaslampi mine is low, i.e. the mine does not restrict forestry or other utilization of nature except in the mining area itself. Although talc is not the only source of contamination, it can be assumed that there is strong correlation between Ni concentration and forest stress level.

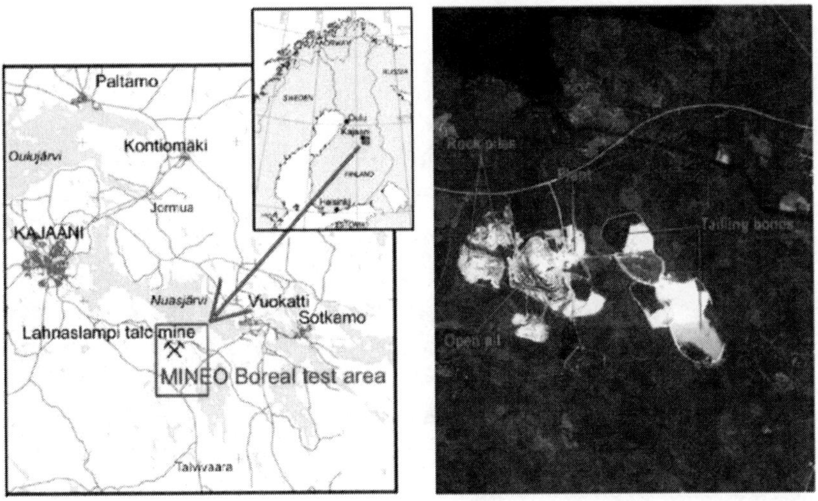


**Figure 1.** Geographic location of the test area (Kuosmanen et al., 2004) (left), true colour image of the test area (right).

## 7.2. Data Acquisition

The data acquisition was part of the EU-funded MINEO project, which aimed at the development advanced methods for the extraction of information and knowledge from earth observation data. This study utilized imagery from the HyMap airborne hyperspectral scanner recorded at 28th of July 2000. During the data acquisition cloud cover was non-existent. The HyMap sensor collects reflected solar radiation in 126 bands covering the wavelengths from 420 to 2480 nm. This includes visible, near infrared and short-wave infrared regions of the electromagnetic spectrum. The ground resolution of one pixel was 5*5 meters (Kuosmanen et al., 2004). Hyperspectral data was atmospherically corrected using ATCOR software. ATCOR uses MODTRAN 4 atmospheric model and parameters describing atmospheric type, solar geometry and hyperspectral sensor.

## 7.3. Methods

In order to remove non-forest pixels from hyperspectral data, a mask image was constructed. Masked pixels were not accounted when test results were calculated. Non-forest pixels were detected by calculating the forest discrimination index (FDI)

$$FDI = \rho_{838} - (\rho_{714} + \rho_{446}) \tag{18}$$

where $\rho_x$ represents reflectance measured in the spectral band centred at *x* nm (Lucas et al., 2008). All pixels whose FDI-value was under certain threshold were removed from hyperspectral data. It is quite difficult to separate different types of green vegetation reliably. Therefore it was necessary to remove pixels representing green agricultural fields and grasslands manually when they were located near the sites where moss samples were collected. High albedo of the ore minerals causes an atmospheric scattering halo effect around the mining area, a halo which is clearly visible in HyMap imagery especially at VNIR-wavelengths. In order to avoid the influence of halo effect on the test results, 50 meters wide buffer zone was masked around bright ore material. The mask image used to remove non-forest pixels is shown on the left panel of Figure 2.

During the flight campaign 51 moss samples were collected from the test area around Lahnaslampi talc mine. Moss samples were collected by hand using contamination-free gloves. The samples were stored at low temperature and the concentrations of heavy metals were measured in the Geolaboratory of the Geological Survey of Finland (Helminen & Räisänen, 2002). Measurements of nickel concentration were utilized in this case study. Sample sites were divided into six classes according to measured nickel concentration as follows:

- Class 1 Ni concentration 0-5 mg/kg
- Class 2 Ni concentration 6-16 mg/kg
- Class 3 Ni concentration 17-30 mg/kg
- Class 4 Ni concentration 31-60 mg/kg
- Class 5 Ni concentration 61-90 mg/kg
- Class 6 Ni concentration 91-130 mg/kg

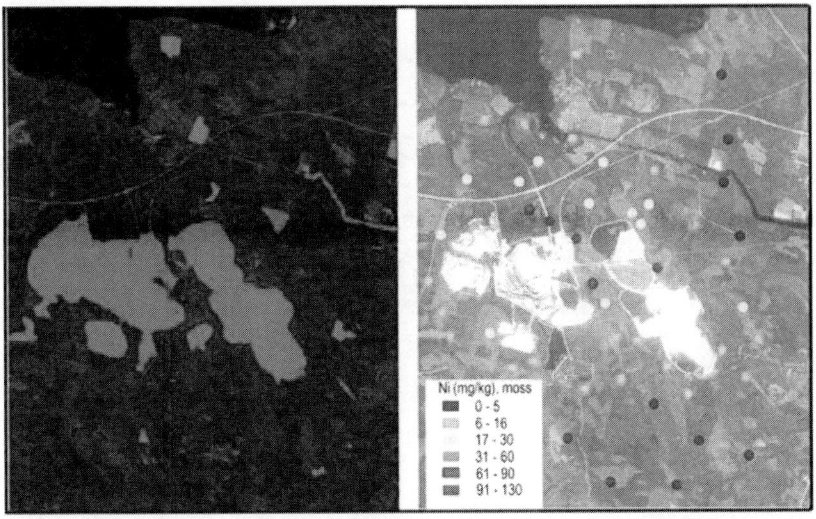


**Figure 2.** Mask image used to eliminate non-forest pixels (left), ground truth map indicating Ni-content (mg/kg) of collected moss samples (right).

It is safe to assume that Ni concentration varies slowly within immediate neighborhood of the sampling site. Circular neighborhoods (radius 50 m) were created around each sampling site. These circular areas were color coded according to their Ni concentration levels. Resulting color coded map was used as a ground truth map in the testing process of Vegetation Indices. Color coded ground truth map is shown on the right panel of Fig 2. Correlation between the values of VIs and the measured Ni concentration as well as separability of the classes were considered as the evaluation criteria for Vegetation Indices. Both measures were calculated for each VI using all pixels situated within the 50 m circular neighborhood around the sample sites. Computation of tested VIs is described in section 5.

## 7.4. Results and Discussion

The coefficients of determination $r^2$ for each VI are shown in table 1. Separability between two classes was calculated using the following simple and robust measure

$$d_{norm} = \frac{|\mu_1 - \mu_2|}{\sigma_1 + \sigma_2} \quad (19)$$

where mu and sigma are the mean and standard deviation of the particular class, respectively (Landgrebe, 2003).

Overall separability between all different class combinations is then

$$S = \frac{1}{15} \sum_{i=1}^{5} \sum_{j=i+1}^{6} \frac{|\mu_i - \mu_j|}{\sigma_i + \sigma_j} \quad (20)$$

where *I* and *j* denote the classes. Separability values for the vegetation indices are shown in table 1.

**Table 1.** Coefficients of determination and separability values for different vegetation indices (N/A means not applicable).

| Vegetation Index | R^2 | S |
|---|---|---|
| Normalized Difference Vegetation Index (NDVI) | 0.9027 | 0.4866 |
| Enhanced Vegetation Index (EVI) | 0.6405 | 0.2953 |
| Red Edge Normalized Difference Vegetation Index (RENDVI) | 0.8135 | 0.3110 |
| Area Normalized to Maximum Band depth (ANMB) | 0.8101 | 0.3036 |
| Water Band Index | 0.4092 | 0.1514 |
| Normalized Difference Water Index (NDWI) | 0.2403 | 0.0925 |
| Moisture Stress Index (MSI) | 0.40141 | 0.2026 |
| Anthocyanin Reflectance Index 700 (ARI_700) | 0.8704 | 0.3242 |
| Anthocyanin Reflectance Index NIR (ARI_NIR) | 0.8098 | 0.1653 |
| Carotenoid Reflectance Index 550 (CRI_550) | 0.7490 | 0.2868 |
| Carotenoid Reflectance Index 700 (CRI_700) | 0.7785 | 0.3223 |
| Normalized Difference Lignin Index (NDLI) | N/A | N/A |
| Cellulose Absorption Index (CAI) | N/A | N/A |
| Photochemical Reflectance Index (PRI) | 0.8141 | 0.2707 |
| Red Green Ratio Index (RGRI) | 0.6258 | 0.1411 |
| Structure Intensive Pigment Index (SIPI) | 0.0023 | 0.1146 |

In general, correlation and separability values obtained for greenness VIs were relatively high. Highest correlation and separability values were obtained for the NDVI index. Narrow band indices (RENDVI, ANMB) produced better results than the broad band EVI. Test results of water content indices were rather poor. Among these indices best separability of classes was achieved by the MSI index. The separability of NDWI index was the worst of all tested VIs. The value of the WBI index is almost the same for all classes; this can clearly be seen from Figure 3 where the median as well as upper and lower quartiles values are shown for each class.

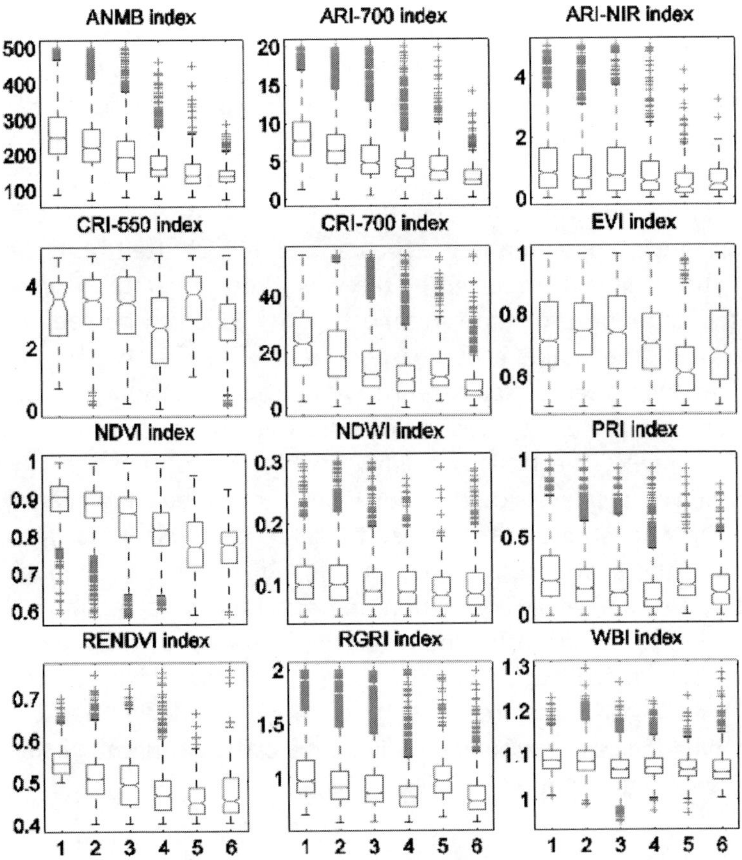


**FIGURE 3.** Median, upper and lower quartile values of classes for tested vegetation indices.

Leaf pigment indices that measure the carotenoids and anthocyanins concentration produced quite good test results. Among these indices the best separability and correlation values were obtained by the ARI_700 index. The separability of classes was poor using the ARI_NIR index. VIs designed to provide an estimate of the amount of carbon, i.e., NDLI and CAI were totally useless in the detection of Ni contaminated forest areas. The test results of VIs used to estimate light use efficiency varied a lot. The PRI index produced good results while the results of the SIPI index were very poor.

## 7.5. CONCLUSIONS

Most of the tested VIs could not detect Ni contaminated forest areas with adequate accuracy. However, three VIs, the NDVI, the ARI_700, and the PRI, produced good results. When the spatial distribution of the ARI_700 index is studied from Figure 4, it is easy to see that the more contaminated areas are stretched into North East direction from the mine. From this we can correctly conclude that North East is the prevailing wind direction in the area as the dust emitted from the talc plant and mine is transported by the wind. Poor separability of classes was the common problem for almost all VIs. This can be seen clearly in Figure 3. The major cause of poor separability is the mixed pixel problem. One pixel is usually representing canopy, shadow and soil. In this Boreal test site the forest is so dense that soil is not present in the pixels normally. The basic problems of the VIs i.e., non-linearity and saturation at high levels can be seen when the test results are analyzed. When Fig 4. is studied, it is easier to detect stressed forest areas by analyzing the spatial distribution of the ARI_700 index, although the NDVI index had better correlation and separability values in the test. This is due to non-linearity and high-end saturation of the NDVI index. The carbon concentration Vis, the CAI and the NDLI failed to produce any meaningful test results. It is understandable because Ni contamination is not assumed to increase carbon content of the tree unless the contamination level is very high.

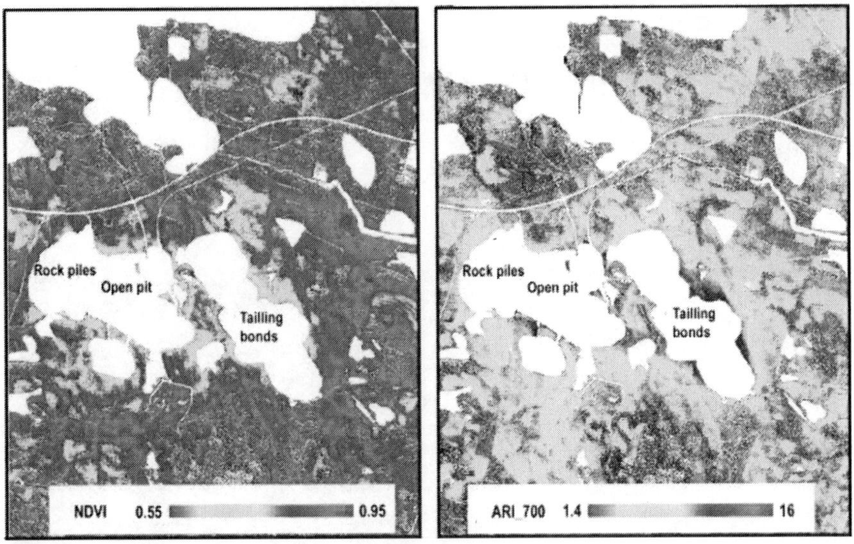


**Figure 4.** Spatial distribution of the NDVI and the ARI_700 indices.

VIs estimating canopy water content had only very weak correlation with Ni concentration. Water absorption regions of hyperspectral data are very difficult to measure as often data in those regions is just noise. Light use efficiency indices produced mixed results. PRI had rather good correlation and separation values, but SIPI did not have any correlation with Ni concentration. A probable cause of this is the high noise level in one channel in the blue region of the spectrum. Test results of SIPI index were not excluded from this study, but they should be interpreted with cautious mind. As an overall conclusion from this study it can be said that the deterioration of forest health due to Ni contamination can be detected using carefully selected VIs.

## 8. CASE STUDY 2: DETECTION OF PEST INFLICTED DEFOLIATION

The objective of this study was to evaluate how well defoliated tree crowns can be separated from healthy tree crowns using different vegetation indices. Pest inflicted stress is globally one of the most important threats to forest health. Pest inflicted damages can be sometimes avoided or reduced with proper countermeasures. Therefore it is important to be able to monitor and detect early stage pest damages in forest areas.

## 8.1. Test Area

The test area constituted a sand soiled boreal forest area in South-West Finland. Forest in the test area is dominated by Pine with some occasional Spruce and Birch trees. In the summer of 2006 a seriously damaged pine forest area was found in South-West Finland.

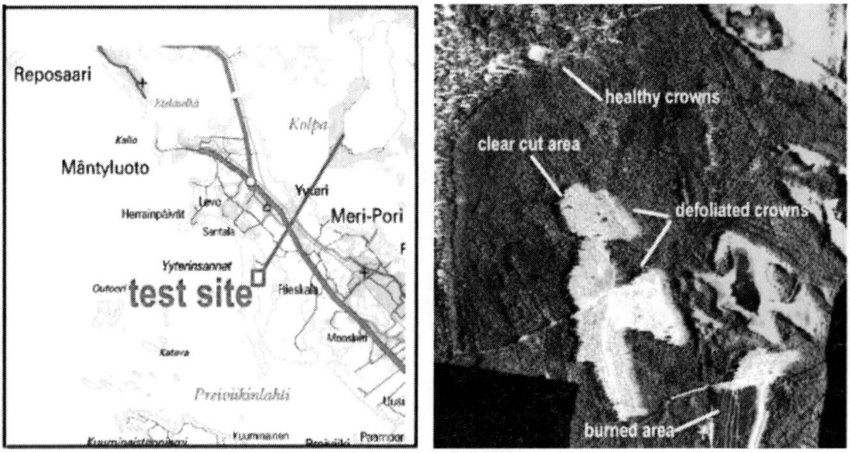


**Figure 5.** Geographic location of the test area(left), true colour image of the test area (right).

Studies showed that the cause of the damage was the great webspinning pine sawfly (acantholyda posticalis). Great webspinning pine sawflies have been found in Finland before, but they have not caused any forest damage before. This time recent warm weathers and sandy soil of the area have made it possible for the sawflies to reproduce and cause damage. It is very possible that forest damage caused by great webspinning pine sawfly will be more frequent due to climate change. The tree mortality in the area was so high that is was necessary to clear cut 30 hectares of forest. 10 hectares were burned in order to decrease future damage. Most of the damaged trees were successfully clear cut, but there were still some left.

## 8.2. Data Acquisition

This study utilized data from the AISA dual airborne hyperspectral scanner recorded at 13th of July 2008. During the acquisition the cloud cover was non existent. The AISA dual spectrometer collects reflected solar radiation in 481 bands covering the wavelengths from 399 to 2452 nm. This includes the visible, near infrared and short-wave infrared regions of the electromagnetic spectrum. The ground resolution of one pixel was 2.5*2.5 meters. Hyperspectral imagery was atmospherically corrected using ATCOR software. This study utilized also LIDAR data acquired on May 1st, 2008, by a small aircraft carrying Leica ALS50-II LIDAR scanner. Acquisition was carried out in early spring, so deciduous trees were still without leafs. The point density of the LIDAR data was 0.76 points per sq. m. Measurement accuracy of elevation data was 0.15 m.

## 8.3. Methods

Ground truth data was collected by doing an extensive field study in the test area. Moderately defoliated tees were located using visual inspection. Heavily defoliated trees were excluded because in this case most of the reflected radiation actually comes from the soil. Geo-coordinates of promising candidates were recorded using a GPS-instrument. Candidates for healthy trees were also collected. Final selection of the trees included in the study was done using LIDAR data.

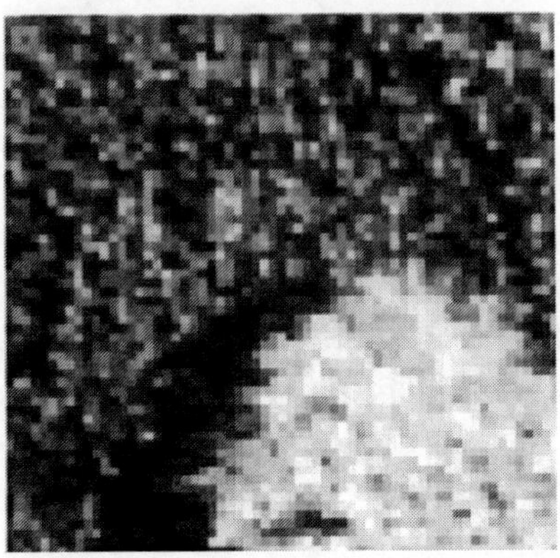

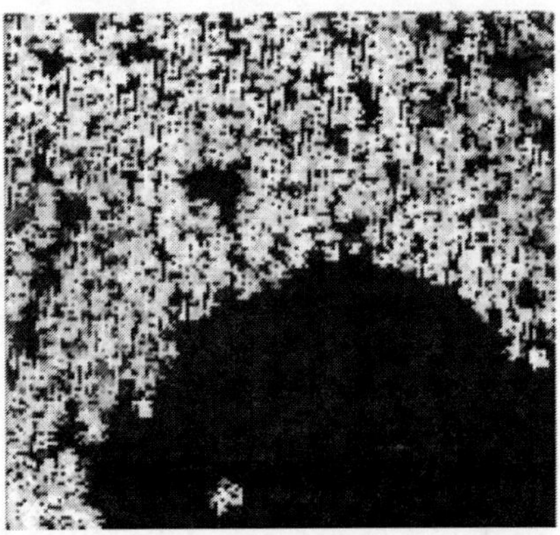

**Figure 6.** True colour image of the hyperspectral data(left), DSM image from the same area(right).

The reason for using LIDAR data is mixed pixel problem. Usually spectral signature imaged from the test area represents the mixture of canopy, soil, shadow and maybe some other materials too. In order to test VIs reliably it was necessary to select final tree crown using digital surface model (DSM). DSM was generated by rasterizing and interpolating raw LIDAR data using ENVI software. DSM was geo-referenced to same coordinate system as the hyperspectral data using polynomial triangulation of ERDAS software. Both data sets were linked in ENVI software and the neighborhood of each tree crown candidate were studied. The tree crowns for which the DSM of the tree top neighborhood showed full canopy coverage in the corresponding hyperspectral pixel were chosen for VI testing. Geo-referenced DSM image is shown on the right panel Fig 7. White pixels show laser beam returns from tree tops. As a result altogether 20 hyperspectral pixels representing 10 healthy tree crowns and 10 partly defoliated tree crowns were selected for the analysis.

## 8.4. Results And Discussion

VIs were calculated for all 20 test pixels. Mean and standard deviation values for both healthy and defoliated classes are shown in table 1. Separability *S* between healthy and defoliated classes was calculated using the formula

$$S = \frac{|\mu_1 - \mu_2|}{\sigma_1 + \sigma_2} \tag{21}$$

where mu and sigma are the mean and standard deviation of the particular class, respectively.

Many of the tested VIs did not produce any meaningful results. PRI used one wavelength channel that appeared to be noisy and therefore the results for this index are not valid. NDLI and CAI probably failed because they use SWIR channels. Most VIs were calculated using ENVI software VI calculator and some using MATLAB codes. Best separability value was obtained using NDVI. NDVI correlates well with reduced chlorophyll content, so its good performance could be expected the result is understandable. ARI_700 and ARI_NIR produced also rather good results. Recently published ANMB produced clearly better separability than the well established EVI and RENDVI VIs.

**TABLE 2.** Mean, standard deviation and separability values for each tested Vis (N/A means not applicable).

| Vegetation index | Healthy | | Defoliated | | |
|---|---|---|---|---|---|
| | Mean | std. | Mean | std. | S |
| NDVI | 0.934 | 0.0219 | 0.791 | 0.035 | 2.513 |
| EVI | N/A | N/A | N/A | N/A | N/A |
| RENDVI | 0.502 | 0.0139 | 0.4823 | 0.01844 | 0.5820 |
| ANMB | 232.7 | 50.69 | 139.8 | 28.79 | 1.168 |
| WBI | N/A | N/A | N/A | N/A | N/A |
| NDWI | N/A | N/A | N/A | N/A | N/A |
| MSI | 0.3865 | 0.0810 | 0.2805 | 0.0872 | 0.6302 |
| ARI_700 | 0.0059 | 0.0011 | 0.0030 | 0.0009 | 1.4500 |
| ARI_NIR | 1.654 | 0.433 | 3.487 | 0.611 | 1.7471 |
| CRI_550 | 0.0101 | 0.0144 | 0.0023 | 0.0031 | 0.7936 |
| CRI_700 | 0.0143 | 0.0021 | 0.0177 | 0.0024 | 0.7348 |
| NDLI | N/A | N/A | N/A | N/A | N/A |
| CAI | N/A | N/A | N/A | N/A | N/A |
| PRI | N/A | N/A | N/A | N/A | N/A |
| RGRI | 0.8574 | 0.1078 | 0.70838 | 0.1540 | 0.5692 |
| SIPI | 1.003 | 0.0174 | 1.0280 | 0.0189 | 0.6887 |

### 8.5. Conclusions

The objective of this study was to determine if defoliated tree crowns can be detected using vegetation indices. Widely used and tested NDVI proved to be the best VI in detection of defoliation. Separability was calculated using standard deviation. The number of tested pixels in each class was rather small for accurate statistical analysis. The results show that some of the VIs can detect defoliated tree areas rather reliably. Results also indicated that reduced chlorophyll content and increased anthocyanoid pigment levels are good indicators of defoliation.

## 9. THE FUTURE OF FOREST HEALTH RELATED REMOTE SENSING

High spatial resolution remote sensing for forestry applications has reached an almost mature phase with wide range of applications available. However, numerous opportunities and challenges remain. The robustness of the data processing methods is one of the issues to be considered. Processing methods currently in use often need extensive calibration and adjustment for each new imaged forest area. Processing methods should be able to handle different types of forest areas in routine fashion despite the variation of climate zone, soil type or forest structure. A lot of research has still to be done to fully utilize the potential of high spatial and spectral resolution data.

In the forest health assessment, most studies are characterized by a limited geographic extent concentrating on test sites were the complexity of forest environment is low. The methods are mostly empirical and require local calibration. Expanding forest health studies over wide geographic regions is a challenge because the added complexity of varying vegetation types can hide the relatively weak signal feature associated with forest stress. One possible solution to overcome this problem is to identify forest health change using data obtained at different times as it may be easier to identify stress from spectral change, rather than from the spectral properties of stress itself (Brandberg & Warner, 2006).

It would be highly desirable to be able to use satellite data, instead of airborne data for remote sensing of individual trees. Airborne remote sensing campaigns are very costly limiting the accessibility of data. Airborne hyperspectral instruments offer superior signal quality and spatial resolution, however, wide coverage multi-temporal forest health monitoring using airborne data is not feasible. Satellite data has the advantage of a relatively uniform illumination and view angle over large regions, thus minimizing problems associated with combining data from individual flight lines. Satellite-borne high spatial resolution hyperspectral data is not available at the moment, but can be anticipated in the future as the development of space technology continues. Spatial resolution of the hyperspectral imager currently in space (EO1 Hyperion), at 30 m, is too coarse for studying individual trees. OrbView-4 which failed to reach orbit after launch in 2001, would have offered 250 band hyperspectral data at 8 m spatial resolution. Specifications of Orbview-4 provide some kind of reference on what can be expected from the capability of future sensors (Castro-Esau & Kalashka, 2008).

Data fusion where several data sources are used together has the potential for revolutionary impact on forest health measurement. For example, with LiDAR data it is possible to directly measure the structural attributes of trees. The data fusion of high spectral resolution LiDAR and high spectral resolution hyperspectral data can raise forest health studies on a new level: precise information on foliar chemistry pin pointed to a single tree. Data fusion of LiDAR and hyperspectral data has already showed promising results (Solberg et al., 2005).

## REFERENCES

1. E. C. Barret, D. E. Curtis, 1997 Introduction to Environmental Remote Sensing, Chapman & Hall, 0-41237-170-7
2. T. Brandtberg, T. Warner, 2006 High-Spatial-Resolution Remote Sensing In: Computer applications in sustainable forest management, Shao, G. & Reynolds, K.M., 19 39 , Springer, 978-1-40204-305-5 Dordrecht
3. K. L. Castro, M. Kalacska, 2008 Tropical Dry Forest Phenology and Discrimination of Tropical Tree Species Using Hyperspectral Data, In: Hyperspectral Remote Sensing of Tropical and Sub-Tropical Forests, Kalacska, M. & Sanchez-Azofeifa, G.A., (1.), 1 25 , CRC Press, 978-1-42005-341-8 Boca Raton
4. P. Ceccato, S. Flasse, S. Tarantola, S. Jacquemoud, J. M. Gregoire, 2001 Vegetation Leaf Water Content Using Reflectance in the Optical Domain. Remote Sensing of Environment, 77 22-33
5. J. Clement, 2004 LiDAR Derived 3D Forest Stand Parameters of Dutch Pine, Wageningen University, Wageningen
6. C. S. T. Daughtry, 2001 Discriminating Crop Residues from Soil by Short-Wave Infrared Reflectance. Agronomy Journal, 93 125-131
7. C. S. T. Daughtry, E. R. Hunt, J. E. McMurtrey, 2004 Assessing Crop Residue Cover Using Shortwave Infrared Reflectance. Remote Sensing of Environment, 90 126 134
8. M. Ferretti, 2001 Forest health assessment and monitoring-Issues for consideration. Environmental monitoring and assessment, 48 45-72
9. T. F. Fourty, S. Baret, Jacquemoud, G. Schmuck, J. Verdebout, 1996 Leaf Optical Properties with Explicit Description of Its Biochemical Composition: Direct and Inverse Problems. Remote Sensing of Environment, 56 104 117

10. J. A. Gamon, J. Penuelas, C. B. Field, 1992 A Narrow-Waveband Spectral Index That Tracks Diurnal Changes in Photosynthetic Efficiency. Remote Sensing of Environment 41 35-44

11. B. C. Gao, 1995 Normalized Difference Water Index for Remote Sensing of Vegetation Liquid Water from Space. Proceedings of SPIE, 2480 225-236

12. A. A. Gitelson, M. N. Merzlyak, O. B. Chivkunova, 2001 Optical Properties and Nondestructive Estimation of Anthocyanin Content in Plant Leaves. Photochemistry and Photobiology, 71 38-45

13. A. A. Gitelson, Y. Zur, O. B. Chivkunova, M. N. Merzlyak, 2002 Assessing Carotenoid Content in Plant Leaves with Reflectance Spectroscopy. Photochemistry and Photobiology, 75 272 281 .

14. T. R. Helminen, M. L. Räisänen, 2002 Regional atmospheric deposition patterns of dust in the vincinity of the Lahnaslampi talc mine, Sotkamo, Finland, as revealed by moss and humus samples, Archive report RS/2002/8, 10 p., 9 appendices. Geological Survey of Finland, Espoo

15. S. Huber, M. Kneubuhler, A. Psomas, K. Itten, N. E. Zimmermann, 1997 Estimating foliar biochemistry from hyperspectral data in mixed forest canopy. Forest Ecology and Management,256 3, 491 501

16. A. R. Huete, H. Liu, K. Batchily, W. van Leeuwen, 1997 A Comparison of Vegetation Indices Over a Global Set of TM Images for EOS-MODIS. Remote Sensing of Environment, 59 440 451

17. J. L. Innes, 1993 Forest health: its assessment and status, CAB international, 0-85198-793-1 Oxon UK

18. Z. Jiang, A. R. Huete, J. Chen, Y. Chen, J. Li, G. Yan, X. Zhang, 2006 Analysis of NDVI and scaled difference vegetation index retrievals of vegetation fraction. Remote Sensing of Environment, 101 366 378

19. T. E. Kolb, M. R. Wagner, W. W. Covington, 1994 Forest health from different perspectives, Proceedings of the 1995 national sivilculture workshop, 5 13 , Gen. Tech. Rep. RM-267, U.S. Department of Agriculture, Fort Collins, Colorado

20. V. Kuosmanen, H. Arkimaa, T. Helminen, E. Hyvönen, E. Kuronen, J. Laitinen, J. Lerssi, M. Middleton, T. Ruohomäki, M. L. Räisänen, J. Saarelainen, R. Sutinen, 2002 MINEO Boreal environment test site, Finland. Contamination/impact mapping and modelling- Final report, Archive report RS/2004/2, 85 p., 6 appendices. Geological Survey of Finland, Espoo

21. D. A. M. Landgrebe, 2003 Signal theory methods in multispectral remote sensing, John Wiley & sons, 047142028-X New Jersey

22. D. G. Leckie, C. Jay, F. A. Gougeon, R. N. Sturrock, D. Paradine, 2004 Detection and assessment of trees with Phellinus weirii(laminated root) using high resolution multi-spectral imagery. International Journal of Remote Sensing, 25 793 818

23. A. Leopold, 1949 A Sand County Almanac and Sketches Here and There, Oxford Univ. Press, 0-19505-305-2 New York

24. R. Lucas, A. Mitchell, P. Bunting, 2008 Hyperspectral Data for Assessing Carbon Dynamics and Biodiversity of Forests, In: Hyperspectral Remote Sensing of Tropical and Sub-Tropical Forests, Kalacska, M. & Sanchez-Azofeifa, G.A., (1.), 50 51 , CRC Press, 978-1-42005-341-8 Boca Raton

25. Z. Malenovsky, C. Ufer, Z. Lhotakova, J. Clevers, M. E. Shaepman, J. Albrechtova, P. Cudlin, 2001 A new hyperspectral index for chlorophyll estimation of forest canopy: area under curve normalized to maximal band depth between 650-725nm. EARSeL eProceedings, 5 2/2006,161-173

26. E. Monnig, J. Byler, 1992 Forest health and ecological integrity in the Northern Rockies, USDA For. Ser. FMP Rep. 92-7

27. B. Moore, G. Allard, 2008 Climate change impacts on forest health, Working Paper FBS/34E02/8, 38 p. FAO, Rome

28. O. Mutanga, A. K. Skidmore, 2004 Narrow band vegetation indices overcome the saturation problem in biomass estimation. International Journal of Remote Sensing, 25 3999 4014

29. J. O'Laughlin, R. Livingston, R. Their, J. Thornton, D. E. Toweill, L. Morelan, 1994 Defining and measuring forest health. Journal of Sustainable Forestry, 2 65 85

30. J. Penuelas, F. Baret, I. Filella, 1995 Semi-Empirical Indices to Assess Carotenoids/Chlorophyll-a Ratio from Leaf Spectral Reflectance. Photosynthetica, 31 221 230

31. J. Penuelas, I. Filella, C. Biel, L. Serrano, R. Save, 1995 The Reflectance at the 950-970 Region as an Indicator of Plant Water Status. International Journal of Remote Sensing, 14 887-1905.

32. L. Serrano, J. Penuelas, S. L. Ustin, 2002 Remote Sensing of Nitrogen and Lignin in Mediterranean Vegetation from AVIRIS Data: Decomposing Biochemical from Structural Signals. Remote Sensing of Environment, 81 55-364

33. D. A. Sims, J. A. Gamon, 2002 Relationships Between Leaf Pigment Content and Spectral Reflectance Across a Wide Range of Species, Leaf Structures and Developmental Stages. Remote Sensing of Environment, 81 337 354.

34. S. Solberg, 1999 Forest health monitoring:Evaluation of methods, trends and causes based on a Norwegian nationwide set of monitoring plots, Norwegian Forest Research Institute, 8-27169-897-4, Oslo

35. S. Solberg, H. Lange, L. Aurdal, R. Solberg, E. Naesset, 2005 Monitoring forest health by remote sensing of canopy chlorophyll: first results from a pilot project in Norway, Proceedings of 31st International Symposium on Remote Sensing of Environment 2005, St. Petersburg, June 20-24 2005

36. C. J. Tucker, 1979 Red and Photographic Infrared Linear Combinations for Monitoring Vegetation. Remote Sensing of the Environment, 8 127-150

37. J. Tuominen, T. Lipping, V. Kuosmanen, 2008 Assesment of ENVI forest health tool in detection of dust and seepage contaminated forest areas. Proceedings of IEEE International Geoscience & Remote Sensing Symposium, 1358 1361, Boston, July 07-11, IEEE

38. A. Van Laar, A. Acka, 2007 Forest mensuration (Managing Forest Ecosystems), Springer, 978402059902, Dordrecht

39. J. E. Vogelmann, B. N. Rock, 1989 Use of the thematic mapper data for the detection of forest damage caused by the pear thrips. 30 217-225

40. Q. Xiao, E. McPherson, 2005 Tree health mapping with multispectral remote sensing data at UC Davis, California. Urban Ecosystems, 8 December 2005, 349 361.

# Chapter 2

# Optical Satellite Volcano Monitoring: A Multi-Sensor Rapid Response System

**Kenneth A. Duda[1], Michael Ramsey[2], Rick Wessels[3] and Jonathan Dehn[4]**

*[1] SGT, contractor to U.S. Geological Survey (USGS) Earth Resources Observation and Science (EROS) Center, Sioux Falls, South Dakota; work performed under USGS contract 08HQCN0005, United States*

*[2] Department of Geology & Planetary Science, University of Pittsburgh, and Advanced Spaceborne Thermal Emission and Reflection Radiometer (ASTER) Science Team, United States*

*[3] USGS Alaska Volcano Observatory, United States*

*[4] University of Alaska Fairbanks and Alaska Volcano Observatory, United States*

## 1. INTRODUCTION

In this chapter, the use of satellite remote sensing to monitor active geological processes is described. Specifically, threats posed by volcanic eruptions are briefly outlined, and essential monitoring requirements are discussed. As an application example, a collaborative, multi-agency operational volcano monitoring system in the north Pacific is highlighted with a focus on the 2007 eruption of Kliuchevskoi volcano, Russia. The data from this system have been used since 2004 to detect the onset of volcanic activity, support the emergency response to large eruptions, and assess the volcanic products produced following the eruption. The overall utility of such integrative assessments is also summarized.

The work described in this chapter was originally funded through two National Aeronautics and Space Administration (NASA) Earth System Science research grants that focused on the Advanced Spaceborne Thermal Emission and Reflection Radiometer (ASTER) instrument. A skilled team of volcanologists, geologists, satellite tasking experts, satellite ground system experts, system engineers and software developers collaborated to accomplish the objectives. The first project, *Automation of the ASTER Emergency Data Acquisition Protocol for Scientific Analysis, Disaster Monitoring, and Preparedness*, established the original collaborative research and monitoring program between the University of Pittsburgh (UP), the Alaska Volcano Observatory (AVO), the NASA Land Processes Distributed Active Archive Center (LP DAAC) at the U.S. Geological Survey (USGS) Earth Resources Observation and Science (EROS) Center, and affiliates on the ASTER Science Team at the Jet Propulsion Laboratory (JPL) as well as associates at the Earth Remote Sensing Data Analysis Center (ERSDAC) in Japan. This grant, completed in 2008, also allowed for detailed volcanic analyses and data validation during three separate summer field campaigns to Kamchatka Russia. The second project, *Expansion and synergistic use of the ASTER Urgent Request Protocol (URP) for natural disaster monitoring and scientific analysis*, has expanded the project to other volcanoes around the world and is in progress through 2011.

The focus on ASTER data is due to the suitability of the sensor for natural disaster monitoring and the availability of data. The instrument has several unique facets that make it especially attractive for volcanic observations (Ramsey and Dehn, 2004). Specifically, ASTER routinely collects data at night, it has the ability to generate digital elevation models using stereo imaging, it can collect data in various gain states to minimize data saturation, it has a cross-track pointing capability for faster targeting, and it collects data up to ±85º latitude for better global coverage. As with any optical imaging-based remote sensing, the viewing conditions can negatively impact the data quality. This impact varies across the optical and thermal infrared wavelengths as well as being a function of the specific atmospheric window within a given wavelength region. Water vapor and cloud formation can obscure surface data in the visible and near infrared (VNIR)/shortwave infrared (SWIR) region due mainly to non-selective scattering of the incident photons. In the longer wavelengths of the thermal infrared (TIR), scattering is less of an issue, but heavy cloud cover can still obscure the ground due to atmospheric absorption. Thin clouds can be optically-transparent in the VNIR and TIR regions, but can cause errors in the extracted surface reflectance or derived surface temperatures. In regions prone to heavy cloud cover, optical remote sensing can be improved through increased temporal resolution. As more images are acquired in a given time period the chances of a clear image improve dramatically. The Advanced Very High Resolution Radiometer (AVHRR) routine monitoring, which commonly collects 4-6 images per day of any north Pacific volcano, takes advantage of this fact. The rapid response program described in this chapter also improves the temporal resolution of the ASTER instrument.

ASTER has been acquiring images of volcanic eruptions since soon after its launch in December 1999. An early example included the observations of the large pyroclastic flow deposit emplaced at Bezymianny volcano in Kamchatka, Russia. The first images in March 2000, just weeks after the eruption, revealed the extent, composition, and cooling history of this large deposit and of the active lava dome (Ramsey and Dehn, 2004). The initial results from these early datasets spurred interest in using ASTER data for expanded volcano monitoring in the north Pacific. It also gave rise to the multi-year NASA-funded programs of rapid response scheduling and imaging throughout the Aleutian, Kamchatka and Kurile arcs. Since the formal establishment of the programs, the data have provided detailed descriptions of the eruptions of Augustine, Bezymianny, Kliuchevskoi and Sheveluch volcanoes over the past nine years (Wessels et al., in press; Carter et al., 2007, 2008; Ramsey et al., 2008; Rose and Ramsey, 2009).

The initial research focus of this rapid response program was specifically on automating the ASTER sensor's ability for targeted observational scheduling using the expedited data system. This urgent request protocol is one of the unique characteristics of ASTER. It provides a limited number of emergency observations, typically at a much-improved temporal resolution and quicker turnaround with data processing in the United States rather than in Japan. This can speed the reception of the processed data by several days to a week. The ongoing multi-agency research and operational collaboration has been highly successful. AVO serves as the primary source for status information on volcanic activity, working closely with the National Weather Service (NWS), Federal Aviation Administration (FAA), military and other state and federal emergency services. Collaboration with the Russian Institute of Volcanology and Seismology (IVS)/Kamchatka Volcanic Eruption Response Team (KVERT) is also maintained. Once a volcano is identified as having increased thermal output, ASTER is automatically tasked and the volcano is targeted at the next available opportunity. After the data are acquired, scientists at all the agencies have access to the images, with the primary science analysis carried out at the University of Pittsburgh and AVO. Results are disseminated to the responsible monitoring agencies and the global community through e-mail mailing lists.

## 2. OVERVIEW

Few natural hazards have the devastating impact of large volcanic eruptions (Figure 1). These events can affect scales from the local citizen to the rare global impact where large Plinian eruptions can alter global climate for years to decades (Yang and Schlesinger, 2002). Eruptions can present significant and varying levels of threat to public safety and health by way of explosive forces that launch rocks and ash, destructive pyroclastic lava flows and gaseous emissions, disruption of transportation and communication, introduction of disease and the loss of life. In addition to threats on the ground, aircraft and passengers are at considerable risk if travelling in affected areas (Miller and Casadevall, 2000). This is a particular concern in the north Pacific where numerous flight routes pass over active volcanoes (Figure 2).

**Figure 1.** An ash-rich volcanic cloud from the 1989-1990 eruption of Redoubt volcano (courtesy J. Warren, April 21, 1990). Large eruption clouds such as these are a major hazard to commercial aviation in the busy north Pacific corridor.

Therefore, advance warning of incipient volcanic activity can lead to better and more accurate eruption prediction and help to save lives. High-quality and detailed data are needed in order to assess the relative risks posed by any one of the potentially active volcanoes along the Aleutian-Kamchatka-Kurile arcs. These data can include seismic monitoring, deformational analysis, studies of the emitted gas, visual and thermal observations, as well as the use of orbital remote sensing. The volcanoes can be monitored using in situ equipment, airborne and satellite imaging systems, or some combination of all of these.

Ground-based seismic and deformation measurements commonly provide the earliest indications of renewed activity and subsurface magma movement at a particular volcano (Stephens and Chouet, 2001;Fournier et al., 2009; Lu et al., 2007). Increased seismic activity and a determination of the frequency, depth, and type of earthquakes below a volcano can be leading indicators of later eruptions. Field- or space-based measurements of volcanically induced deformation together with increased levels of heat and gas emissions can further verify the seismic results and lead to a more complete picture of the changes with time. Changes in the alignment of surface features are noted through precise measurements and the collection of spectral, thermal, and GPS data can serve to validate the satellite-based observations.

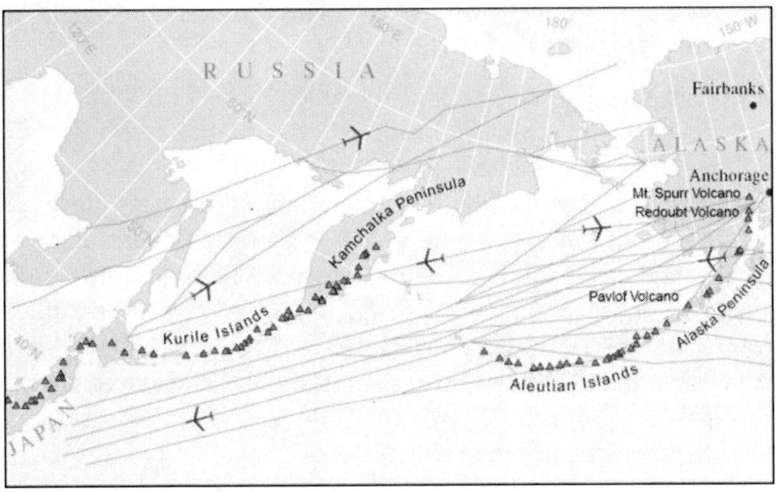


**Figure 2.** North Pacific flight route map and the locations of the active volcanoes of the Alaska Peninsula, the Aleutian Islands, the Kamchatka Peninsula, the Kurile Islands, and into Japan (Courtesy USGS Fact Sheet 030-97).

In addition to ground-based measurements, satellite-based observations of volcanoes are of value. Satellite-based observations are useful in the northern Pacific and at many other active volcanoes around the world, where limited resources do not allow extensive ground-based monitoring. Large regions of the world and widely distributed and commonly remote targets can be monitored frequently and economically using satellite data. These images can reveal changes in thermal or gas emission, in the composition of the surface rocks, and in the deformation over time. They can thus provide additional insight into the evolving status of volcanoes. Fixed-wing aircraft and helicopters can also be used where resources permit to carry sensors such as thermal cameras, gas sensors, and cameras, which obtain higher resolution data of the volcanic activity. The integrated analysis of all available data offers an enhanced perspective on the subtle and unique differences among volcanoes and on the likelihood of impending eruptive activity at a given site.

The scientists and engineers assembled to carry out the previously described NASA-funded research programs have the goal of improving volcanic monitoring in the north Pacific region in order to minimize the subsequent risks and increase scientific knowledge of these dynamic geologic processes. Incorporating the finer spatial resolution, multispectral ASTER data in the analysis process allowed this possibility and added greater clarity to the characterization of many remote volcanoes in the north Pacific region.

The work involved developing and implementing algorithms and tools to detect new activity, establish protocols for escalating response activity, create system linkages between AVO, the LP DAAC and JPL to enable semi-automated transmission of satellite tasking requests, develop the tools and procedures to identify ASTER overpass opportunities and control scheduling requests, secure sensor tasking authorization from the ASTER Science Team, employ existing systems to capture, downlink and process ASTER data, and finally create new data distribution mechanisms to ensure the timely availability of the acquired data. New data analysis procedures are then employed to assess current conditions and issue alerts when needed.

## 3. SENSORS

The critical imaging requirements (i.e., optimum temporal frequency, spatial resolution, wavelengths, etc.) must be understood in order to use sensors on Earth-orbiting spacecraft to monitor volcanoes. The most commonly used instruments for volcano monitoring in the north Pacific include AVHRR, Geostationary Operational Environmental Satellite (GOES), Multifunctional Transport Satellites (MTSAT-1R), Moderate Resolution Imaging Spectroradiometer (MODIS), the Landsat Enhanced Thematic Mapper Plus (ETM+), and ASTER (Table 1). Sub-meter visible data from several commercial satellites (e.g. QuickBird, WorldView, IKONOS and GeoEye) have also recently become integrated as another useful tool for volcano monitoring.

Current sensors such as AVHRR, MODIS, MTSAT, and GOES provide the frequency necessary to detect the onset of large thermal anomalies in near real-time (Dehn et al, 2000; Wright et al., 2002,Schneider et al., 2000). In comparison, high spatial resolution instruments such as ASTER and Landsat ETM+ provide data at a much improved spatial scale ideal for scientific analysis, damage assessment, and smaller scale monitoring, but at the expense of rapid repeat times (Harris et al., 1998; Ramsey and Dehn, 2004). In order to obtain frequent status updates where attempting to initially identify thermal anomalies, sensors offering quick revisits (i.e., GOES, AVHRR, MODIS) are used (Figure 3). The disadvantage is that such data are collected at a coarse spatial resolution allowing only very large or very hot anomalies to be detected, whereas the non-eruptive or small-scale activity is missed completely at many of the remote volcanoes. Higher spatial resolution data (i.e., ASTER, ETM+) are employed in order to obtain more detailed information of lava flows, thermal anomalies, and gas emissions. However, nominally ASTER can only revisit a site once every 16 days at the equator viewing nadir. This temporal frequency is improved with off-nadir pointing and/or at higher latitudes. For example, at the higher latitudes of Kamchatka the temporal frequency can be shortened to 13 hours with off nadir pointing.

The ASTER URP has been used in support of North Pacific volcano monitoring to improve the collection probability and the speed of ASTER data availability. Through this approach, data are typically available within six hours after acquisition. This combination of frequent coarse spatial resolution change detection and less frequent detailed higher resolution imaging has proven very valuable scientifically and operationally. The three bands of ASTER VNIR data span from 0.52 to 0.86 micrometers at a ground resolution of 15 m. These data have been used to generate digital elevation models, map the eruption deposits and changing surface characteristics and to detect ground temperatures in excess of 800 C. Thermal anomaly and gas emissions identification is accomplished primarily using sensor bands in the SWIR and TIR wavelengths. The six ASTER SWIR bands cover wavelengths from 1.6 to 2.43 micrometers at 30-m ground resolution. Unfortunately, the ASTER SWIR subsystem is no longer operational after 2008, but the data were commonly used to detect alteration minerals on the surface and temperatures between 100 C and 460 C. The ASTER TIR subsystem has five TIR bands from 8.125 to 11.65 micrometers at 90-m ground resolution. These data have been used to extract temperatures less than 100 C, map silicate and carbonate minerals, model the vesicle content of lavas, and estimate the thermal inertia of the surfaces. A brief history of the ASTER mission, including key orbital and data characteristics, is provided in the text box entitled The ASTER Mission.

**Table 1.** Comparison of key characteristics for sensors used in the operational north Pacific monitoring system.

| Criteria | GOES/MTSAT | AVHRR | MODIS | LANDSAT ETM+ | ASTER |
|---|---|---|---|---|---|
| Revisit frequency | Geostationary | 2 passes per day | 1-2 days | 16 days | 16 days, less off-nadir |
| Ground resolution | At 60°N latitude: 2 km VIS 8 km IR | 1.1 km (LAC) | 250m (B1-2), 500m (B3-7), 1000m (B8-36) | 30m (B1-5, 7) 60m (B6) 15m (B8) | 15m (B1-3), 30m (B4-9), 90m (B10-14) |
| Spectral coverage | 5 bands: 0.55 to 12.5 μm | 6 bands: 0.58 to 12.4 μm | 36 bands: 0.405 to 14.385 μm | 8 bands: 0.45 to 12.5 μm | 14 bands: 0.52 to 11.65 μm |
| Swath Width | Full Earth view | 2,399 km | 2,330 km | 185 km | 60 km |
| Orbit Altitude | 35,800 km | 833 km | 705 km | 705 km | 705 km |

## 4. GROUND SYSTEMS AND DATA ACQUISITION

The infrastructure needed to support ASTER urgent request tasking, data collection, processing and eventual distribution is summarized in Figure 4. The five key steps of the volcano monitoring sequence are shown in the center of the figure. Corresponding contributions by participating groups are shown in the outer circle, occurring in a sequence that commences at the arrow and proceeds clockwise. The existing ASTER mission procedures and systems were used for final tasking, collection, and data processing.

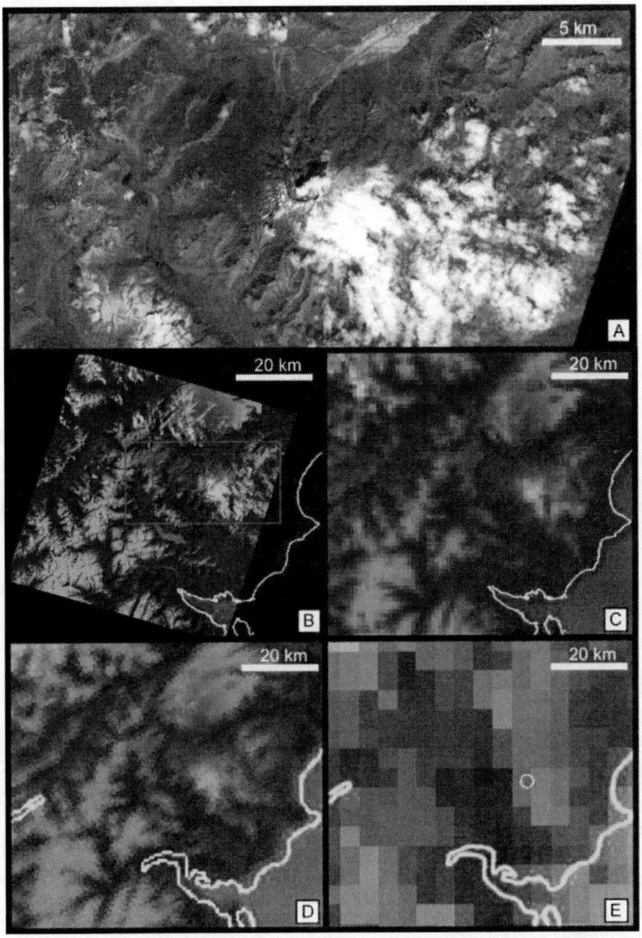


**Figure 3.** Comparison of satellite data at several spatial resolutions over Redoubt volcano, Alaska during daylight hours between 21:30 and 22:01 UTC on June 6, 2009. (A) False-color subset of ASTER 15 m VNIR bands 3,2,1 as R,G,B. Figures B-E show TIR bands (centered on about 11 micrometers) from four different sensors: (B) ASTER 90 m TIR band 14, (C) MODIS 1 km TIR band 31, (D) AVHRR 1.1km TIR band 4, and (E) GOES 8 km TIR band 4.

**Figure 4.** Volcano monitoring actions and participant involvement. Size of each portion is not scaled to the actual work level involved. Acronyms: Alaska Volcano Observatory (AVO); Advanced Spaceborne Thermal Emission and Reflection Radiometer (ASTER); Jet Propulsion Laboratory (JPL); Earth Remote Sensing Data Analysis Center (ERSDAC); Ground Data System (GDS); Goddard Space Flight Center (GSFC); Flight Operations Team (FOT); Earth Observing System Data and Information System (EOSDIS) Data and Operations System (EDOS); Land Processes Distributed Active Archive Center (LP DAAC); University of Pittsburgh (U. Pitt).

Initial thermal anomaly event detection in AVHRR or MODIS data triggers the ASTER ground systems, scheduling, and eventual data acquisition. The methods used to detect anomalous events include a detailed set of screening algorithms designed to minimize false positives prior to tasking ASTER. Once ASTER is tasked, a series of procedures are used to determine timing, scheduling support, and the eventual distribution of ASTER data. This entire system of cross-satellite integration and supporting systems is located at AVO; LP DAAC; JPL; ERSDAC; Goddard Space Flight Center (GSFC); White Sands, New Mexico; and includes the ASTER instrument on the Terra spacecraft. The work of this project has contributed new techniques in event detection and tasking of a complex instrument.

## 4.1. Event Detection

The core of the triggering mechanism for the ASTER emergency acquisition requests is based on the Okmok algorithm (Dean et al., 1998). This algorithm uses a time series of AVHRR data to detect thermal anomalies above an expected average seasonal background temperature. Deviations in temperature may signal increased thermal emission and an impending eruption at a particular volcano. The algorithm scanned a small subsector of data over each volcano for the warmest IR band 3 (3.5-3.9 micrometers) radiant temperature. This value along with all other values in the data subset are recorded and tested based on a simple criterion: if the AVHRR band 3 temperature was greater than 35 C, then an e-mail was generated and sent to all AVO staff. This process worked well, however its overall effectiveness was limited due to the poor geolocation accuracy of the AVHRR instrument and the generation of numerous false alarms due to noise.

In conjunction with the NASA funded projects, the Okmok algorithm has been significantly refined and applied to other sensors, such as MODIS. It has also been revised to better constrain the AVHRR data stream. The new algorithm, called Okmok II, focuses on decreasing the number of false alarms through a series of additional logic tests (Dehn and Harris, in press). However, it does not sacrifice the sensitivity to low grade thermal anomalies, which are notoriously hard to detect. Okmok II is built around seven levels of data processing and is interfaced with a new Web-based visualization tool (Figure 5). If triggered, an e-mail alert is sent directly to the ASTER Emergency Scheduling Interface and Control System (AESICS) database to immediately initiate the ASTER scheduling process.

## 4.2. Aster Tasking And Collection

In support of the ASTER project, and for use with other applications, the LP DAAC created the ASTER Overpass Predictor to simplify the determination of future ASTER tasking opportunities. The data entry page includes the scene center latitude and longitude, and desired forecast window (Figure 6). Returned results identify possible dates, day or night opportunities, and which ASTER subsystems would be available. In addition to supporting the north Pacific volcano monitoring applications, this tool is now routinely used in support of other emergency response and research activities.

After an anomalous thermal event is detected, AVO initiates an ASTER tasking request and notifies the LP DAAC. These incoming requests are logged in a database included in AESICS. AESICS is a Web-based tool that was created to simplify tasking request submission and to control these requests (Figure 7). New tasking requests can be entered either manually or through the automated procedure, which begins through processes in place at AVO. After new requests are logged in the AESICS database, e-mail notifications are immediately forwarded to the ASTER team at JPL for the actual scheduling using pre-existing mission protocols. Approvals to task the sensor occur at each stage of the process to ensure the best use of available resources and compliance with mission requirements and international agreements. For example, if there are multiple AVHRR alerts over a particular volcano in a given day, the system of checks guarantees that the ASTER scheduling system is not overloaded.

Following final approvals, and after confirming the absence of scheduling conflicts, the ASTER JPL team accesses the ASTER scheduling system at ERSDAC in Japan and the tasking request is uploaded. The final schedule is determined at ERSDAC and provided to the GSFC Flight Operations Team (FOT) for uplink to the Terra spacecraft. This entire process can be as quick as several hours. Once the data are acquired, they are stored in the Terra solid state recorder until downlink to receivers at White Sands, New Mexico. Raw data are transferred from White Sands via network to GSFC EDOS for initial processing from the raw image data format and then sent to the LP DAAC.

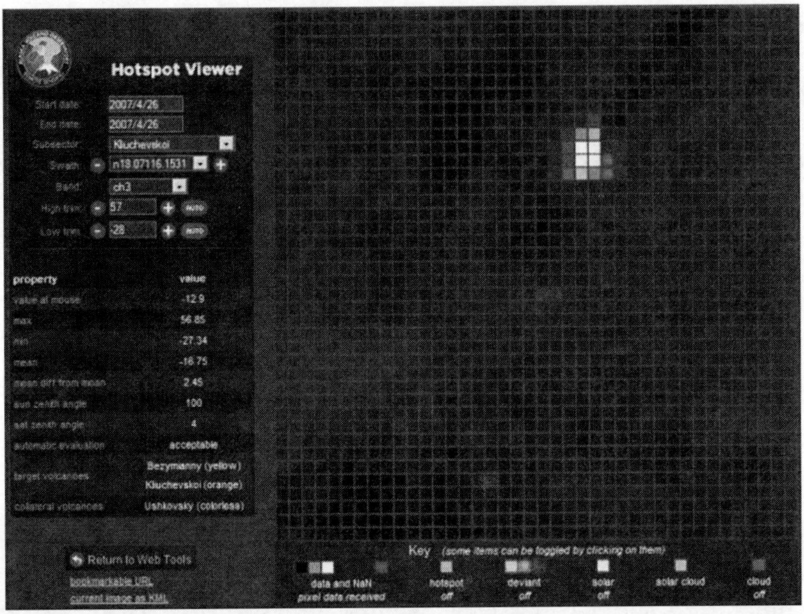


**Figure 5.** AVO Hotspot Viewer Web interface for detection of AVHRR thermal anomalies. The AVHRR pixels are reformatted into 1-km grid cells with user-selectable tools for data enhancement. This tool is used to screen for actual anomalies and trigger an ASTER observation. Shown here are the data collected on April 26, 2007 from AVHRR on NOAA-18 for Kliuchevskoi volcano. This data viewer is also used by AVO in their routine volcano monitoring operations. (http://avo-animate.images.alaska.edu/auto_obs_viewer.php).

### *The ASTER Mission*

ASTER is a joint endeavor involving NASA, Japan's Ministry of Economy, Trade and Industry, and other organizations. The ASTER sensor was launched in 1999 on the Terra spacecraft as part of NASA's Earth Observing System with the goal of conducting a global land mapping mission. ASTER has been accomplishing this goal very successfully, having already acquired over 1.5 million scenes. Though the instrument design life has been exceeded, ASTER continues to acquire approximately 450 new 60 by 60 km images of Earth's land surfaces daily, providing Earth land surface information useful for a wide variety of applications.

Terra orbits Earth once every 98.88 minutes at an elevation of 705 km in a sun-synchronous orbit at an inclination of 98.3 degrees. The descending orbit has a 10:30 AM equatorial crossing time and the revisit time is 16 days at the equator (less at higher latitudes or when off-nadir pointing is used). ASTER is tasked through scheduled observations at an 8% duty cycle, and has three imaging subsystems: VNIR, SWIR and TIR. Ground resolution is 15m for VNIR, 30m for SWIR, and 90m for TIR. ASTER has 14 spectral bands ranging from 0.52 micrometers to 11.65 micrometers, including a back-looking VNIR band 3 that enables the generation of digital elevation models. Data are distributed in the HDF-EOS format by the LP DAAC in the USA and by ERSDAC in Japan. More information is contained in Yamaguchi et al. (1998).

Level-1A reconstructed unprocessed instrument data are archived, and other products are generated from these data at the request of customers. Additional products include registered radiance at the sensor, surface reflectance, brightness temperature, surface kinetic temperature, surface emissivity, decorrelation stretch, polar surface and cloud classification, orthorectified, and digital elevation model (Abrams, 2000). The file size of the registered radiance product is approximately 118 MB, and contains data from each subsystem that are geometrically co-registered and radiometrically calibrated.

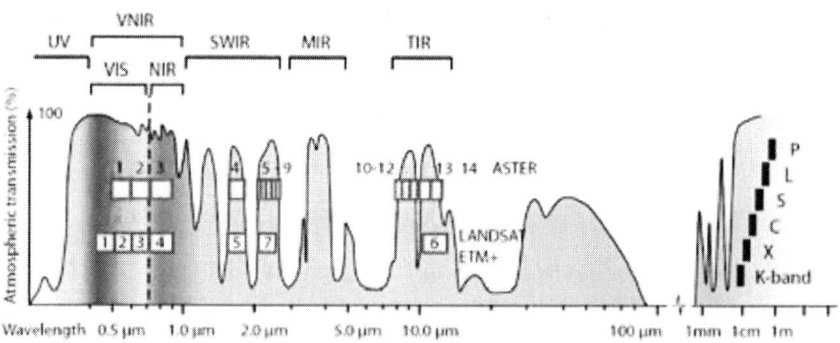


**The Earth's atmospheric windows with the spectral coverage of various instruments shown.** ASTER (shown in red), Landsat ETM+ (shown in blue), and SAR (shown in black). (Kaab, 2005)

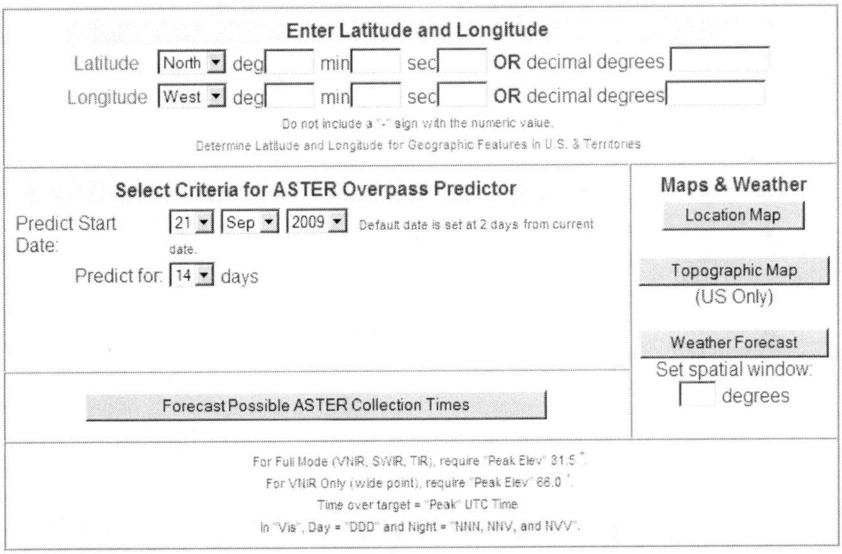


**Figure 6.** ASTER Overpass Predictor Web page. (http:// igskmncnwb 001.cr.usgs.gov/aster/estimator/reference_info.asp)

## 4.3. Aster Data Processing and Distribution

Following standard ASTER mission protocols, raw ASTER data from the Terra spacecraft are processed by EDOS to Level-0. Level-0 data are then transferred via network to LP DAAC. LP DAAC receives the Level-0 data and processes it to Level-1 using executable code and observation schedule information received from ERSDAC. Upon completion of Level-1 processing, data are staged for retrieval via FTP. LP DAAC created a Recent Expedited Production Web site for this project to simplify and speed initial data assessment and downloading of ASTER expedited data. Data are also made available via AESICS, an FTP site, and through the Warehouse Inventory Search Tool (WIST) data search and order mechanism.

Employing the ASTER expedited approach greatly assists in the timely availability of image data for analysis. ASTER expedited data are typically available within six hours after collection, whereas the standard data product availability takes several days after acquisition. This lag time was actually several weeks when the URP volcano monitoring program was first conceived.

ASTER expedited products are very similar to standard products but there are some minor differences. Expedited data do not contain the back-looking band 3, so stereo data are not initially available for digital elevation model (DEM) generation. DEMs can be created later once the standard Level-1 products become available.

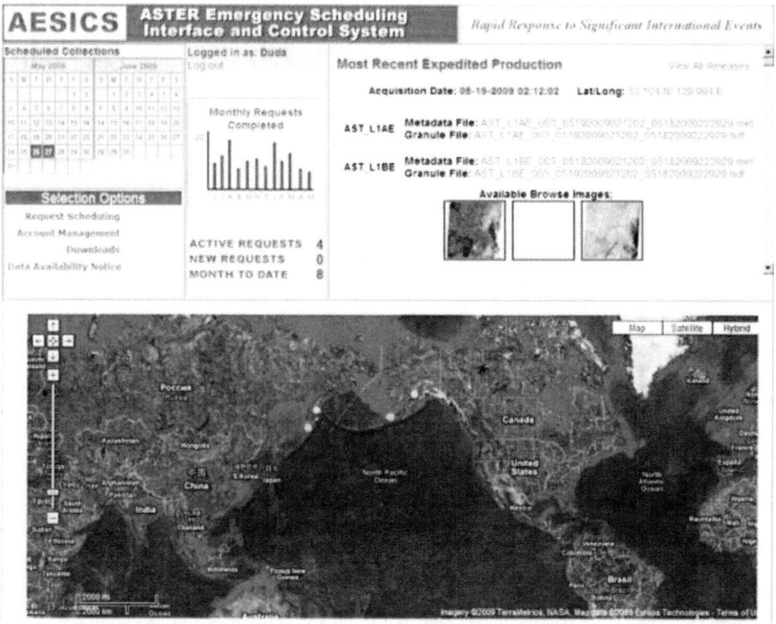


**Figure 7.** AESICS Web-based interface. Users can rapidly access statistics on the number of ASTER urgent requests in the last month and year, thumbnail images of the most recent acquisitions, and a map-based display of all ASTER tasking requests in the past 30 days. Based on the Google Earth application, this interface allows users to see which ASTER tasks are new (red dot), approved (yellow dot), and completed (green dot). Each target can also be queried for more information.

In addition, short-term calibration for the TIR is not available so long-term calibration is used. As a result, expedited TIR data quality is expected to be somewhat reduced. Inter-telescope registration quality is also slightly lower since numerous adjacent scenes are not available as is normally the case. Finally, expedited processing uses raw spacecraft ephemeris data, so the geometry is slightly different than for the data produced using standard processing, which uses refined (post-processed) ephemeris data. Even with these caveats, the image data are still of excellent quality and commonly used to report the latest activity and thermal output values.

### 4.4. Aster Data Analysis and Alert Status Updates

After ASTER Level-1 data are processed, image analysts and volcanologists at AVO and the University of Pittsburgh obtain the images for inspection to gain further insight on the current state of the volcano. This may include a visual assessment of the available bands, a characterization of temperature conditions, areal extent of the anomaly, time series analysis, documentation of flow patterns, and plume extent and content. New data are compared with historical satellite data and other information to identify trends. If warranted, volcano status alerts are updated to notify government authorities and other interested individuals. The volcano monitoring plan is also then updated to ensure that continued observations occur as needed.

## 5. THE 2007 ERUPTION OF KLIUCHEVSKOI VOLCANO, RUSSIA

One of the most volcanically active regions in the world is Kamchatka, Russia, which is located in the northwestern region of the Pacific Ocean. Several of the more than two dozen active volcanoes on this peninsula are commonly erupting at any given time. These eruptions can produce hazards for the sparse local population living near the volcanoes. But of far more consequence are the eruptions that produce larger ash columns, which are carried by the easterly winds into the routes of approximately 200 aircraft and 20,000 people overflying the region each day (Miller and Casadevall, 2000).

Kliuchevskoi is the highest (> 4800 m) and one of the most active volcanoes in Kamchatka. In the last century, summit eruptions have increased in frequency averaging one every 1–2 years. These eruptions are commonly caused by lava interaction with melting snow/ice and start with increased fumarolic activity within the summit crater. This thermal activity is typically followed by Strombolian explosions, the effusion of blocky lava flows, and the generation of hot avalanches and lahars. Less common are the paroxysmal eruptions, which last occurred in 1994. That eruption included a large convecting column, pyroclastic flows, lahars and lava flows. The largest explosive eruption reached 18 km above sea level and travelled approximately 1,000 km southeast into the north Pacific air traffic routes (Miller et al., 1994).

Ground-based techniques for monitoring the remote volcanoes of Kamchatka include seismic and visual observations. In 2007, there were 33 seismic stations deployed in Kamchatka (Chebrov, 2008). Human observations, reports, and a Web-based video camera system are also used. The Web camera has been installed in the town of Klyuchi 30 km north of Kliuchevskoi. Using the Web-based system, the height of the eruption column has been correlated with the level of seismic activity, thereby allowing seismic signals to predict the height of the eruption plume at night or in times of bad weather (McNutt, 1994; Roach et al., 2004). In addition to these data, high temporal/low spatial resolution orbital remote sensing monitoring of these remote volcanoes has been used for nearly two decades by volcanologists in Russia and the United States. ASTER has been an integral part of this monitoring since it was launched in 1999. Some of the first scientific images collected by ASTER in early 2000 were of the large eruption deposits of Bezymianny volcano (Ramsey and Dehn, 2004). Beginning in 2004, the ASTER rapid response/urgent request system has been linked to the routine remote sensing monitoring done by AVO and KVERT. The ASTER data from this collaborative program have provided the basis for enhanced monitoring efforts, new discoveries, and numerous scientific results (Carter et al., 2007, 2008; Ramsey et al., 2004; Rose and Ramsey, 2009).

During the period between 2000 and 2008, Bezymianny volcano, 10 km south of Kliuchevskoi, was nearly continuously active with approximately two large eruptions per year (Ramsey and Dehn, 2004;Carter et al., 2007). Therefore, this volcano produced numerous AVHRR thermal alerts and subsequent ASTER urgent request images (after the URP system was active). When both Bezymianny and Kliuchevskoi are active, the poor geolocation accuracy of AVHRR makes it difficult to discern which volcano is responsible for producing the thermal anomaly (Dehn et al, 2000). Similarly, both volcanoes commonly appear in one ASTER 60-km scene. It is therefore common for ASTER observations targeting one volcano to capture activity at the other. Typically, low-level thermal activity is seen long before visual or even AVHRR spaceborne observations detect that activity. The better radiometric accuracy, higher spatial resolution, and more precise geolocation of ASTER compared to AVHRR makes the TIR data ideal for detection of the very early stages of new activity at a volcano. However, the poorer temporal frequency commonly limits this important aspect of ASTER. Future TIR instruments with a temporal frequency of hours to days will provide a critically important new dataset for volcano monitoring and eruption prediction.

The detection of renewed activity at Kliuchevskoi first occurred in 2005 in an ASTER TIR scene collected to observe the waning stages of an eruption at Bezymianny (Rose and Ramsey, 2009). A similar situation occurred nearly two years later when observations of the 2006 eruptive activity at Bezymianny also showed a very slight increase in thermal output (~ 5 C above background) at the summit of Kliuchevskoi as early as November 2, 2006. This activity was noted but did not raise concern due to its very low level and no other detectable activity from visual or seismic observations. On November 27, 2006 the activity remained unchanged. It was not until the first detection by AVHRR on December 14, 2006 that a rapid increase in activity was noted. Two AVHRR pixels between 10 and 30 C above the average background temperature were detected at the summit, which likely meant that very hot gases and/or small amounts of lava had reached the surface. By December 22, 2006 the activity had further increased enough to trigger an ASTER urgent request image, which was scheduled for January 4-5, 2007 (Figure 8). No activity was detected in the subsequent ASTER VNIR data although a pixel-integrated brightness temperature of 332 C was extracted from the SWIR data. By January 12, 2007 the color-code for Kliuchevskoi was raised from green to yellow indicating that heightened activity was taking place. However, this activity was low-grade enough that the first ground-based visible observations of a vigorous steam plume and the presence of lava at the summit were not confirmed until February 16, 2007.

Commonly, the winter weather in Kamchatka is clear and cold. Minimal to no cloud cover can be the norm for long stretches between December and May in this region, thereby making optical remote sensing an excellent tool. In the spring and summer months, low-level thicker clouds typically form after 9am and can persist the entire day. These clouds can hinder ground-based visual observations and make visits to the summits of the volcanoes difficult. However, the clouds typically do not extend high enough to obscure the summits of the taller volcanoes such as Kliuchevskoi. For these periods, it is not uncommon that optical remote sensing is the only form of monitoring possible. Higher altitude thin clouds, thin volcanic plumes, and jet contrails can also be problematic for optical remote sensing. These clouds are visually hard to detect, but can negatively-impact the extraction of accurate surface composition and temperature.

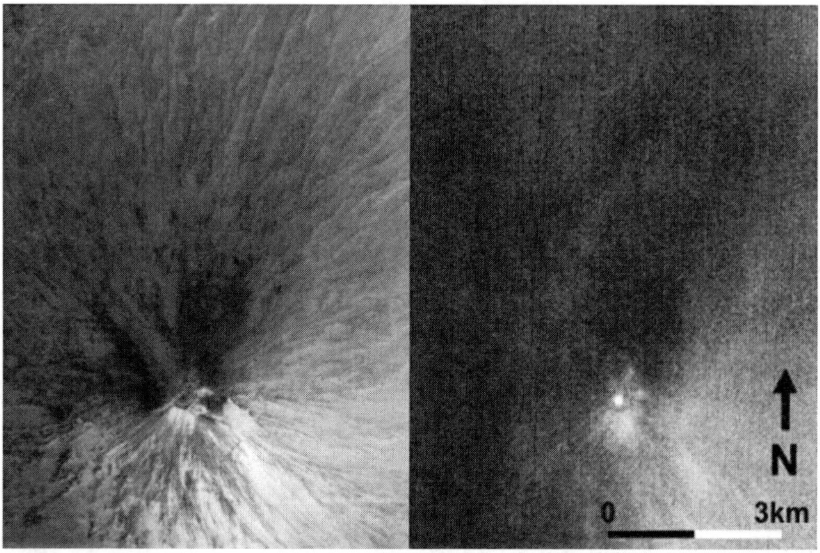


**Figure 8.** ASTER data collected on January 4, 2007 and centered on the summit of Kliuchevskoi volcano. The VNIR color composite image (left) shows almost no signs of activity other than a small darker spot in the center of the summit crater (indicating snowmelt) and a minor amount of steam. The band 9 SWIR image (right) however shows a distinct thermal anomaly centered over the dark spot in the summit crater. The maximum integrated brightness temperature derived from the SWIR data was 332 C indicating the presence of hot gases and/or magma very close to the summit.

For example, this can be seen in the lower surface temperatures derived during periods of thin cloud cover for Kliuchevskoi (Figure 9). From mid-January through mid-April, ASTER continued to collect regularly scheduled and urgent request data of the volcanic activity at Kliuchevskoi. During this period, ground-based observations were hindered at times by the presence of low-level cloud cover. However, these clouds were commonly lower than the summit of Kliuchevskoi allowing the activity to be observed with ASTER and AVHRR. The summit activity continued to increase during this time, producing more thermally elevated pixels and raising the brightness temperature enough to saturate the ASTER SWIR data by early April (Figure 9). This was caused by the presence of a large amount of non-crusted lava in the summit crater from either a small actively overturning lava lake or very vigorous Strombolian eruption activity. On April 9, 2007 the clouds had diminished and photographs from the ground confirmed the Strombolian explosions and the presence of lava that was flowing down the northern slope at a location nearly identical to the 2005 flow (Rose and Ramsey, 2009).

ASTER captured another day/night pair of urgent request images on April 26-27, 2007 (Figure 9). By this time, the active lava flow had been present for over a month and the SWIR and TIR data were commonly saturated in many locations due to the high temperatures. These lava flows were producing numerous lahars that were emplaced 10-15 km further down the northern slope from the base of the lava flow. During this period, larger Vulcanian-style eruptions were also common at the summit crater. These short-duration explosions produced ash columns 5-15 km above the summit, which then commonly drifted S-SE over the peninsula and into the northern Pacific Ocean.

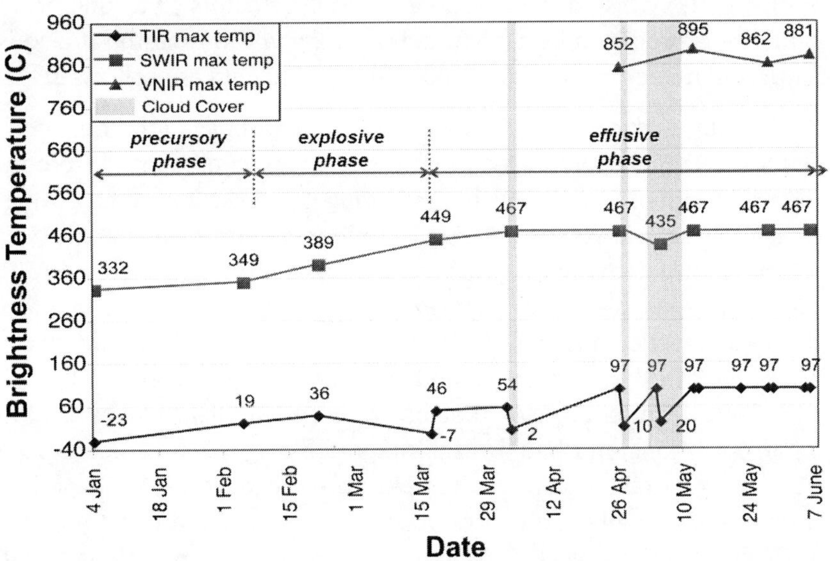


**Figure 9.** Time-temperature plot of the 2007 eruption of Kliuchevskoi volcano showing the extracted pixel-integrated brightness temperatures for the three wavelength regions and the three phases of eruptive activity.

The plumes also produced proximal airfall deposits easily seen in contrast to the underlying snow (Figure 10). The April 26-27, 2007 ASTER image pair was unique in several other ways. The data were collected several hours before an AVHRR overpass, which had triggered the next ASTER urgent request (later acquired on May 4-5, 2007). However, because the AVHRR data are displayed on the AVO Web site (see Figure 5) within minutes to hours after collection, this image was seen before the ASTER data and initial descriptions of the activity on April 26 were based on the AVHRR data. The collection of both these datasets within hours of each other provided a rare opportunity to compare detailed ground data/features at the 15- to 90-m scale to what was imaged by the AVHRR instrument at 1-km spatial scale. AVHRR had 23 thermally-elevated pixels, five of which were saturated. The poor spatial resolution did not allow the discrimination of the two lava flows; however, the wide area of hot AVHRR pixels in conjunction with two recovery pixels at the far eastern edge of the thermally elevated area indicated that a new active lava flow was likely present, which was then later confirmed in the ASTER data. Recovery pixels are defined as areas with anomalously low temperatures that commonly occur on the down-scan margin of very high temperature thermal features (Higgins and Harris, 1997). This adverse response of the detectors to a zone of high radiance is produced by a slight delay during which time the previously saturated detector elements equilibrate and no data are collected.

A cold plume extending to the NE in the AVHRR image (and not seen in the ASTER image) indicated the volcano was continuing to produce larger eruptions every few hours during the lava flow emplacement phase. These plumes were carried in different directions (SE in ASTER, NE in AVHRR) depending on the local wind at the time. Once the ASTER data were available, it was confirmed that lava was now being emplaced in a new direction and that the previous lava flow was beginning to cool. This new lava flow was active and large enough to be seen in the 15m VNIR data (Figure 10C). At that point the open channel was 15-30 m wide, 3 km long and had a pixel-integrated brightness temperature of 852 C. To extract brightness temperatures in daytime VNIR or SWIR data, the solar reflected contribution in each pixel had to be removed. This was done by calculating that amount in each wavelength band using non-thermally-elevated pixels from a nearby region under similar lighting conditions (Rose and Ramsey, 2009). The flow was being emplaced to the north, in the large volcano shadow because of the low sun angle at this latitude and this time of year. Therefore, the solar correction was minimal and the extracted brightness temperatures from the VNIR and SWIR data were much more accurate.

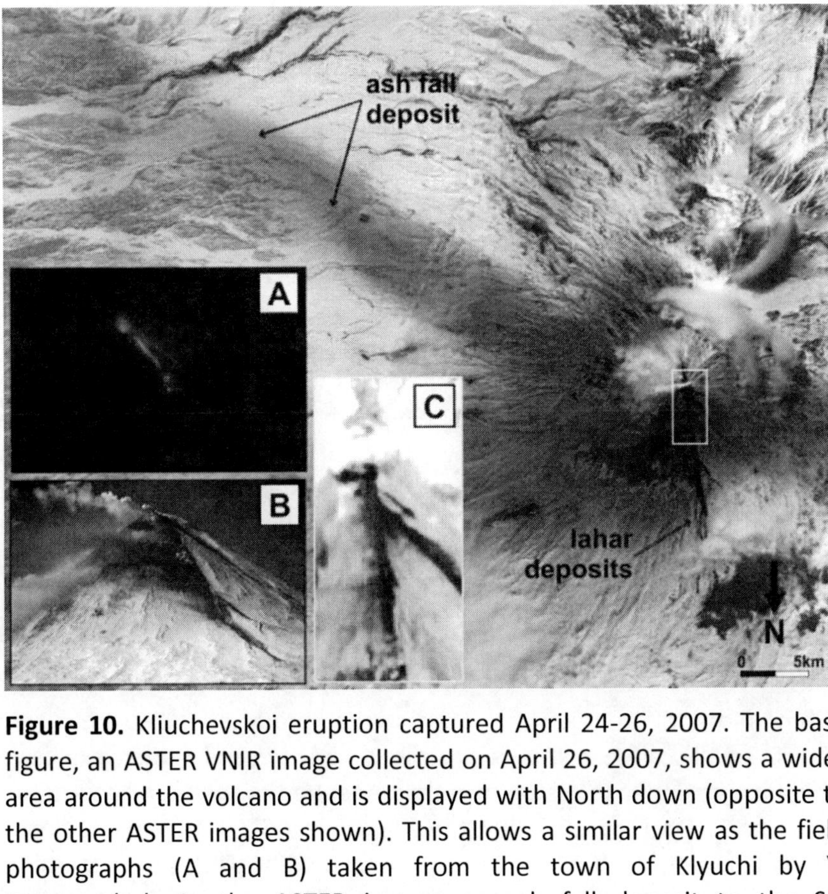


**Figure 10.** Kliuchevskoi eruption captured April 24-26, 2007. The base figure, an ASTER VNIR image collected on April 26, 2007, shows a wider area around the volcano and is displayed with North down (opposite to the other ASTER images shown). This allows a similar view as the field photographs (A and B) taken from the town of Klyuchi by Y. Demyanchuk. In the ASTER image, an ash fall deposit to the SE, numerous lahar deposits to the N, and a vigorous steam plume at the summit are all visible. (A) Nighttime photograph of the summit taken on April 24, 2007 showing the Strombolian activity and two distinct lava flows. (B) Daytime photograph taken one day later (and one day before the ASTER image) showing both lava flows and the dendritic patterns of the lahar deposits all of which are clearly visible in the ASTER image. (C) Enhanced linear stretch of the ASTER VNIR data of the summit region (denoted by the yellow rectangle). The active incandescent lava flows are seen and the maximum brightness temperature extract from these data was 852 C.

This dataset also marked the start of approximately two months of nearly cloud-free ASTER data where the locations of the lava flows and their VNIR-derived temperatures were tracked and used for detailed monitoring of the eruption (Figures 9-11). The extracted pixel-integrated brightness temperatures were divided into three phases of activity, which also correlated with visible observations and seismic data. In the precursory phase (November 2006 to February 2007), TIR temperatures began to rise above the average background temperature (approximately -40 C in January) and became hot enough to be detected by the SWIR data. This phase was dominated by fumarolic degassing and minor Strombolian activity at the summit. In the explosive phase (February 2007 to March 2007), degassing became more intense and Strombolian explosions at the summit were nearly constant. However, the amount of lava was not large enough to saturate either the TIR or SWIR data. Beginning in mid-March 2007 and continuing until mid-June 2007, the effusive phase was ongoing with the emplacement of three large basaltic andesite lava flows. Each of these flows contained an active center channel for long periods of time resulting in the saturation of both the TIR and SWIR data and allowing temperatures to be extracted from the 15-m VNIR data. These pixel-integrated temperatures are slightly lower than similarly derived temperatures for Hawaiian lava flows indicating that the actual temperature of the lavas were between 1050 and 1100 C (Ramsey and Wessels, 2007).

From mid-April to mid-June, Kliuchevskoi produced three new lava flows, small-scale Strombolian summit activity, and larger explosive eruptions resulting in ash plumes that extended to the east hundreds of kilometers. On May 29, 2007 the nighttime ASTER TIR image clearly showed the three N-NW lava flows and a weak TIR signal extending approximately 400 m to the SE of the summit crater. This linear thermal anomaly was predicted to be the start of new lava flow direction but was not confirmed until the next ASTER image pair on June 6-7, 2007 (Figure 11). The prediction of this new flow direction was initially discounted by most observers/scientists because of the high level of activity ongoing to the north and lack of historical lava flows emanating from the summit in this direction. The data collected on June 6, 2007, verified that the northern lava flows were no longer active. Their temperatures (between 10-20 C above the background temperature) had cooled well below the detection threshold for ASTER VNIR and SWIR but could still be discerned in the TIR image. The most obvious change was the presence of two new SE trending active lava flows in the exact direction as the linear thermal anomaly seen in the May 20, 2007, image. The larger flow was 3.1 km long and was emplaced at an average rate of 16 m/hr, which was nearly the same flow rate as the northern flows. However, the flow rate increased significantly near the end of the effusive phase of the 2007 eruption. The flow rate was calculated by examining changes in the small flow south of the larger SE flow (Figure 11B). ASTER collected a nighttime image 13 hours after the data shown in Figure 11and the advancement of this flow was easily seen. Using the digital elevation model derived from the VNIR image (Figure 11A), the slope in this region was calculated and used to derive a flow rate of 26 m/hr. This flow rate continued for the rest of June, after which the effusive phase ended. The lava flows continued to cool over the next several months and no new explosive activity was observed. The eruption was officially declared over in July 2007.

During the six month long 2007 eruption of Kliuchevskoi, ASTER provided unprecedented views of the summit activity. The eruption began in the winter, making ground or helicopter observations nearly impossible because of the harsh conditions. As the weather warmed, cloud cover further limited ground views of the summit. Thirty-two ASTER images were acquired during the day and night over this time period, which was an average of one scene every six days, although many images were day/night pairs collected 13 hours apart. Of this total, three day/night pairs (January 4-5, March 17-18, May 4-5) were collected as a result of the automatic triggering of the ASTER urgent request system. An additional seven image pairs were also acquired using the manual tasking of this urgent request system. The manual tasking was used during cloudy periods where AVHRR data were limited or to augment the volume of automatically tasked images. The ASTER URP program therefore produced a 63% increase in the data volume during the eruption of Kliuchevskoi, making it an invaluable tool for monitoring. The twelve additional ASTER images collected during this time period were part of the routine volcano monitoring performed by ASTER for all active volcanoes around the world (Pieri and Abrams, 2004).

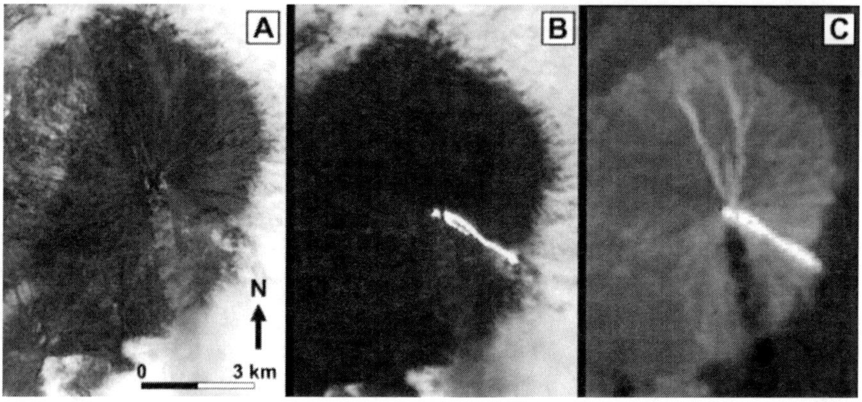


**Figure 11.** ASTER urgent request data collected on June 6, 2007 over the summit region of Kliuchevskoi volcano. (A) VNIR image now showing the lack of snow on the upper flanks of the volcano and an ash-rich plume from the summit crater and extending southward. Note the low level cloud deck surrounding the volcano at ~2500 m. Such clouds are common starting in the spring and completely obscure ground observations. ASTER provided the only detailed record of eruptive activity during this period. (B) SWIR band 4 image covering the same area. The summit crater and SE trending lava flow are clearly visible, as is a secondary breakout flow to the south of the larger flow. Using the ASTER-derived DEM of the summit and the difference in the flow lengths between the day and nighttime images ($\Delta T$ = 13 hours), the flow velocity was calculated to be 26 m/hour. This was significantly faster than the previous lava flows to the N and NW. (C) TIR band 10 image showing the active lava flow, the colder plume, and the two previous lava flows (N and NW) that were still cooling but no longer visible in the VNIR or SWIR data.

The numerous datasets provided by ASTER as part of the collaborative rapid-response program in conjunction with the relatively clear weather and the summit elevation and high latitude of Kliuchevskoi resulted in the largest and most comprehensive multispectral/multispatial high resolution dataset of a volcanic eruption. These data provide another means to monitor and characterize eruptive activity of this region of the globe. Precursors of eruption onset and new behavior can be better recognized using high resolution image data across the wavelength region and thus minimize future risks.

# 6. CONCLUSIONS

The value of using remote sensing assets in geoscience applications has been demonstrated by many authors and detailed here for the north Pacific volcano monitoring program. Numerous successes have been realized using this programmatic approach for capturing higher temporal frequency data, and a more rapid dissemination of critical information has been established for integration with future higher resolution datasets. The work described enabled the incorporation of higher spatial resolution ASTER data into an existing coarser resolution volcano monitoring initiative. New techniques were developed for event detection, satellite tasking, data distribution, and data analysis. These resulted in more rapid data availability, more detailed information, greater scientific insight on geologic processes, and more reliable alerts to the communities involved.

Specifically for Kamchatka, several beneficial factors have combined resulting in nearly 1,400 ASTER images of the five most thermally active Kamchatka volcanoes (Bezymianny, Karimsky, Kliuchevskoi, Sheveluch and Tolbachik). These factors include the orbital alignment of Terra, the high latitude of the peninsula, and the persistent activity in this region. From the inception of the automated rapid response program in 2004, an additional 350 scenes have been acquired over these volcanoes, many soon after larger eruptions. These data have produced valuable quantitative information on the small-scale activity and larger eruptions. A detailed example of the 2007 eruption of Kliuchevskoi described here enabled fundamental lava flow parameters to be determined. Numerous eruptions have been observed in Kamchatka by ASTER, which have displayed varying volcanic styles including basaltic lava flow emplacement, silicic lava dome growth, pyroclastic flow emplacement, volcanic ash plume production, fumarolic activity, and geothermal emission. The high spatial resolution and moderate spectral resolution of the data are ideal for deriving the energy flux from both high and low temperature systems, mapping chemical and textural changes of the volcanic products, and for imaging and understanding recent volcanic deposits.

The international collaboration developed for this work created professional relationships and infrastructure that will prove valuable in future work. The focus of further research will be on specific eruptions of the Kamchatka and Alaska volcanoes, the science results stemming from those data, expansion plans for global ASTER urgent request data and support for other types of events. The current ASTER rapid response program in Kamchatka and Alaska has produced a large archive of data, which has only been sampled to a small degree. These data offer a source for both the timely completion of current studies and new scientific analysis. It has also improved the timeliness and reliability of resulting hazard notifications to responsible authorities and affected communities. However, it should be noted that the ASTER instrument has long exceeded its initial design life. The SWIR subsystem has now failed and the sensor could suffer further catastrophic losses at any time. Therefore, there is a critical need for a continuing series of sensors with similar characteristics to ASTER in order to support such geoscience applications.

## 7. ACKNOWLEDGEMENTS

The URP project described here was made possible by the efforts and support of the ASTER Science Team and through NASA funding to M. Ramsey (grants: NNG04GO69G and NNX08AJ91G).The authors would like to thank S. Rose for her detailed research on the Kliuchevskoi ASTER data and helping to create Figure 9. The authors greatly appreciate technical review comments received from Zhong Lu and Russell Rykhus, and editorial assistance from Tom Adamson and Aleksandar Lazinica. ASTER data are courtesy of NASA, GSFC, Japan's Ministry of Economy and Industry, ERSDAC, Japan Resources Observation System and Space Utilization Organization (JAROS), the U.S./Japan ASTER Science Team, and LP DAAC at the USGS EROS Center.

## REFERENCES

1. M. Abrams, 2000 The Advanced Spaceborne Thermal Emission and Reflectance Radiometer (ASTER): data products for the high spatial resolution imager on NASA's Terra platform. International Journal of Remote Sensing, 21, 847–859.
2. A. J. Carter, M. S. Ramsey, A. B. Belousov, 2007 Recent crater formation at Bezymianny Volcano lava dome: Significant changes observed in satellite and field data, Bull. Volc.,
3. A. J. Carter, O. Girina, M. S. Ramsey, Y. V. Demyanchuk, 2008 ASTER and field observations of the 24 December 2006 eruption

of Bezymianny Volcano, Russia, Rem. Sens. Environ., 112 2569 2577

4. V. Chebrov, 2008 Complex seismological and geophysical investigation of Kamchatka and Commander Islands. Annual report of KB GS RAS. Ed. V. Chebrov. 268 (in Russian).

5. K. Dean, M. Servilla, A. Roach, B. Foster, K. Engle, 1998 Satellite monitoring of remote volcanoes improves study efforts in Alaska, EOS, Trans. Amer. Geophys. Union, 79 413 422 423.

6. J. Dehn, K. G. Dean, K. Engle, 2000 Thermal monitoring of north Pacific volcanoes from space, Geology, 28 755 758.

7. J. Dehn, A. J. L. Harris, (in press) Thermal Anomalies at Volcanoes. in: Monitoring Volcanoes in the North Pacific: Observations from Space, Dean K.G. and Dehn J. editors, Praxis/Springer.

8. T. Fournier, J. Freymueller, P. Cervelli, 2009 Tracking magma volume recovery at Okmok volcano using GPS and an unscented Kalman filter, J. Geophys. Res., 114, B02405

9. A. J. L. Harris, L. P. Flynn, L. Keszthelyi, P. J. Mouginis-Mark, S. K. Rowland, J. A. Resing, 1998 Calculation of lava effusion rates from Landsat TM data, Bull. Volc., 60 52 71 .

10. J. Higgins, A. Harris, 1997 VAST: A program to locate and analyse volcanic thermal anomalies automatically from remotely sensed data, Computers & Geosci., 23 n. 6, 621 645 .

11. A. Kaab, 2005 Remote sensing of mountain glaciers and permafrost creep, Physical Geography Series, Zurich, 48, 266 pages, 3-85543-244-9

12. Z. Lu, D. Dzurisin, C. Wicks, J. Power, O. Kwoun, R. Rykhus, 2007 Diverse deformation patterns of Aleutian volcanoes from satellite interferometric synthetic aperture radar (InSAR), in Volcanism and Subduction: The Kamchatka Region (edited by J. Eichelberger et al.), American Geophysical Union Geophysical Monograph Series 172 249 261 .

13. S. R. McNutt, 1994 Volcanic tremor amplitude correlated with the Volcanic Explosivity Index and its potential use in determining ash hazards to aviation. Acta Vulcanol. 5. 193 196.

14. T. P. Miller, V. Y. Kirianov, H. L. Kelley, 1994 Klyuchevskoy Fact Sheet. U.S. Geological Survey Fact Sheet. 94-067, 4 Also online (http://eq.giseis.alaska.edu/volcanoes/klyu/klyufact.html).

15. T. P. Miller, T. J. Casadevall, 2000 Volcanic Ash Hazards to Aviation. In: Sigurdsson, H., Houghton, B., McNutt, S.R., Rymer, H. and Stix, J., Editors, 2000. Encyclopedia of Volcanoes, Academic Press, San Diego, CA, 915 930.

16. D. C. Pieri, M. J. Abrams, 2004 ASTER watches the world's volcanoes: a new paradigm for volcanological observations from orbit. Journal of Volcanology and Geothermal Research. 135 13 28.

17. M. S. Ramsey, J. Dehn, 2004 Spaceborne observations of the 2000 Bezymianny, Kamchatka eruption: The integration of high-resolution ASTER data into near real-time monitoring using AVHRR, J. Volc. Geotherm. Res., 135, 1-2, 127 146.

18. M. S. Ramsey, J. Dehn, R. Wessels, J. Byrnes, K. Duda, L. Maldonado, J. Dwyer, 2004 The ASTER emergency scheduling system: A new project linking near-real-time satellite monitoring of disasters to the acquisition of high-resolution remote sensing data, Eos Trans. AGU, 85(47), Fall Meet. Suppl., Abstract SF23A-0026.

19. M. S. Ramsey, R. L. Wessels, 2007 Monitoring changing eruption styles of Kilauea Volcano over the summer of 2007 with spaceborne infrared data, Eos Trans. AGU, 88(52): Fall Meet. Suppl., Abstract 51H-07 .

20. M. S. Ramsey, S. Anderson, R. Wessels, 2008 Active dome and pyroclastic flow deposits of Sheveluch Volcano, Kamchatka: Unique thermal infrared and morphologic field observations, IAVCEI, General Assem. Prog., 67

21. A. L. Roach, J. P. Benoit, K. G. Dean, S. R. McNutt, 2004 The combined use of satellite and seismic monitoring during the 1996 eruption of Pavlof volcano, Alaska. Bulletin of Volcanology. 62 (6-7): 385 399.

22. S. R. Rose, M. S. Ramsey, 2009 The 2005 eruption of Kliuchevskoi volcano: Chronology and processes derived from ASTER spaceborne and field-based data, J. Volc. Geotherm. Res.,

23. D. Schneider, K. G. Dean, J. Dehn, T. P. Miller, V. Yu. Kirianov, 2000 Monitoring and analyses of volcanic activity using remote sensing at the Alaska Volcano Observatory: case study for Kamchatka, Russia, December 1997. Remote Sensing of Active Volcanism, Geophysical Monograph 116 65 85.

24. C. D. Stephens, B. A. Chouet, 2001 Evolution of the December 14, 1989 precursory long-period event swarm at Redoubt Volcano, Alaska: Journal of Volcanology and Geothermal Research, 109 n. 1, 133 148 .

25. R. L. Wessels, D. J. Schneider, M. L. Coombs, J. Dehn, M. S. Ramsey, (in press) 2006 High-resolution Satellite and Airborne Thermal Infrared Imaging of the 2006 Eruption of Augustine Volcano, Alaska, in Power, J.A., Coombs, M.L., and Freymueller, J.T., eds., Investigations of Augustine Volcano, Alaska after the 2006 eruption, U.S. Geological Survey, Professional Paper.

26. R. Wright, L. Flynn, H. Garbeil, A. Harris, E. Pilger, 2002 Automated volcanic eruption detection using MODIS, Rem. Sens. Environ., 82 135 155 .

27. Y. Yamaguchi, A. B. Kahle, H. Tsu, T. Kawakami, M. Pniel, 1998 Overview of Advanced Spaceborne Thermal Emission and Reflection Radiometer (ASTER). IEEE Transactions on Geoscience and Remote Sensing, 36, 1062–1071.

28. F. Yang, M. E. Schlesinger, 2002 On the surface and atmospheric temperature changes following the 1991 Pinatubo volcanic eruption; a GCM study, J. Geophys. Research, 107; 7 8.

# Chapter 3

## Habitat Suitability Modeling of Endangered Primates in Nigeria: Integrating Satellite Remote Sensing and Spatial Modeling Techniques

**Alex O. Onojeghuo[1,2*], Alan G. Blackburn[1], Francis Okeke[3], Ajoke R. Onojeghuo[2,4]**

*Alex O. Onojeghuo[1,2*], Alan G. Blackburn[1], Francis Okeke[3], Ajoke R. Onojeghuo[2,4]*

## ABSTRACT

This paper investigates the impacts of forest cover and spatial structure changes on the forest landscape across Afi-Mbe-Okwangwo protected area of Cross River State, Nigeria and its corresponding implication on two endangered primates (Cross River Gorilla and Nigeria-Cameroon Chimpanzee) habitat using satellite remote sensing and modeling techniques. Using remote sensing change detection analysis, the spatial extent and annual rate of deforestation for the study area was determined as 34,620 hectares and 1.5% respectively (from 2000 to 2014). The protected areas with highest annual deforestation rates were Afi Mountain Wildlife Sanctuary (2.6%) and Mbe Mountains (2.2%), both prominent for gorilla and chimpanzee sightings and nests. Further investigations on changes to the forest landscape structure revealed high levels of forest fragmentation across the study area for the 14-year period investigated. As a means of further understanding effects of forest landscapes changes across the study area, a 14-year forward simulation was performed using the

Markov model as to determine the spatial extent of futuristic forest cover changes. The results showed that if this current trend of forest cover change continued, 28,121 hectares of forests would be lost to deforestation in 2028 (approximately 16% of the total landmass of the entire study area). Using Maxent modeling, suitable primate habitats were predicted and the total coverage determined as 30,940 hectares (54.4% situated in CRNP—Okwangwo division, 29.4% in AMWS, 14.3% in Mbe Mountains and 1.9% in ARFR). Further analysis revealed 6468 hectares of predicted primate habitat were affected by deforestation in 2014 (21% of the predicted primate habitats). These results indicate that suitable primate habitats (particularly for gorillas and chimpanzees) are under immense pressure from deforestation and forest fragmentation. This paper presents a cost effective and time saving approach for determining suitable primate habitats and understanding the effects of forest transition on primate habitat suitability.

**Keywords:** Maxent, Remote Sensing, Primates, GIS, Deforestation, Markov, FRAGSTATS

## 1. INTRODUCTION

The negative impacts of forest loss and hunting activities across the world's tropical forests are sources of concern, as it affects the continued existence of endangered primates [1] -[3]. The Afi-Mbe-Okwangwo forest landscape is a biodiversity hotspot with known sightings of four Red List endangered primates namely: the Nigeria-Cameroon Chimpanzee Pan Troglodytes ellioti, Cross River Gorilla Gorilla gorilla diehli, Preuss's guenon Cercopithecus preussi and drill Mandrillus leucophaeus [4] -[6]. The area is also known to have suitable roosting sites for migratory birds [4]. However, this region of great ecological importance is faced with a range of human induced threats to the existing wildlife and forest ecosystem at large. Furthermore, the large and growing population of Cross River State depends greatly on subsistence agriculture, which has over the years resulted in high rates of forest loss and over-exploitation of natural resources [1]. This once undisturbed stretch of tropical forests in Cross River State is faced with diverse threats of deforestation in forms of large commercial plantations of plantain, rubber, cashew and pineapples [7] -[9] . In Africa, primates population have been most affected by loss and modification of forest habitats resulting from factors like intense logging, shifting cultivation, bush burning, over harvesting of fuel wood for commercial and household use, creation of grazing land and clearance for plantation farming [7] [10] -[12]. The interaction of high population growth rates and deforestation rates has caused a significant loss in the spatial coverage of remaining tropical forest landscapes across Africa [13] [14]. The study by [4] noted that primates, such as gorillas, are particularly vulnerable given their reliance on tropical forest habitats, slow rates of reproduction and massive body sizes. Hence, a better understanding of environmental factors such as forest cover distribution, pattern of forest landscape dynamics and modeling of species distribution would prove vital in enhancing conservation and proper management of forest protected areas across Africa.

The combined use of satellite remote sensing and ground truth data has shown to be an effective method of understanding changes in forest landscapes across the world [4] [15] -[18]. Using outputs from remote sensing in form of classified images, a better understanding of forest structure changes is possible, providing valuable results that are crucial to formulating strategies which assist in forest conservation and preservation of existing wildlife. A major problem identified in a number of conservation related studies is the negative impact of forest fragmentation on existing wildlife that relies on forested habitats for survival [19]-[21]. The study [20], assessed levels of forest fragmentation in the conservation of priority landscapes across parts of India using FRAGSTATS analysis. The investigation as to changes in the spatial structure of forest landscape provided a better understanding of forest fragmentation and its impact on protected areas across the study area.

An important component of this study was to perform a forward simulation of forest landscape transition in the study area as a means to create futuristic scenarios based on existing trends of forest cover dynamics. This would provide a direct representation as to the spatial and structural changes of forest landscape in the future, provided the current trend of forest transition continues. With such outputs, it would be particularly useful for wildlife conservation and decision makers involving in the day to day running and management of protected areas across the country. To this end, the Markovian process which had capabilities to model future state of landscapes using the immediate preceding state as a basis for forward prediction [22] was investigated in the study. [23] proposed making use of the cellular automata-Markov Chain model to simulate a counterfactual scenario of evaluating the role of forest conservation and investigate degrees of forest fragmentation protected areas in part of India.

Having a good understanding of wildlife species distribution and identifying factors that influence habitat selection would greatly assist wildlife conservation and management. Within Afi-Mbe-Okwangwo protected area landscape, delineating the current and potential distribution of endangered species, such as the Cross River Gorillas and Chimpanzees primates are critical to conservation management and planning. Part of this study shall investigate the use of presence-only data and other environmental variables for mapping suitable habitats for endangered primates in the study area. The Maximum Entropy modeling (Maxent), allows for suitable habitat distribution through combined use of presence data and environmental variables [24]-[28]. The advantages of Maxent modeling include high levels of relative insensitivity to positional errors of species location data, non-dependence on large datasets for predictive modeling and better performance levels when compared to other predictive modeling techniques [27]. Although we acknowledge the existence of other predictive models, emphasis of this study focuses on use of Maxent model in primate habitat distribution across parts of Nigeria. Since the Maxent modeling technique relies greatly on species presence data, collection of such information should be unbiased in nature and great care must be taken to avoid data duplication [25].

Taking into consideration the afore-mentioned, this study aimed to investigate impacts of forest cover and spatial structure changes on the forest landscape of Afi-Mbe-Okwangwo protected area in Cross River State, Nigeria particularly its corresponding implication on two endangered primates (Cross River Gorilla and Nigeria-Cameroon Chimpanzee) habitat using satellite remote sensing and modeling techniques. The key objectives of the study were: 1) to examine the changes in forest cover and spatial structure across the study area using satellite remote sensing and ground data; 2) to perform predicted forward simulation of forest cover changes across the study area as a means of projecting possible impacts of deforestation on existing forest landscapes; 3) to model habitat suitability of endangered primates using a combination of species presence data and other key environmental variables; and 4) to estimate the spatial extent and effects of deforestation on predicted primate habitats.

# 2. MATERIALS AND METHODS

## 2.1. Study Area

The study area, Afi-Mbe-Okwangwo forest landscape, is composed of four protected areas namely Afi Mountain Wildlife Sanctuary (AMWS), Afi River Forest Reserve (ARFR), Mbe Mountains and Cross River National Park (CRNP) (Okwangwo division) all situated in the northern part of Cross River State (Figure 1). The entire landscape covers an approximate area of 176,835 hectares and is geographically located between latitudes 6°04'19" to 6°28'35" and longitudes 8°49'55" to 9°21'47".

The AMWS is the westernmost protected area that serves as home to estimated 25 - 30 Cross River Gorilla sightings. The sanctuary created in 2000 was established to protect wildlife in approximately 100 km2 of lowland and hill forests in the north-western corner of the 380 $km^2$ Afi River Forest Reserve (ARFR) [3]. The protected area is currently managed by the Cross River State Forestry Commission with assistance from the environmental NGO Wildlife Conservation Society (WCS). Other NGOs that have been involved region include Fauna and Flora International, Nigerian Conservation Foundation and Pandrillus. The AMWS is dominated by rocky hills with sparse vegetation and large grasslands making it a suitable roost for migratory European barn swallows (Hirundo rustica). [4] estimates 20 million winter birds, particularly swallows, roost in the suitable grasslands of the AMWS each year. The sanctuary is surrounded by sixteen villages and farms which has led to cases of encroachment beyond its boundaries, scattered logging in the lowland areas, occurrence of forest fires during farm clearing by farmers in farming seasons, poaching of endangered wildlife (such as the Cross River Gorilla, chimpanzee etc.) and over exploitation of non-timber forest products such as herbs for commercial and household use [1] [3] [4] [29].

To the east of the Afi Mountains and adjourning the ARFR is the Mbe Mountains protected area, a forest dominated hilly landscape with sightings of gorillas since 1983 was established a wildlife sanctuary in 2000 [30]. It covers an estimate area of 100 $km^2$ and has hills rising to elevation of 900 metres. To the northern, western and southern parts of Mbe Mountains are nine villages (namely, Kanyang 1, Kanyang 2, Bamba, Abo Mkpang, Abo Obisu, Abo Ogbagante, Bokalom, Wula 1 and Wula 2) which have led to problems of encroachment by surrounding

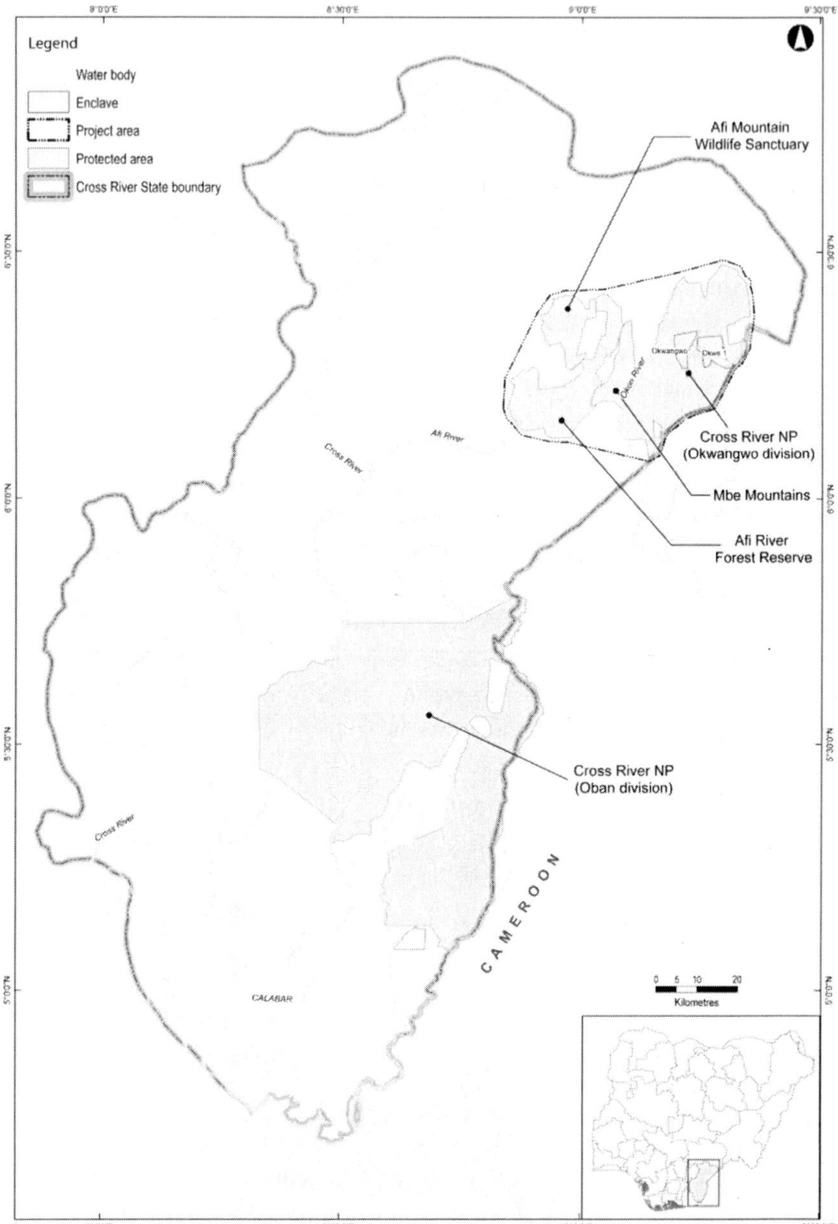


**Figure 1**. Map of study area showing protected areas investigated and an insert map of Nigeria showing Cross Rivers State.

communities. The Mbe Mountains is an important corridor between AMWS and east of the CRNP (Okwangwo Division). The Mbe Mountains has been managed as a community wildlife sanctuary by the Conservation Association of the Mbe Mountains (CAMM) established by the earlier mentioned nine communities. The community wildlife sanctuary is managed by CAMM and supported by the CRS Forestry Commission and Wildlife Conservation Society (WCS). As is the case with AMWS, similar challenges from agricultural activeties, poaching, bush burning and fragmentation of existing forest landscapes are key issues of concern in the protected area. [30] confirmed the severity of threats from hunting, farm encroachment, fire and logging as threats to the ecological integrity of the landscape which left unchecked could affect the continued existence of the Cross River Gorilla and other primates present.

Further to the east of Mbe Mountains, divided by the Okon River is CRNP Okwangwo Division. The CRNP located in south eastern Nigeria consists of two divisions, the Oban division to the south of Cross River state and the Okwangwo division to the north as shown in Figure 1. The Okwangwo division of the CRNP, created in 1991, covers an approximate area of 4000 $km^2$ and comprises of the former Boshi, Okwangwo, Boshi Extension Forest reserves and some inhabited enclaves (Okwangwo and Okwe 1) [2] (Figure 1). The CRNP Okwangwo division landscape, dominated by mixture of primary and secondary high forests and savanna woodland formations, has a high biodiversity particularly with respect to primates like the Cross River Gorilla [31]. Although the wildlife in the park are in theory under protected status, hunting in the park is widespread and cases of fire in the montane forests on the edge of the Obudu Plateau a reoccurring event [3]. The gorilla habitats in the CRNP (Okwangwo division) are under immense threat from farming encroachment as the population within both enclaves has greatly increased over the years.

## 2.2. Field Data Collection

The field data on endangered primates (gorillas and chimpanzees) presence across the study area was collected using cyber-tracking technology employed by patrol officers of WCS and CRSFC [32]. The signs of gorilla and chimpanzee presence were based on vocalisation; presence of nests, food remains, dung and direct encounters. The use of cyber tracking based systems was introduced in the Mbe Mountains in

2009 and has since allowed for ease in tracking gorillas across the protected area since [32]. Other sources of endangered primate sightings were obtained from the WCS annual reports available on the official website (http://www.wcsnigeria.org/). The gorilla and chimpanzee presence data from 2011 to 2014 was used for this study. Past studies on gorilla distribution in Cross River observed that combining earlier surveys did not alter the current distribution of the primates, as recent surveys showed that all earlier known areas were still occupied [4].

## 2.3. Forest Cover Mapping Using Remote Sensing

The change in the forest landscape across the study area was performed using satellite remote sensing data for two epochs (2000 and 2014). For this study Landsat 7 Enhanced Thematic Mapper (ETM+) with World Reference System path 187, row 056 (10 December 2010) and UK-DMC satellite data (7 January 2014) were used. The Landsat data was downloaded from the United States Geological Survey GLOVIS web-page (http://glovis.usgs.gov/) while the UK-DMC satellite data was supplied by Nigeria Space Agency, NASRA (National Space Research and Development Agency), based on sharing agreement with UK partners (Disaster Monitoring Constellation Company, DMCii). The Landsat image was processed to Level 1 Terrain-corrected, implying it was already ortho-rectified making it a suitable product for direct image-to-image comparison in change detection analysis. Similarly the UK-DMC satellite image was radiometrically and geometrically corrected. Since the images were already radiometrically and geometrically corrected on delivery, the Landsat and UK-DMC images were atmospherically corrected [33] and the digital numbers converted to ToA (i.e. Top of Atmosphere Reflectance) [34] [35] . This conversion process was done to remove all forms of noise present in the satellite images (such as instrument error, errors from view changes during data acquisition and effects of illumination geometry) [34] [36] [37]. A similar approach was implemented on the UK-DMC image used in the study. Before image classification, Landsat bands with same spectral ranges as the UK-DMC image were selected for input during image classification (Figure 2).

For this study, the supervised maximum likelihood classifier (MLC) was applied to each satellite image. A total of six broad classes were used in the study, based on the Intergovernmental Panel on Climate Change land use classification scheme [38]. The six broad classes comprised of forestland, cropland, grassland, wetlands, settlements and other land classes. For purpose of this study theses classes were further re-categorized into two distinct classes namely Forest and Non-forest. The process of image classification and accuracy assessment was performed using independent training and testing data. The training and ground truth data were obtained using a variety of sources namely Google Earth, GPS (global positioning system) data over the study area, historic/recent aerial photographs and visual interpretation of the satellite images.

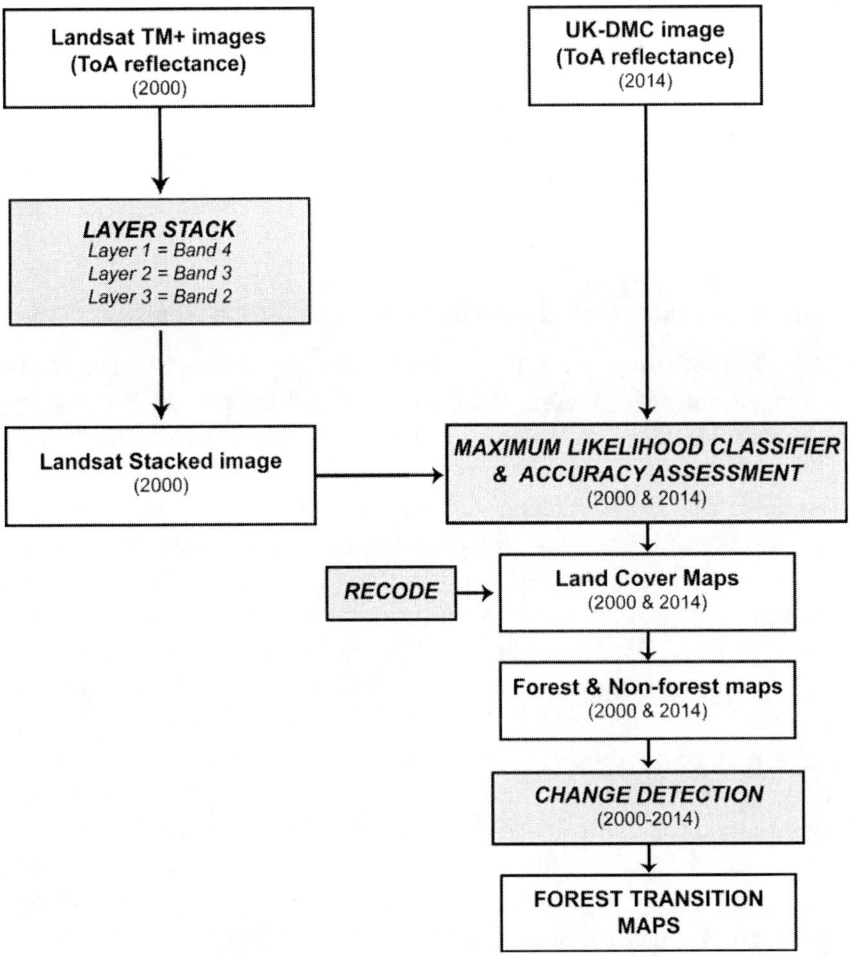


**Figure 2**. Procedure for post classification comparison using Landsat and UK-DMC satellite images.

## 2.4. Forest Cover Change Analysis

In this study, the IDRISI Selva Land Cover Modeler was used to analyses forest cover change over the study area. Using results of the forest cover maps for 2000 and 2014, forest cover change maps across the protected areas were determined for 2000 to 2014. The forest cover change maps for both time intervals showed deforested, unchanged forest and afforested landscapes classes. In order to calculate, the annual deforestation and forest changes for the two time periods were used.

## 2.5. Spatial Dynamics of Forest Landscape and Predictive Modeling

The open source software, FRASTATS (version 4.3.1.603), [39] was used to calculate the landscape and class metrics needed to better understand the spatial structure of forest cover classifications. The class metrics used in this study included the number of patches (NP), mean patch area (MPA) and patch density (PD) (which is the number of patches of a patch type divided by the total landscape area).

An important component of this study, with particular emphasis on the management of protected areas and eventual benefits to endangered primate habitats, was predicting the future status forest landscapes across the study area using current trends of forest cover transition. To this end, the Markov Chain Analysis and Cellular Automata Analysis (CA_MARKOV) module in IDRISI Selva was used to model forest cover scenario over a 14-year future projection (i.e. from 2014 to 2028). The Markovian process models the future state of any chosen landscape using the immediate preceding state as the basis for forward prediction [22]. For this study, the forest cover maps for 2000 and 2014 were used to model future predictions of forest cover over a 14-year period. Using the forest cover maps for 2000 and 2014 as inputs, the MARKOV module was used to produce the transition probability matrices and conditional probability images. To implement the CA_MARKOV module, the transition area text file, transition suitability image and 2014 forest map were used to generate a 14-year forward simulation of forest cover. The Cellular Automata iteration was specified as 14, allowing the model to be based on 1-year increments. The default cellular Automata filter of 5 × 5 was used for the study.

## 2.6. Maximum Entropy Modeling of Endangered Primates in Cross River State

The Maximum Entropy Distribution (Maxent) [26] [27] was used to model the presence of endangered gorilla and chimpanzee primates across the study area. Maxent is a machine learning programme that utilises maximum entropy to estimate species distribution probabilities' using the presence only data and environmental variables (categorical or continuous) [26]. For this study endangered primates' presence data across the study area and environmental variable (continuous and categorical) layers were used as inputs into the modeling process. The open source software Maxent version 3.3.3k (http://www.cs.princeton.edu/~schapire/maxent/) was used to generate estimates of probabilities for primates presence varying from 0 to 1 (where 0 is the lowest and 1 the highest probability respectively). The presence data, Global Positioning System (GPS) positions of gorilla and chimpanzee nests and sightings for 2009 and 2013 were sample layer inputs used in Maxent to model potential distribution of environmental suitability for the endangered primates. Topographical data used in modeling was taken from the Shuttle Radar Topography Mission Digital Elevation Model (SRTM DEM). The SRTM DEM was downloaded from the Global Land Cover Facility (GLCF) website (http://glcfapp.glcf.umd.edu:8080/esdi/).

The modeling of environmental suitability for endangered primates using the presence data of gorilla and chimpanzee sightings and nests and three key environmental variables namely, DEM (for terrain information), Slope map (derived from the DEM) and forest cover map of the study area. The recommended default values for maximum iteration numbers (50), convergence threshold (100,000) and regularization values (10,000) were used. The random test percentage of value used was 30 (meaning that 30% of the presence data was used for validation while 70% for training of the model). For analysis a threshold was specified to help perform binary predictions of habitat suitability conditions with suitable habitats predicted above such threshold and unsuitable indicated as below. In this study a convergence threshold of 0.00001 was used with maximum iterations of 500. The ASCII raster outputs of predicted primate habitats were further analyzed using ESRI ArcGIS software version 10.2 [40] .

# 3. RESULTS AND DISCUSSIONS

## 3.1. Classification Accuracy Results

The overall land cover classification accuracy results were approximately 89 percent (2000) and 91 percent (2014) respectively. Table 1 presents the classification accuracy reports for the classifications performed. The classification accuracy results were within VCS REDD Methodologies specification [41].

**Table 1**. Accuracy assessment results showing user accuracy (UA), producer accuracy (PA) and overall accuracy (OA) generated for classified images.

| Land Use | 2000 (ETM+) | | 2014 (UK-DMC) | |
|---|---|---|---|---|
| | UA (%) | PA (%) | UA (%) | PA (%) |
| Forest | 87.1 | 95.7 | 91.8 | 88.2 |
| Farmland | 88.6 | 87.8 | 90.9 | 86.2 |
| Grassland | 93.6 | 96.7 | 91.7 | 94.3 |
| Wetland | 86.4 | 88.4 | 90 | 87.8 |
| Settlement | 100 | 54.6 | 93.3 | 96.6 |
| Other land | 100 | 100 | 86.7 | 100 |
| *OA (%)* | *89.4* | | *90.8* | |

## 3.2. Changes in Forest Cover and Spatial Structure

The classification result shows that total forest cover for the entire study area was 152,509 hectares in 2000 and 129,339 hectares in 2014 (Table 2). The total area of forest cover across the project area landscape declined from 86% in 2000 to 73% in 2014. Overall, the four protected areas experienced a general decline in the spatial cover of forest landscape for the 14-year period investigated (Table 2).

The forest cover transition results indicated that overall deforestation for the study area from 2000 to 2014 was 34,620 hectares (Table 3). The total forest landscape that remained unchanged and afforested for the same period were 117,730 hectares and 11,574 hectares respectively. In terms of percentage landmass, 13.3% of the total forest landscape was deforested over the entire study area. Amongst the four protected areas, the most affected by deforestation were AMWS and Mbe Mountains with annual deforestation rates of 2.6% and 2.2%, followed by CRNP (1.1%) and ARFR (1.5%). The AMWS and Mbe Mountains, both known to have gorilla and chimpanzee sightings, are under immense pressure from illegal logging, encroachment farming and bush burning. The recent WCS annual report (2014) on AMWS and Mbe Mountains identified rise in the number of hunting incidents in the area, a major threat to the existence of endangered primate population. Furthermore, the close proximity of surrounding communities to the primate habitats has further heightened the pressures of farming encroachment and hunting activities, (particularly in forms of wire snares across the sanctuary). These factors have greatly contributed to the rise in deforestation and consequent disturbance of the gorilla and chimpanzee habitats across the protected area.

Using spatial metrics measures (NP, PD and MPA) over time, the changes in spatial structure over the 14-year period was investigated for the study area The NPs for the entire study area increased from 1255 in 2000 to 2843 in 2014. Similarly, the PD of the entire study area increased from 0.48 in 2000 to 1.12 in 2014 while MPA decreased from 122 hectares (in 2000) to 46 hectares (in 2014) (Table 2). The implication of these results was that the forest landscape across the study area experienced high levels of forest fragmentation over the 14-year period. Results of spatial structure dynamics for all four protected areas indicated that there were significant rises in NP and PD and a general decline in MPA (particularly for Mbe Mountain which experienced a sharp MPA decline from 4472 hectares in 2000 to 48 hectares in 2014). The effects of large decline in forest cover and increased levels of forest fragmentation over the 14-year period are indications of habitat loss, decreased connectivity of remaining vegetation (which prevents migration and movement of primates) and reduction in forest patch sizes (unsuitable for primate habitation). Other key deforestation drivers that have ravaged the Afi-Mbe- Okwangwo landscape include over-exploitation of non-timber forest products and occurrence of forest fires (a common practice in seasonal farming activities in the lowland forests) [1] [3] [4]. The combined effects of the afore-mentioned drivers of deforestation are serious threats to the conservation of the forest ecosystem, upon which existing primates depend for survival. Hence, the major challenge of the Cross River State Government and Nigeria at large is the establishment of suitable measures to combat or possibly mitigate these effects of de forestation across existing protected areas.

**Table 2**. Results of spatial structure changes for forest landscape in Afi-Mbe-Okwabgwo protected areas from 2000 to 2014.

| Zone | Forest Cover (ha) | | Number of Patches | | Patch Density (no. per 100 ha.) | | Mean Patch Area (ha) | |
|---|---|---|---|---|---|---|---|---|
| | 2000 | 2014 | 2000 | 2014 | 2000 | 2014 | 2000 | 2014 |
| Project area | 152,509 | 129,339 | 1255 | 2843 | 0.48 | 1.12 | 121.52 | 45.49 |
| AMWS | 9715 | 7418 | 76 | 474 | 0.35 | 2.18 | 127.83 | 15.65 |
| ARFR | 27,033 | 22,315 | 113 | 338 | 0.13 | 0.39 | 239.23 | 66.02 |
| Mbe Mountains | 8944 | 6728 | 2 | 141 | 0.01 | 0.67 | 4475.24 | 47.72 |
| CRNP (Okwangwo) | 61,828 | 53,614 | 80 | 496 | 0.06 | 0.37 | 772.85 | 108.09 |

**Table 3**. Results of forest cover transition for the Afi-Mbe-Okwangwo forest landscape from 2000 to 2014.

| Zone | Deforested (ha) | Unchanged forest (ha) | Afforested (ha) | Deforested (%) | Unchanged forest (%) | Afforested (%) | Annual deforestation rate (%) | Annual rate of change in forest cover (%) |
|---|---|---|---|---|---|---|---|---|
| Project area | 34,620 | 117,730 | 11,574 | 13.26 | 45.09 | 4.46 | 1.48 | 0.51 |
| AMWS | 3229 | 6486 | 933 | 14.84 | 29.80 | 4.29 | 2.61 | 0.84 |
| ARFR | 5559 | 21,474 | 830 | 6.43 | 24.85 | 0.96 | 1.48 | 0.59 |
| Mbe Mountains | 2311 | 6634 | 94 | 10.92 | 31.35 | 0.45 | 2.19 | 0.88 |
| CRNP (Okwangwo) | 9623 | 52,196 | 1388 | 7.17 | 38.89 | 1.03 | 1.06 | 0.44 |

To this end, the Nigerian Government is actively involved in the United Nations REDD+ (Reducing Emissions from Deforestation and Forest Degradation) programme aimed at preserving existing forest landscape across the country, particularly forest dependent communities [42]. As part of the REDD+ programme in Nigeria, portions of Afi-Mbe-Okwangwo protected areas have been assigned as one of the three main REDD pilot sites for Cross River State. The implementation of the REDD+ programme across the state would facilitate sustainable management of forest landscape and its resources whist promoting wildlife conservation. Other approaches include creating public awareness on the need to refrain from killing endangered primates and avoidance of human activities such as agricultural encroachment, forest fires, illegal logging and over-exploitation of non-timber forest products. Figure 3 shows the forest cover (2000 and 2014) and forest transition (2000-2014) maps of the entire study area.

### 3.3. Forest Cover Forward Simulation

Figure 4(a) shows the spatial extent of forest cover across the entire study area, including the four protected areas from 2000 to 2028. The results show that for the entire study area, forest landscape experienced a gradual decline over the 28 year period. The spatial extent of forest decline from 2000 to 2014 and 2014 to 2028 were 23,170 hectares and 41,338 hectares respectively. The results of forest landscape decline were representative of approximately 15% and 27% of the earlier forest cover in 2000. Figure 4(c) and Figure 4(d) show the 2014 forest cover map and 2028 forward simulated forest cover map of the study area. At a glance there's an obvious decline in the spatial coverage of forest cover in 2028.

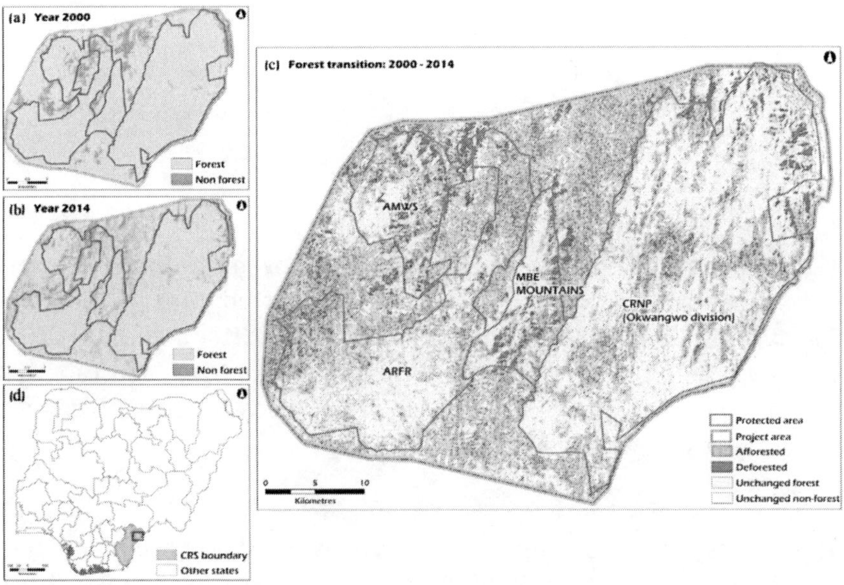


**Figure 3**. Map of study area showing: (a) Forest cover for 2000; (b) Forest cover for 2014; (c) Forest transition for 2000 to 2014 indicating deforested, unchanged forests, afforested and unchanged non forest landscapes across Afi-Mbe-Okwangwo protected areas in Cross River State; and (d) Insert map of Nigeria showing the study area.

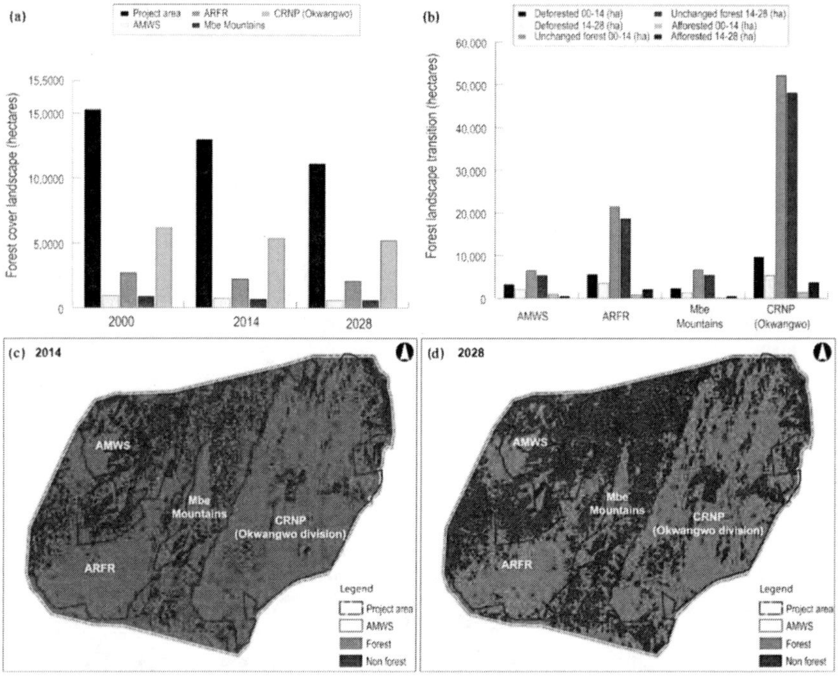


**Figure 4**. Results of forest transition analysis showing (a) Graph of forest cover landscape for 2000, 2014 and 2028 across study area; (b) Graph of forest transition across study area for 2000-2014 and 2014-2028, indicating deforested, unchanged forest and afforested landscape; (c) Forest cover map of 2014; and (d) 2028 predicted forest cover map generated using Markov module.

The results of forest cover change analysis for the entire study area based on the 2014 and 2028 predicted forest maps showed that 28,121 hectares of forest cover would be deforested. The spatial extent of unchanged forest and afforested landscape between 2014 and 2028 would be 101,183 and 9988 hectares respectively. These predicted results are based on the current trend of forest cover change driven by current forest management practices and pressures from human activities (such as farming encroachment, bush burning and illegal logging). Figure 4(c) shows forest landscape transition highlighting deforested, unchanged and afforested landscapes (for 2000 to 2014 and 2014 to 2028) across the protected areas. Overall, the results show that forest cover transition across the study area is prone to high levels of deforestation particularly outside and within the protected areas

(Figure 4(d)). Given the current trend of forest transition, measures should be put in place to mitigate associated threats to endangered primates and human population.

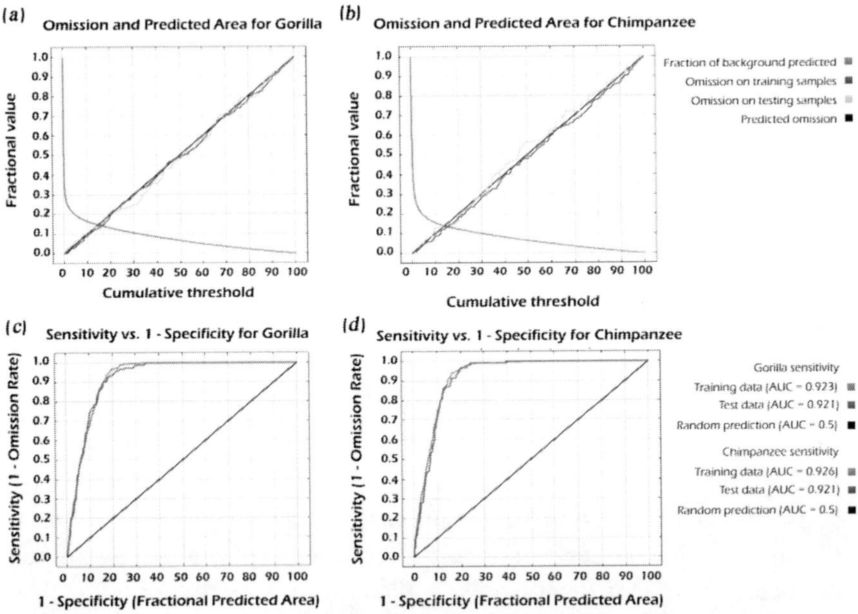


**Figure 5**. Results of statistical analysis of Maxent modeling for endangered species (Cross River Gorilla and Nigeria-Cameroon Chimpanzee) habitat suitability results of forest transition analysis showing (a) Graph of forest cover landscape for 2000, 2014 and 2028 across study area; (b) Graph of forest transition across study area for 2000-2014 and 2014-2028, indicating deforested, unchanged forest and afforested landscape; (c) Forest cover map of 2014; and (d) 2028 predicted forest cover map generated using Markov module.

### 3.4. Habitat Suitability Modeling of Endangered Primates Using Maxent

The omission and predicted area plots of the endangered primates (Figure 5) show how testing and training omissions and predicted areas vary with the choice of cumulative threshold. Figure 5(a) and Figure 5(b) showed that the omission on test samples (for both primates) was a very good match to the predicted omission rate. Figure 5(c) and Figure 5(d) showed the receiver operating curve (ROC) for both training and test data used in the modeling process. The red (training) line shows the "fit" of the model to the training data while the blue (testing) indicated the fit of the model to the testing data (note, this is the real test of the models predictive power). Given that the blue line in graph 5(c) and graph 5(d) were towards the top left of the graph, it was an indication that the model was better at predicting the presence contained in the test sample of the data [43] . The accuracy of the predictive modeling for both primates was determined by the test area under the ROC (AUC).

The AUC and standard deviation for the modeled prediction of gorillas and chimpanzees habitats were 0.921 (standard deviation 0.006) and 0.921 (standard deviation 0.007) respectively. The gorilla suitability habitat modeling was performed using 359 presence records for training and 153 for testing. For the chimpanzee habitat suitability modeling 217 presence records was used for training and 93 for testing. The environmental variables used in the model included the study area DEM, classified forest map and slope map datasets.

During the training phase of species prediction modeling Maxent software keeps track of environmental variables that contribute to fitting the model. Consistent with ecological knowledge the model predicted that Cross River Gorilla and chimpanzees are found mostly in forested areas (particularly tropical high and montane forests) in mountainous locations across the site. Figure 6(a) and Figure 6(b) show the predicted habitat suitability map outputs of endangered gorilla and chimpanzee primates generated using presence data and key environmental variables in Maxent modeling software. The most important variables in order of importance for both primates as indicated by the Maxent model were DEM, Slope derived map and Forest map of study area.

### 3.5. Effect of Deforestation on Predicted Primates Habitat

The raster datasets showing predicted primate habitats generated using Maxent model were imported into ArcGIS software for further analysis and visualization. The raster outputs were reclassified to binary images showing only predicted primate habitats with possible probability of occurrence (i.e. 0.4 - 1.0) (Figure 6(c)). The binary images were digitized and compared with current and historic primate presence data. The results indicated that the total coverage of predicted suitable habitat for endangered primates over the entire study area was 30,940 hectares (54.4% situated in CRNP—Okwangwo division, 29.4% in AMWS, 14.3% in Mbe Mountains and 1.9% in ARFR). The spatial coverage of modeled primate habitats for both gorillas and chimpanzees for

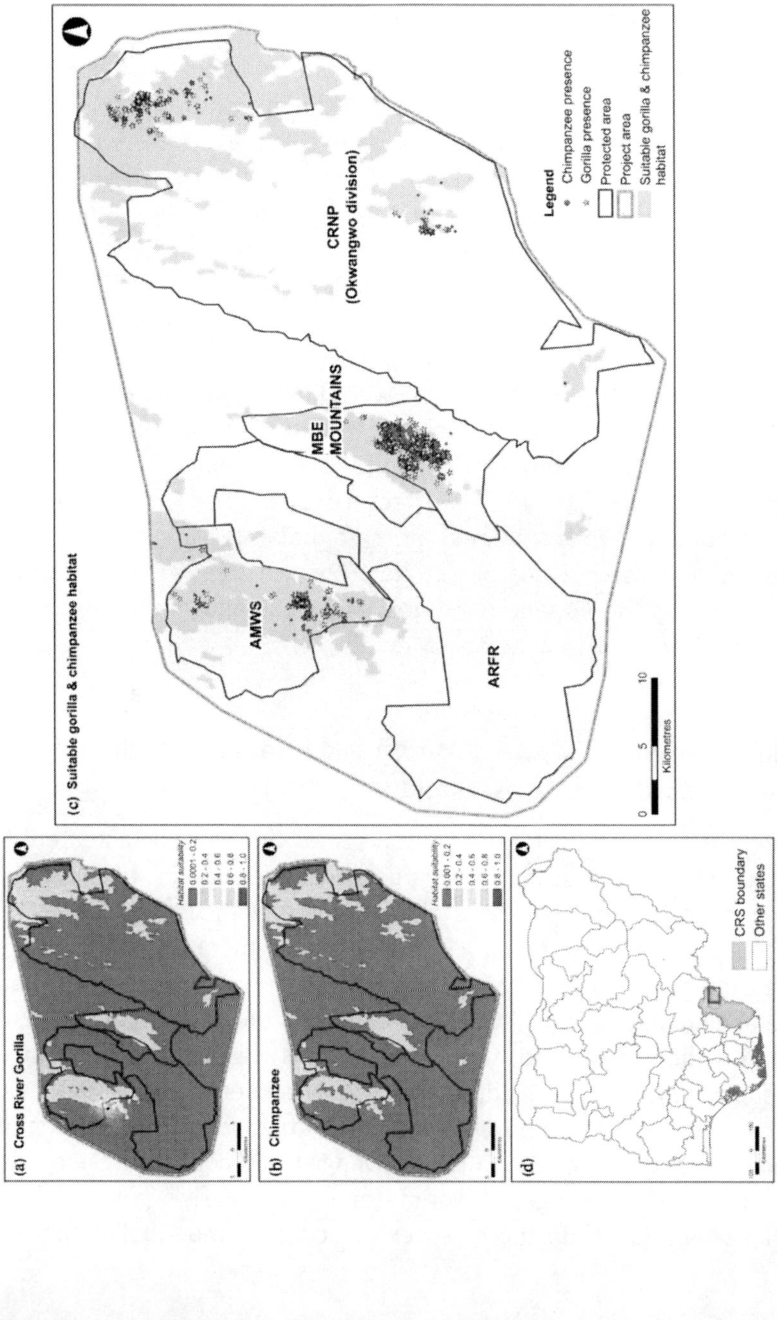


Figure 6. Map of study area showing Maxent predicted habitat suitability maps for (a) Cross River Gorilla and (b) Nigeria-Cameroun Chimpanzee; (c) Combined suitable gorilla and chimpanzee habitats and primate presence data from 2011 to 2014 across protected areas in the Afi-Mbe-Okwangwo forest landscape and (d) Insert map of Nigeria showing location of study area.

AMWS, ARFR, Mbe Mountains and CRNP (Okwangwo division) were approximately 9094 hectares, 573 hectares, 4428 hectares and 16,846 hectares respectively. These results would prove vital in the effective management and development of conservation plans for endangered primate habitats across the study area required by wildlife conservation organizations and the State government.

Following the modeling of suitable gorilla and chimpanzee habitats, the authors investigated the effects of deforestation trends (2000-2014) on the spatial coverage of predicted primate habitats across the study area. Using spatial analysis techniques in ArcGIS, the forest transition map (Figure 3(c)) and predicted primate habitat map (Figure 6(c)) were reclassified to binary images, each having two classes (deforestation map with 1 = deforested landscape & 0 = others landscape and primate habitat map with 1 = primate habitats & 0 = others habitats). Using the afore-mentioned binary raster, a map showing deforestation hotspots across the predicted primate habitat were generated (Figure 6(c)). The results indicated that approximately 6468 hectares of predicted primate habitats were affected by deforestation (21% of the predicted forested primate habitat) in 2014. The effects of deforestation and forest fragmentation on the predicted primate habitats reflects the poor management of forest landscape in the protected areas across the study area, justifying the urgent need for existing management practices across the protected areas to be reviewed

## 3.6. Implication of Forest Transition and Human Activities on Primates Habitat and Conservation

The high levels of forest fragmentation across the four protected areas investigated in this study are largely dependent on the current prevailing agricultural practices across the region. The presence of intensified pressure from agricultural encroachment by surrounding villages and communities are key catalysts to the disturbance of existing gorilla and chimpanzee habitats. Over the years the combined effects of forest fragmentation and deforestation has resulted in extensive restriction of primates to isolated and highly fragmented forest patches. This pattern of forest transition combined with uncontrolled hunting in protected areas such as the CRNP (Okwangwo division) has resulted in rapid loss of endangered primates (particularly gorillas and chimpanzees) and in drastic cases extinction of some wildlife species such as the grey-cheeked mangabey (Lophocebus albigena) and crowned guenon (Cercopithecus pogonias) [2] [44] . The problems of hunting and agricultural encroachment are further heightened by the presence of large enclaves in CRNP (Okwangwo division). Similar occurrences are observed in AWFS, ARFR and Mbe Mountains.

In the Mbe Mountains, a key catalyst asides from the above mentioned factors, is the presence of the Ikom—Obudu highway which traverses across Ago Ogbagante, Kayang-1, Kayang-2, Wula-1 and Wula-2 communities. Results of past studies [2] [4] and current interviews have noted a distinct pattern in forest clearance along the Ikom—Obudu highway for agricultural practice by surrounding communities.

The devastating impact of deforestation and other human related activities on remaining forest landscape and wildlife (in particular endangered primates) justifies the need for Cross River State (CRS) Government (in collaboration with the National Government) to put in place mitigative measures and policies/laws that would protect all wildlife and forests. In this regard, the CRS government on February 2014 established a mobile court under the CRSFC charged with the responsibility of prosecuting any individual found guilty of killing or hunting any animal (or endangered species) as listed in the CRS Forestry Commission Law No. 3 of 2010 [45] . Another avenue explored by CRS is the implementation of the UN-REDD+ programme. In 2010 CRS Government demonstrated its commitment to the REDD+ programme by enacting a deforestation moratorium, which is still in force till date, banning all forms of logging in the state [46] [47] .

## 4. CONCLUSIONS

The study has demonstrated the use of remote sensing data and techniques combined with various modeling approaches to understand effects of forest cover transitions on the spatial structure and prediction of suitable primate habitats. The results show that suitable habitats for gorillas and chimpanzees are under immense pressure from deforestation and high levels of forest fragmentation. Forest change analysis results showed that the overall extent of deforestation across the study area (2000 to 2014) was 34,620 hectares with annual deforestation rate of 1.5%. The protected area with the highest annual deforestation rates were AMWS (2.6%) and Mbe Mountains (2.2%), both prominent for gorilla and chimpanzee sightings and nests. The key drivers of deforestation across the study area were identified as illegal logging, over exploitation of non-timber forest products, hunting, farm encroachment, and forest fires. Further investigations as to changes in forest landscape structure revealed presence of high forest fragmentation levels, factors that had adverse negative effects on endangered primate and other wildlife habitat. As a means of further understanding the effects of the current forest cover transition trends, a forward simulation of forest landscape was performed for a 14-year forward prediction. The results showed that if the current trend of forest cover changes continued, a total of 28,121 hectares of forest cover would be deforested in the 2028 (approximately 16% of the total landmass of the entire study area). Using Maxent, suitable primate habitats were modeled and the total coverage of predicted habitat was estimated to be 30,940 hectares (54.4% situated in CRNP—Okwangwo division, 29.4% in AMWS, 14.3% in Mbe Mountains and 1.9% in ARFR). Further analysis as to the extent of deforestation in the predicted primates habitat revealed that 6468 hectares of forest cover had been deforested in 2014 (an estimated 21% of the predicted primate habitats). These results show that suitable primate habitats (particularly for gorillas and chimpanzees) are under immense pressure from deforestation and forest fragmentation.

This study has demonstrated that using remote sensing analysis combined with predictive modeling tools (such as Markovian and Maxent models) provides better understanding of forest transition and corresponding effects on primate habitats across parts of Africa. The approach presented in this study is a cost effective and time saving method of accurately predicting endangered primate habitats and understanding the dynamics of forest change and structure in the area. The information derived using this approach would assist conservation organizations, wildlife managers and the Government of Nigeria in making decisions and implementing policies that would help protect the forests and conserve endangered primates and other wildlife that are forest dependent for their existence.

In conclusion, the continued conservation of forests across African countries is paramount, as gorillas and chimpanzees alike depend on forest for survival. Hence, the only way to ensure their survival is to fight for the conservation of remaining forests. Such efforts will require support from government of African nations, Conservation Organizations, World Donor Agencies and most important the assistance of communities where such forests and endangered primates are found. The place of financial independence of forest protected areas can be promoted through ecotourism and infrastructure development [48] . Furthermore, implementing simple non-cap- ital intensive and sustainable projects on a local scale can promote self-sufficiency across forest dependent communities in Nigeria.

## ACKNOWLEDGEMENTS

The authors would like to appreciate Wildlife Conservation Society (Cross River) for providing the relevant field data on endangered primates across the study area and National Space Research and Development Agency (NASRDA) for making available the UK-DMC satellite image of Cross River State for image analysis.

## CITE THIS PAPER

Alex O.Onojeghuo,Alan G.Blackburn,FrancisOkeke,Ajoke R.Onojeghuo, (2015) Habitat Suitability Modeling of Endangered Primates in Nigeria: Integrating Satellite Remote Sensing and Spatial Modeling Techniques. *Journal of Geoscience and Environment Protection*, **03**,23-38.

## REFERENCES

1. Al-Quraishi, A.M.F., Guang, D.H. and Jian, G.C. (2004) Land Degradation Detection, Mapping, and Monitoring in the Northwestern Part of Hebei Province, China, Using RS and GIS Technologies. Map Asia Conference 2004, Beijing, 26-29 August 2004.
2. Angelsen, A. and Kaimowitz, D. (1999) Rethinking the Causes of Deforestation: Lessons from Economic Models. The World Bank Research Observer, 14, 73-98.
3. Baldwin, R.A. (2009) Use of Maximum Entropy Modeling in Wildlife Research. Entropy, 11, 854-66.
4. Bergl, R.A., Brenda, J.B, Nsubuga, A. and Vigilant, L. (2008) Effects of Habitat Fragmentation, Population Size and Demographic History on Genetic Diversity: The Cross River

Gorilla in a Comparative Context. American Journal of Primatology, 70, 848-59.

5. Bergl, R.A., Ymke, W., Aaron, N., Dunn, A., Inaoyom, I., Jacqueline, L. Sunderland-Groves and Oates, J.F. (2012) Remote Sensing Analysis Reveals Habitat, Dispersal Corridors and Expanded Distribution for the Critically Endangered Cross River Gorilla Gorilla gorilla diehli. Oryx, 46, 278-89.

6. Boyd, D.S., Foody, G.M. and Ripple, W.J. (2002) Evaluation of Approaches for Forest Cover Estimation in the Pacific Northwest, USA, Using Remote Sensing. Applied Geography, 22, 375-92.

7. Chander, G., Markham, B.L. and Helder, D.L (2009) Summary of Current Radiometric Calibration Coefficients for Landsat MSS, TM, ETM+, and EO-1 ALI Sensors. Remote sensing of Environment, 113, 893-903.

8. Cowlishaw, G. and Dunbar, R.I.M (2000) Primate Conservation Biology. University of Chicago Press, Chicago.

9. Elith, J., Graham, C.H., Anderson, R.P., Dudík, M., Ferrier, S., Guisan, A., Hijmans, R.J., Huettmann, F., Leathwick, J.R., Lehmann, A., Li, J., Lohmann, L.G., Loiselle, B.A., Manion, G., Moritz, C., Nakamura, M., Nakazawa, Y., Mc Overton, J., Peterson, A.T. and Phillips, S.J. (2006) Novel Methods Improve Prediction of Species' Distributions from Occurrence Data. Ecography, 29, 129-51.

10. ESRI (2014) ArcGIS Desktop: Release 10.2. Environmental Systems Research Institute.http://www.esri.com/

11. FAO (2010) Global Forest Resources Assessment 2010. Food and Agriculture Organization of the United Nations, Rome.

12. Geist, H.J. and Lambin, E.F. (2001) What Drives Tropical Deforestation. LUCC Report Series No. 4, 116 p.

13. Harcourt, A.H., Stewart, K.J. and Inahoro, I.M. (1989) Gorilla Quest in Nigeria. Oryx, 23, 7-13.

14. Harcourt, A.H. (1996) Is the Gorilla a Threatened Species? How Should We Judge? Biological Conservation, 75, 165-176.

15. Harcourt, A.H. and Doherty, D.A. (2005) Species-Area Relationships of Primates in Tropical Forest Fragments: A Global Analysis. Journal of Applied Ecology, 42, 630-637.

16. Huang, C., Wylie, B., Yang, L., Homer, C. and Zylstra, G. (2002) Derivation of a Tasselled Cap Transformation Based on Landsat 7 At-Satellite Reflectance. International Journal of Remote Sensing, 23, 1741-1748.

17. Imong, I., Eban, J. and Mengnjo, C. (2014) Using Technology to Save Gorillas in the Mbe Mountains. Gorilla Journal, 48, 15-17.

18. Iqbal, M. (2012) An Introduction to Solar Radiation. Elsevier, Amsterdam.

19. Irish, R.R. (2000) Landsat 7 Science Data User's Handbook. National Aeronautics and Space Administration, Report 430-15-01-003-0.

20. Ite, U.E. and Adams, W.M. (1998) Forest Conversion, Conservation and Forestry in Cross River State, Nigeria. Applied Geography, 18, 301-314.

21. Kehinde, A.L., Akande, J.A. and Ntabe, E. (2009) Forest Stakeholder's Awareness of Reduced Impact Logging (RIL) in Nigeria and Cameroon. Journal of Horticulture and Forestry, 1, 176-181.

22. Lu, D., Mausel, P., Brondizio, E. and Moran, E. (2002) Assessment of Atmospheric Correction Methods for Landsat TM Data Applicable to Amazon Basin LBA Research. International Journal of Remote Sensing, 23, 2651-2671.

23. Mambo, J. and Archer, E. (2007) An Assessment of Land Degradation in the Save Catchment of Zimbabwe. Area, 39, 380-391.

24. McGarigal, K., Cushman, S.A., Neel, M.C. and Ene, E. (2002) FRAGSTATS: Spatial Pattern Analysis Program for Categorical Maps.

25. Mengnjo, C. (2015) Wildlife Conservation Society Annual Report 2014—Cross River National Park (Okwangwo Divison). Wildlife Conservation Society.

26. Midha, N. and Mathur, P.K. (2010) Assessment of Forest Fragmentation in the Conservation Priority Dudhwa Landscape, India Using FRAGSTATS Computed Class Level Metrics. Journal of the Indian Society of Remote Sensing, 38, 487-500.

27. Mondal, P. and Southworth, J. (2010) Evaluation of Conservation Interventions Using a Cellular Automata-Markov Model. Forest Ecology and Management, 260, 1716-1725.

28. Morgan, B.J., Adeleke, A., Bassey, T., Bergl, R., Dunn, A., Fotso, R., Gadsby, E., Gonder, K., Greengrass, E., Koulagna, D.K., Mbah, G., Nicholas, A., Oates, J., Omeni, F., Saidu, Y., Sommer, V., Sunderland-Groves, J., Tiebou, J. and Williamson, E. (2011) Regional Action Plan for the Conservation of the Nigeria-Cameroon Chimpanzee (Pan troglodytes ellioti). IUCN/SSC Primate Specialist Group and Zoological Society of San Diego.

29. Myers, N., Mittermeier, R.A., Mittermeier, C.G., Da Fonseca, G.A.B. and Kent, J. (2000) Biodiversity Hotspots for Conservation Priorities. Nature, 403, 853-858.

30. Oates, J.F., Bergl, R.A. and Linder, J.M. (2004) Africa's Gulf of Guinea Forests: Biodiversity Patterns and Conservation Priorities. Advances in Applied Biodiversity Science, Vol. 6, Conservation International, Washington DC.

31. Oates, J.F., White, D., Gadsby, E.L. and Bisong, P.O. (1990) Conservation of Gorillas and Other Species. In: Caldecott, J.O., Oates, J.F. and Ruitenbeek, H.J., Eds., Cross River National Park (Okwangwo Division): Plan for Developing the Park and Its Support Zone, Appendix 1, WWF-UK, Goldalming.

32. Oates, J.F., Sunderland-Groves, J., Bergl, R., Dunn, A., Nicholas, A., Takang, E., Omeni, F., Imong, I., Fotso, R. and Nkembi, L. (2007) Regional Action Plan for the Conservation of the Cross River Gorilla (Gorilla gorilla diehli). IUCN/SSC Primate Specialist Group and Conservation International, Washington DC.

33. Oyebo, M., Bisong, F. and Morakinyo, T. (2015) A Preliminary Assessment of the Context for REDD in Nigeria.

34. Phillips, S.J., Anderson, R.P. and Schapire, R.E. (2006) Maximum Entropy Modeling of Species Geographic Distributions. Ecological Modelling, 190, 231-259.

35. Phillips, S.J., Dudík, M. and Schapire, R.E. (2004) A Maximum Entropy Approach to Species Distribution Modeling. Proceedings of the 21st International Conference on Machine Learning, Banff, 4-8 July 2004, 83.

36. Plumptre, A.J., McNeilage, A., Hall, J.S. and Williamson, E.A. (2002) The Current Status of Gorillas and Threats to Their Existence at the Beginning of a New Millennium. In: Taylor, A.B. and Goldsmith, M.L., Eds., Gorilla Biology: A Multidisciplinary Perspective, Cambridge University Press, Cambridge, 414-431.

37. Radespiel, U. and Bruford, M.W. (2014) Fragmentation Genetics of Rainforest Animals: Insights from Recent Studies. Conservation Genetics, 15, 245-260.

38. Ravilious, C., Kapos, V., Osti, M., Bertzky, M., Bayliss, J.L., Dahiru, S. and Dickson, B. (2010) Carbon, Biodiversity and Ecosystem Services: Exploring Co-Benefits. Nigeria: Preliminary Results. UNEP-WCMC, Cambridge, UK.

39. Saatchi, S., Buermann, W., ter Steege, H., Mori, S. and Smith, T.B. (2008) Modeling Distribution of Amazonian Tree Species and Diversity Using Remote Sensing Measurements. Remote Sensing of Environment, 112, 2000-2017.

40. Schoneveld, G.C. (2014) The Politics of the Forest Frontier: Negotiating between Conservation, Development, and Indigenous Rights in Cross River State, Nigeria. Land Use Policy, 38, 147-162.

41. Shoch, D., Eaton, J. and Settelmyer, S. (2011) Project Developer's Guidebook to VCS REDD Methodologies. Conservation International, Arlington.

42. Smith, P., Bustamante, M., Ahammad, H., Clark, H., Dong, H., Elsiddig, E.A., Haberl, H., Harper, R., House, J., Jafari, M., Masera, O., Mbow, C., Ravindranath, N.H., Rice, C.W., Robledo Abad, C., Romanovskaya, A., Sperling, F. and Tubiello, F. (2014) Agriculture, Forestry and Other Land Use (AFOLU). In: Edenhofer, O., Pichs-Madruga, R., Sokona, Y., Farahani, E., Kadner, S., Seyboth, K., Adler, A., et al., Eds., Climate Change 2014: Mitigation of Climate Change. Contribution of Working Group III to the Fifth Assessment Report of the Intergovernmental Panel on Climate Change, Cambridge

University Press, Cambridge, United Kingdom and New York, 811-922.

43. Fielding, A.H. and Bell, J.F. (1997) A Review of Methods for the Assessment of Prediction Errors in Conservation Presence/Absence Models. Environmental Conservation, 24, 38-49.

44. Souza, C., Firestone, L., Silva, L.M. and Roberts, D. (2003) Mapping Forest Degradation in the Eastern Amazon from SPOT 4 through Spectral Mixture Models. Remote Sensing of Environment, 87, 494-506.

45. CRS (2014) Cross River State of Nigeria Gazette. C.R.M. Court, Ed., No. 1, Ministry of Information, Calabar.

46. Suter, J. and Oates, J. (2000) Sanctuary in Nigeria for Possible Fourth Subspecies of Gorilla. Oryx, 34, 71-71.

47. Takada, T., Miyamoto, A. and Hasegawa, S.F. (2010) Derivation of a Yearly Transition Probability Matrix for Land-Use Dynamics and Its Applications. Landscape Ecology, 25, 561-572.

**48.** Ezebilo, E.E. and Mattsson, L. (2010) Socio-Economic Benefits of Protected Areas as Perceived by Local People around Cross River National Park, Nigeria. Forest Policy and Economics, 12, 189-193.

# Chapter 4

# Soil Salinity Mapping and Monitoring in Arid and Semi-Arid Regions Using Remote Sensing Technology: A Review

**Amal Allbed, Lalit Kumar**

*School of Environmental and Rural Science, University of New England, Armidale, Australia*

## ABSTRACT

Soil salinity is a serious environmental problem especially in arid and semiarid areas. It either occurs naturally or is human-induced. High levels of soil salinity negatively affect crop growth and productivity leading land degradation ultimately. Thus, it is important to monitor and map soil salinity at an early stage to enact effective soil reclamation program that helps lessen or prevent future increase in soil salinity. Remote sensing has outperformed the traditional method for assessing soil salinity offering more informative and professional rapid assessment techniques for monitoring and mapping soil salinity. Soil salinity can be identified from remote sensing data obtained by different sensors by way of direct indicators that refer to salt features that are visible at the soil surface as well as indirect indicators such as the presence of halophytic plant and assessing the performance level of salt-tolerant crops. The purposes of this paper are to 1) discuss some soil salinity indicators; 2) review the satellite sensors and methods used for remote monitoring, detecting and mapping of soil salinity, particularly in arid and semi-arid regions; 3) review various spectral vegetation and salinity indices that have been developed and proposed for soil salinity detection and mapping, with an emphasis on soil salinity mapping and assessment in arid and semi-arid regions; and 4) highlight the most important issues limiting the use of remote sensing for soil salinity mapping, particularly in arid and semi-arid regions.

**Keywords:** Soil Salinity; Remote Sensing; Halophytic Plant; Salinity Index

## 1. INTRODUCTION

According to the US Salinity Staff Laboratory, soils with conductivity of the saturation extract (EC) > 4 deciSiemens per meter (dS/m) at 25°C, Exchangeable Sodium Percentage (ESP) < 15 and pH (soil reaction) < 8.5 are referred to saline soils [1]. Salt in the soil mostly derives from the weathering of rocks and primary minerals, which formed in situ or transported by water or wind [2]. Other causes of soil salinity are topography, irrigation and dryland salinity, which occur due to forest clearance, overgrazing, and cutting bushes that cause water-tables to rise and bring saline groundwater close to the land surface. Thus, soil salinity categories are either primary salinity which is naturally occurring or secondary salinity which is human-induced.

Soil salinity is a prevalent environmental hazard in arid and semiarid regions around the world [3]. The United Nations Food and Agriculture Organization (FAO) has estimated that saline soil covered 397 million hectares of the total land area of the world [4]. Africa, Asia, Australia, Europe, Latin America, Near East and North America are the most affected areas [4]. According to the most recent estimates from a survey of farmers, about 2 million hectares and 20,000 farms across Australia alone showed some signs of salinity [5]. A recent study conducted by McFarlane et al. [6] has estimated that, in Western Australia alone, secondary salinity covers about 10% of the land, and could reach 23% if action is not taken to stop its spread.

Soil salinity adversely affects plant growth, crop production, soil and water quality, and it eventually results in soil erosion and land degradation [7-9]. Soil salinity impacts are not limited only to the environment but also extend to the economy. For instance, the economic losses due to secondary salinisation in Batinah region in Oman have been estimated at US$ 1604 $ha^{-1}$ (28%) when the salinity increases from low to medium level and US$ 4352 $ha^{-1}$ (76%) if it jumps from low to high level [10].

Basically, soil salinity is a dynamic process with severe consequences for the soil, hydrological, climatic, geochemical, agricultural, social, and economic aspects. Therefore, for greater development and implementation of sufficient soil reclamation programs and preventing any further salinization to sustain agricultural lands and natural ecosystems, information on the spatial extent, nature and distribution of soil salinity is becoming very essential. Thus, timely detection of soil salinity, monitoring and assessment of its severity level and extent become very important in its beginning at local and regional scales.

Conventionally, soil salinity has been measured by collecting in situ soil samples and analyzing those samples in the laboratory to determine their solute concentrations or electrical conductivity. However, these methods are time-consuming and costly since dense sampling is required to adequately characterize the spatial variability of an area [11-14]. Ghabour & Daels [15] agreed that detection soil salinity traditionally is time consuming, but remote sensing data and techniques offer more efficiently and economically rapid tools and techniques for monitoring and mapping soil salinity.

Remote sensing data and techniques have been progressively applied to monitor and map soil salinity since 1960s when black-and-white and color aerial photographs are used to delineate salt-affected soils [16]. Multispectral data such as Landsat, Satellite Pour l'Observation de la Terre (SPOT), IKONOS, QuickBird and the Indian Remote Sensing (IRS) series of satellites, as well as hyperspectrl data such as EO-1 Hyperion and HyMap, have been found to be useful in detecting, mapping, and monitoring soil salinity [17-23]. Generally, remote sensing uses the electromagnetic energy reflected from targets to obtain information about the Earth's surface with different levels of detail. So based on this concept, the spectral reflectance of the salt features at the soil surface has been widely studied using remote sensing and used as a direct indicator for soil salinity detection and mapping. However, when the soil moisture is high or the crust salt is invisible on the soil surface or mixed with other soil constituents, this direct approach becomes complicated and may yield unreliable results since these factors influence the soil spectral reflectance. But, the present scattered vegetation or halophytes on the soil surfaces can serve as a sign of the salinity problem, making it possible to indirectly detect and map areas that are affected by soil salinity using the reflectance from vegetation. Normally, unhealthy vegetation has a lower photosynthetic activity, causing increased visible reflectance and the reduced near-infrared reflectance (NIR) from the vegetation [24]. This pattern has been found in various plants subjected to salinity stress [25]. Therefore, based on this finding, several vegetation indices (VIs) such as Normalized Differential Vegetation Index (NDVI) and Soil Adjusted Vegetation Index (SAVI) have been used as indirect indicators assess and map soil salinity. Similarly, a number of researchers have developed different salinity indices to detect and map soil salinity such as Normalized Difference Salinity Index (NDSI) and Salinity Index (SI).

This review concentrates on the problem of soil salinity and discusses some soil salinity indicators and how remote sensing data and technologies are used for monitoring and mapping soil salinity. Additionally, it will discuss the most current vegetation and salinity indices used for soil salinity detecting and mapping, and it highlights some of the limitations and problems of using remote sensing for monitoring and mapping this hazard with an emphasis on soil salinity mapping and monitoring techniques for arid and semi-arid regions.

# 2. SOIL SALINITY AND REMOTE SENSING

## 2.1. Soil Salinity Symptoms

Soil salinity can be detected directly from remotely sensed data through salt features that are visible at the soil surface, such as bare soil with white salt crusts on the surface [19,26] or indirectly from indicators such as the presence of halophytic plant, the performance level of salt-tolerant crops [27-30].

### 2.1.1. Salt Features at the Soil Surface

The dynamic processes at the surface of saline soil limit the monitoring and assessment of the salinization process because they influence the spectral, spatial and temporal behavior of the salt features [31]. Via the physicochemical properties of soil such as soil moisture content, organic matter, soil texture, types of clay color and surface roughness soil spectral reflectance is determined [32-36]. Due to salinity these soil properties change which affect the spectral reflectance of features that occur at the soil surface, including salt crusts and efflorescence besides variations in surface texture and structure [37,38]. For example, Schmid et al. [37]found that crusted saline soil reflects strongly in the visible and near-infrared (NIR) bands; moreover, Rao Singh and Sirohi [39] noted that a crusted saline soil surface is generally smoother than a non-saline surface and exhibits high reflectance in the visible and NIR bands, which has been confirmed by Rao et al. [40]. On the other hand, Metternicht and Zinck [41] found that the reflectance in the visible and NIR bands is highly affected by both the crust color and surface roughness factors.

Despite the effects of salt features on the soil surface on the spectral reflectance, they have been considered good direct indicators of soil salinity. For example, Fernandez-Buces et al. [42] used surface features to predict soil salinity. They found that the correlation coefficient between surface colors, EC and the sodium adsorption ratio (SAR) were statistically significant, which suggested that efflorescence color is a promising surface indicator with which to estimate soil salinity.

#### 2.1.2. Presence of Halophytic Plants

Halophytic plants (salt-tolerant plants) are plants that tolerate high salt concentrations of the soil and can be grown on a salt affected land [43]. Although halophytic plants are common in saline areas, not all have been found to be good remote sensing indicators of soil salinity.For instance, Metternicht [44] found the spectral reflectance curve with high absorption in the visible range and high reflectance in the NIR range of halophyte Chenopodiaceae in Bolivia, to be equivalent to that of chlorophyll-rich vegetation. In contrast, due to lower chlorophyll content the spectral reflectance curve of C. dactylon, also a halophyte, increased continuously in the visible and NIR bands. This study concluded that halophytic plants were promise indicator to distinguish saline areas from non-affected ones.

#### 2.1.3. Crop Performance

The performance of some crops that can be grown on saline soils, such as alfalfa, barley, and cotton, reflect the severity of soil salinity. Cotton is largely cultivated on irrigated land, is therefore considered an ideal indirect indicator for soil salinity, so it has been used as salinity indicators in a variety of studies [31]. For example, based on the high correlations between the Normalized Difference Vegetation Index

(NDVI) values of cotton, sugarcane crops and the EC, Wiegand et al. [45,46] successfully assessed the severity and extent of soil salinity in terms of the economic impact on crop production and also distinguished saline soils from non-affected soils. This strong relationship most likely exists only where salinity is the major factor that causes crop yield variability; lands that suffer from soil salinity are likely to have other factors that affect yields as much or more than salinity, such as high or low temperatures, topography and land management. Therefore, deduction of the relationship between a number of saline fields and an entire landscape is likely to result in large errors, so the possible use of this indicator for determining the level of soil salinity must be checked carefully. To overcome this issue, some researchers have proposed using average crop production over a series of years to mask out the noise from non-soil factors that differ from year to year. For example, Lobell et al. [47,48] warn of the use of inter-annual changes on crop yield as an indirect indicator of soil salinity at regional scales because yield mapping that is limited to only one year does not always give a reliable estimator of soil salinity, particularly when strongly saline soils are scarce. Thus, through the use of a 6-year temporal series of satellite images of yield, they obtained a strong correlation between yield losses and soil salinity; they highlighted that yield loss in agricultural regions could be primarily due to several factors, including soil salinity.

## 3. MAPPING AND MONITORING SOIL SALINITY USING REMOTE SENSING DATA

Advantages of using remote sensing technology include saving time, wide coverage (satellite remote sensing provides the only source when data is required over large areas or regions), are faster than ground methods, and facilitate long term monitoring. These techniques provide multispectral image with resolutions that can be ranged from medium to high, as well as hyperspctral image. These remotely sensed data have been successfully used for monitoring and mapping soil salinity for decades with mixed results. Many researchers have used different techniques to monitor and map soil salinity using remote sensing data, as discussed below.

## 3.1. Multispectral Satellite Sensors for Mapping and Monitoring Soil Salinity

Extensive research using satellite imagery for mapping and monitoring soil salinity has been conducted over the last three decades, mostly with multispectral sensors. These include Landsat Thematic Mapper (TM), Landsat Multispectral Scanner System (MSS), Landsat Enhanced Thematic Mapper Plus (ETM+), SPOT, Advanced Spaceborne Thermal Emission and Reflection Radiometer (Terra-ASTER), Linear imaging self-scanning sensor (LISS-III) and IKONOS [23,49,50]. For example, in the United States of America (USA), Elnaggar and Noller [51] used Landsat TM imagery integrated with decision-tree analysis (DTA) to map soil salinity in central Malheur County. They found that there was a significant relationship between EC values and reflectance in Landsat bands 1, 2, 3 and 4 as well as the Brightness (BI) and Wetness (WI) indices. Maximum likelihood supervised classification was used to classify the image into nonsaline soils (EC < 4 dS/m) and saline soils, with accuracies of 97% and 60% respectively, whereas DTA predicted five classes of soil salinity with an overall accuracy of approximately 99%. Their results indicated that the use of Landsat TM imagery effectively identified bare soils that were characterized by high spectral reflectance due to a high salt content on the surface, and the approach of integrating DTA with remote sensing data was more accurate and effective compared to using remote sensing analysis alone.

Many researchers, including Katawatin and Kotrapat [52], Mehrjardi [53] and Yu et al. [54] have investigated the utility and effectiveness of ETM+ data for soil salinity mapping and monitoring. For example, in Thailand, Katawatin and Kotrapat [52] investigated the use of Landsat-7 ETM+ with different combinations of three sources of ancillary data (topography, geology, and underground water quality) for soil salinity mapping. A maximum likelihood classification method was employed in this study. Their results showed that the use of Landsat ETM+ data bands 4, 5 and 7 in combination with all three types of ancillary data yielded the most accurate soil salinity map, with 83.6% overall accuracy. Additionally, Douaoui et al. [55], Farifteh et al. [56] and Eldeiry and Garcia[57] agreed that an integrated approach using remote sensing techniques in addition to ancillary data such as field data, topography and spatial models geophysical surveys can improve the development of high quality soil salinity maps.

Using multispectral sensors for soil salinity research has also been studied by Goossens et al. [58]. Their study examined and compared the accuracy of Landsat TM, MSS, and SPOT XS imagery for soil salinity mapping. They found that Landsat TM was optimal for soil salinity mapping. Another comparative assessment of the suitability of multisensor data for soil salinity studies was conducted in Pakistan by Ahmed and Andrianasolo[59]. They compared the performance of Landsat TM and SPOT XS in mapping salinity at a semi-detailed level. Their results were completely opposite to that of Goossens et al. [58]. They found that the SPOT XS data were more helpful than Landsat TM as it provided finer details of various thematic variables.

Thermal band has proven to be a useful tool in soil salinity studies. It has a key role in differentiating saline soils, especially in areas of bare soil or sparse and similar vegetation. For example, Verma et al. [50] demonstrated that the addition of the thermal band of Landsat TM to the visible-NIR bands helped overcome spectral similarity issues with saline soils. Furthermore, thermal band was used to discriminate saltand sodium-affected soils by Metternicht and Zinck [41]. They found that the incorporation of the thermal band allowed for better salt and sodium detection. Furthermore, in a case study in Iran, AlaviPanah and Goossens [60] found that the addition of the thermal band to the best Landsat TM visible-NIR band combination had great potential for separating saline soil from gypsiferous soil. This study confirmed the result obtained by Goossens et al. [61], who reported the key role of the TM thermal band in separating gypsiferous soil from saline soils. In China, Huang et al. [62] used Terra ASTER imagery to identified saline areas dominated by sodium chlorides and sodium sulfates. Their results showed a good correlation between surface salt concentrations and band 1 of the ASTER sensor, followed by bands 2 and 3.

It cannot be denied that medium or low spatial resolution of the satellite images can limit the mapping and detection of saline regions, particularly when the affected areas are smaller than the pixel size. Thus, high resolution multispectral sensors with pixel size of less than 5 m are becoming an essential for soil salinity studies [23]. However, only limited attempts have been made to identify and map soil salinity problems using fine spatial resolution (0.6 m - 4 m) images that are available from IKONOS (4 bands) and Quickbird (4 bands) satellites, as well as WorldView-2, which has 8 multispectral bands at 1.84 m spatial resolution and one panchromatic band at 0.5 m spatial resolution [63].

This is most likely due to the higher cost of this higher-resolution imagery and these sensors being more recently developed systems. Elhaddad and Garcia [64] have used IKONOS satellite imagery and crop reflectance to identify the severity level of soil salinity and its effect on crop yield in Arkansas River Basin, Colorado. Image enhancement was used to separate the crop condition into several classes, and supervised classification was applied to delineate the different levels of soil salinity. They concluded that the superior effectiveness of their approach was primarily due to the use of high spatial resolution imagery. On the other hand, Dwivedi et al. [23] conducted a comparative study on the performance of IKONOS imagery and imagery from the IRS-ID LISS-III sensor for mapping salt-affected soils. Different image classification and transformation techniques were used in their study, and an overall accuracy of 92.4% was gained when using IKONOS data compared to an overall accuracy of 78.4% and 84.3% obtained when using the IRS-ID LISS-III multispectral sensor, which indicates the great potential of high spatial resolution IKONOS images for soil salinity mapping and detection. In South Australia, Setia et al. [22] detected and mapped soil salinity in an agricultural area using QuickBird imagery. They found that 99% of the variation in spectral values occurred in bands 2 and 4. Furthermore, they found that by dividing the image into hundreds of paddocks (small fields) and performing an unsupervised classification using a paddock-by-paddock approach, which is a procedure where each individual paddock is classified and mapped separately, they were able to map soil salinity severity levels more accurately. Thus, they concluded that a paddock-by-paddock classification approach for QuickBird imagery is a promising method for detecting degrees of salinity severity at a farm level. However, in spite of these promising results, it should be noted that it would extremely difficult to assess the validity and reliability of this approach at regional scales because it is a time-consuming and labor-intensive procedure. Thus, further investigations will help to elucidate whether and how a paddock-by-paddock classification approach will work at regional scales.

## 3.2. Limitations of Multispectral Satellite Sensors in Soil Salinity Mapping and Monitoring

Multispectral satellite sensors have been and still are the preferred method for mapping and monitoring soil salinity. This is primarily due to the low cost of the imagery (e.g., Landsat, SPOT) and the ability to map extreme surface expressions of salinity. Nevertheless, multispectral data has limited diagnostic capability because of its coarse spatial and spectral resolutions [65]. For example, Furby et al. [66] and Howari [67] reported that direct mapping of soil salinity with multispectral imagery had major limitations that arise, especially where there are no salt features on the soil surface and where saline soils are dominated by halophyte plants. Additionally, Furby et al. [66] stated that multispectral satellite sensors caused confusing reflectance, as they found that non-saline soils were confused with bare, extremely saline areas. In the case of Landsat imagery, Fraser and Joseph [68] reported that the spectral resolution of Landsat was insufficient due to the difference between the spectra of saline land and waterlogged land not being sufficient to allow spectral separation, as well as the variable spectral response of saline soil. Additionally, Hick & Russell [69] state that the ability of discriminating plant species and plant health conditions is challenging with Landsat imagery due to the absence of narrow bands in the range of 700 - 730 nm, 730 - 760 nm and 900 - 1100 nm.

## 3.3. Hyperspectral Remote Sensing Data

Most of the studies discussed above have attempted to map and monitor soil salinity using different multispectral satellite sensors. However, the use of such sensors is restrictive, as their spectral resolution influences the quality and quantity of the information they provide. The development of airborne and satellite-based hyperspectral sensors has overcome some of the spatial and spectral limitations of multispectral satellite imagery for monitoring and mapping soil salinity, both regionally and locally. Hyperspectral sensors offer a large number of spectral bands with high spatial resolution that allow the discrimination of halophyte plants from non-halophyte plants as well as the identification of surface salt features in more detail than the multispectral sensors [70,71]. Taylor et al. [72] demonstrated the possibility of using airborne hyperspectral data to map salinity in the soil. They described the use of visible-NIR and shortwave infrared (SWIR) hyperspectral data that were collected with airborne Geoscan sensor to map soil salinity at Pyramid Hill, Victoria, Australia. The found that differentiation of salt-affected soils based on the mapping of halophytic plants simply achieved via the employing of the principal component analysis of Geoscan imagery. The potential of the HyMap airborne hyperspectral sensor, which captures images within a spectral range of 450 - 2500 nm in 128 bands, for soil salinity studies has been tested by Dehaan and Taylor [73,74]. They concluded that HyMap has considerable potential for mapping saline areas that characterize the variety of salinity levels and scattered halophyte plants. Likewise, Farifteh et al. [75] measured reflectance spectra from multiple sources (experimental, field, and airborne datasets) to predict salt concentrations with Partial Least Squares Regression (PLSR) and Artificial Neural Networks (ANN) and maped soil salinity using airborne hyperspectral data acquired with HyMap. Their results indicated that both PLSR and ANN enabled good mapping of soil salinity.

After the year 2000, two experimental hyperspectral sensors were launched, CHRIS and Hyperion [76]. Compared to airborne hyperspectral remote sensing, very limited numbers of studies on soil salinity have been conducted with these space-borne hyperspectral data. In Turkey, soil salinity was assessed by Satir et al. [77] using CHRIS PROBA sensor imagery and the spectral characteristics of indicator crops. Their study showed a strong correlation between the spectral wavebands of CHRIS PROBA and the reflected signals of cotton and wheat for detecting soil salinity, which confirmed that the condition of cotton and wheat crops is a good indirect indicator of soil salinity. On the other hand, a study by Dutkiewicz [71] evaluated the performance of Hyperion imagery for mapping surface symptoms of dryland salinity using mixture-tuned matched filtering in southern Australia. She found that the hyperspectral imagery wasunable to distinguish halophytic samphire vegetation at slight or moderate levels of salinity; however, it could be used to map high to very high and extremely high salinity. In China, Weng et al. [18] have investigated the potential of data from the spaceborne Earth Observing 1 (EO-1) Hyperion sensor for the prediction of soil salinity. Partial Least-Squares Regression (PLSR) and Stepwise Linear Regression (SWR) were used as prediction models. Their results indicated that the PLSR model produced more accurate estimations of soil salt content than SWR and could overcome the difficulties that were caused by high dimensionality and strong correlation among input variables as well as noisy data. Thus, they concluded that PLSR was a promising approach for the quantitative mapping of soil salinity with Hyperion data over a large area. These results suggest that further studies need to be conducted to examine prediction models based on nonlinear regression methods, and the spectral noise of hyperspectral data should be taken into consideration to increase the accuracy of soil salinity mapping. In addition, a more recent study was conducted in Iran by Hamzeh et al. [78] to investigate the ability of Hyperion spaceborne hyperspectral data for mapping salinity stress in sugarcane fields. Different classifications such as Support Vector Machine (SVM), Spectral Angle Mapper (SAM), Minimum Distance (MD) and Maximum Likelihood (ML) were used with different band combinations to classify soil salinity into three classes (low, moderate and high salinity). Their results indicated that SVM classification using all bands as input data yielded a salinity map with good accuracy, with an overall accuracy and kappa coefficient of 78.7% and 0.68, respectively.

# 4. USING VEGETATION AND SOIL INDICES IN SOIL SALINITY STUDIES

## 4.1. Vegetation and Soil Indices

As mentioned previously, halophytic plants grow naturally in saline soil, and can be adapted to high soil salinity. Therefore, vegetation has been used as an indirect indicator to predict and map soil salinity. Accordingly, numerous researchers have conducted studies on the mapping and delineation of soil salinity using different Spectral Vegetation Indices (SVI). Among the vegetation indices, NDVI, SAVI, Ratio Vegetation Index (RVI) and Tasseled Cap Transformation that consisted of the Soil Brightness Index (SBI), the Green Vegetation Index (GVI), and the Wetness Index (WI) have been used in soil salinity studies [26,29,30,57,79-81].

Due to absorption in the visible range and high reflectance in the NIR range of the electromagnetic spectrum, the NDVI (**Table 1**) has been widely used to map soil salinity by monitoring halophytic plants [42,51,79]. The difference in reflectance between the visible and NIR bands is divided by the sum of the two bands' reflectance (**Table 1**). This normalizes differences in the amount of incoming light and produces a number from −1 to 1; the range of actual values is approximately 0.1 for bare soils to 0.9 for healthy vegetation [82]. In Mexico, FernandezBuces et al. [42] found a significant correlation between NDVI, EC and SAR. Moreover, Pérez González et al. [83] have correlated the NDVI of halophytic vegetation with the spatial variability of the chemical and physical properties of a transect to identify saline hydromorphic soils. Their results showed the NDVI to be very proper in detecting halophytic plant and relating it to saline soils. Additionally, Bannari et al. [84] have stated that because plant growth declines due to soil salinity, salt stress could be predicted using the NDVI. However, researchers such as Metternicht and Zinck [76] and Zhang, et al. [29] argue that detecting soil salinity using the NDVI is challenging because the presence of vegetation could cause spectral confusion with the reflectance properties of salt and also because the NDVI is considered an unreliable indicator, as it is also correlated to other yield variables such as chlorophyll content, biomass and leaf area. Liu and Huete [85] have developed a modification of the NDVI to reduce the atmospheric and canopy background noise, the enhanced soil and atmosphere resistant vegetation index (EVI) (**Table 1**). A comparison study of the efficiency of the EVI and NDVI calculated from Multiyear Moderate Resolution Imaging Spectroradiometer (MODIS) imagery for assessing soil salinity in the Red River Valley, United States has been conducted by Lobell et al. [48]. They found that the EVI is a more reliable indicator of salinity than the NDVI. The use of the RVI (**Table 1**) to measure the spectral reflectance of soybean canopy and elephant grass under different salinity and irrigation treatments was investigated by [80] and [86]. In both studies, the results showed that the canopy spectral reflectance in the NIR region was reduced as salinity level increase.

Furthermore, SAVI (**Table 1**) was developed by Huete [87] to eliminate soil-induced variation and for use in areas where soil backgrounds differed and the low canopy cover was present. Depending on the crop density, the L factor generally ranges from 0 for higher densities and 1 for lower densities [87]. The resulting SAVI values in the classified image are either positive, negative or zero. A positive SAVI value indicates that there is a decrease in the vegetation, while a negative value indicates an increase in vegetation. A zero value indicates no change in vegetation. The effectiveness of this index for soil salinity detection and mapping has been studied by several researchers, including Zhang et al. [29], Alhammadi [88], Koshal [89], Elnaggar & Noller [51] and Masoud & Koike [90]. For example, in the United Arab Emirates (UAE) Alhammadi and Glenn [27] used the SAVI index for detecting date palm health under soil salinity. They found that the SAVI values decreased with increasing soil salinity; for instance, the SAVI value was 0.155 at the lowest salinity level of 6900 parts per million (ppm), whereas the value decreased to 0.104 at the very high salinity of 41,000 ppm. These results showed the potential of using the date palm, which is a halophytic plant, as an indirect indicator of soil salinity as well as the effectiveness of the SAVI in detecting plant stress related to severe salinity and thus permitting the identification and mapping of saline areas indirectly. Recently, Zhang et al. [29] have proposed four Soil-adjusted Salinity Indices (SASIs) through the most sensitive bands in a SAVI form. For halophyte plants, SASIs produced better results compared to other selected vegetation indices such as the NDVI and SAVI. These results indicate that in highly saline areas that are covered mainly by halophyte plants, SASIs would give superior results, whereas VIs such as NDVI and SAVI would only be proper to assess salinity in low saline areas covered by salt-sensitive plants.

**Table 1**. Vegetation and soil salinity indices that have been proposed and used for soil salinity monitoring and mapping.

| | Indices | Equation | References |
|---|---|---|---|
| 1 | Normalized Differential Vegetation Index | $NDVI = (NIR - R)/(NIR + R)$ | [82] |
| 2 | Enhanced Vegetation Index | $EVI = 2.5(NIR - R)/(NIR + 6R - 7.5BLUE + 1)$ | [85] |
| 3 | Soil Adjusted Vegetation Index | $SAVI = (NIR - R)/(NIR + R + L) \times (1 + L)$ | [87] |
| 4 | Ratio Vegetation Index | $RVI = NIR/R$ | [91] |
| 5 | Normalized Differential Salinity Index | $NDSI = (R - NIR)/(R + NIR)$ | |
| 6 | Brightness Index | $BI = \sqrt{(R^2 + NIR^2)}$ | [92] |
| 7 | Salinity Index | $SI = \sqrt{BLUE \times R}$ | |
| 8 | Salinity Index | $SI1 = \sqrt{G \times R}$ | |
| 9 | Salinity Index | $SI2 = \sqrt{G^2 + R^2 + NIR^2}$ | [55] |
| 10 | Salinity Index | $SI3 = \sqrt{G^2 + R^2}$ | |
| 11 | Salinity Index | $SI\text{-}1 = ALI9/ALI10$ | |
| 12 | Salinity Index | $SI\text{-}2 = (ALI6 - ALI9)/(ALI6 - ALI9)$ | |
| 13 | Salinity Index | $SI\text{-}3 = (ALI9 - ALI10)/(ALI9 + ALI10)$ | [84] |
| 14 | Soil Salinity and Sodicity Indices | $SSSI\text{-}1 = (ALI9 - ALI10)$ | |
| 15 | Soil Salinity and Sodicity Indices | $SSSI\text{-}2 = (ALI9 \times ALI10 - ALI10 \times ALI10)/ALI9$ | |
| 16 | Salinity Index | $S_1 = Blue/R$ | |
| 17 | Salinity Index | $S_2 = (Blue - R)/(Blue + R)$ | |
| 19 | Salinity Index | $S_3 = (G \times R)/Blue$ | [93] |
| 20 | Salinity Index | $S_4 = \sqrt{Blue \times R}$ | |
| 21 | Salinity Index | $S_5 = (Blue \times R)/G$ | |
| 22 | Salinity Index | $S_6 = (R \times NIR)/G$ | |

Similarly, various spectral salinity indices have been developed for salt mineral detection and mapping. Douaoui et al. [55] have proposed three salinity indices (**Table 1**) produced from SPOT XS imagery to detect and map soil salinity hazards in a semi-arid environment in Algeria. They found that those indices were strongly correlated with measured values, but considerably underestimated the salinity of areas with high levels of surface salt. Besides, Khan et al. [92] have proposed three spectral salinity indices: the Brightness index (BI), Normalized Difference Salinity Index (NDSI) and Salinity Index (SI) (**Table 1**) from the LISS-II sensor of the IRS-1B satellite to assess hydrosalinized land degradation in Pakistan. Among these indices, they found that NDSI yielded the most acceptable results in identifying different salt classes. Another study conducted by Vidal et al. [94] and Vincent et al. [95] looked at salinity by differenting vegetated from non-vegetated areas using NDVI; then the BI was computed to identify the moisture and salinity status of fallow land and deserted fields. Furthermore, three different salinity indices, SI-1, SI-2 and SI-3 (**Table 1**) from the EO-1 ALI spectral bands, have been proposed by Bannari et al. [84] to discriminate slight and moderate soil salinity and sodicity in Morocco. Although the results showed that SI-3 had the highest correlation (46.9%), the result from this index was not adequate to provide precise information. Therefore, they devised another two Soils Salinity and Sodicity Indices (SSSI) (**Table 1**). Their results indicated that these SSSI indices were likely to increase the identification accuracy in areas with low and medium salinity because they offered the most significant correlation (52.9%) with the ground EC measurement. In Pakistan, Abbas and Khan [93] have suggested an integrated approach based on the spatial analysis of both ground and satellite data to assess soil salinity. Remotely sensed data-based salinity indices and a Principal Components Analysis (PCA) were developed to detect soil salinity. Their result showed that out of the six salinity indices (**Table 1**) S3 produced the most promising result compared to ground measurements. Moreover, they concluded that PCA and salinity indices are promising techniques for soil salinity prediction based on satellite images.

Looking at these vegetation and soil salinity indices in the literature, a number of results stand out. Utilizing vegetation indices in the assessment and mapping of soil salinity in areas of densely vegetated soils will yield promising results, whereas on bare soils, the identification of salt based on vegetation indices will not work. Thus, soil salinity indices will be the appropriate method in the case of bare soils or soils with very low scattered vegetation cover, providing super results. These observations are in agreement with Bouaziz et al. [96] and Fan et al. [97]. Bouaziz et al. [96] found that vegetation indices such as SAVI, NDVI and EVI had a low correlation with EC due to an insufficient density of vegetation cover, whereas soil salinity indices indices exhibited higher correlations with EC. Additionally, Fan et al. [97] found that NDVI values had a significant negative relationship with soil salinity in soils covered by vegetation, whereas this relationship was not clear on bare soil.

## 5. ISSUES IN MAPPING SOIL SALINITY IN ARID AND SEMI-ARID REGIONS USING REMOTE SENSING

The spatial, temporal and vertical variability in the soil profile are the limiting factors in the assessment and mapping of soil salinity using remote sensing data because the spectral reflectance is unable to provide information on the whole soil profile, as it only observes the soil surface [98,99]. Moreover, the surface characteristics in many cases may not be representative of the deeper soil profile [100]. However, combining remote sensing data with geophysical surveys and simulation models can be an alternative option [56].

The direct detection of soil salinity becomes applicable and much easier for bare soils and/or whenever salt-related symptoms (e.g., crusts) and scattered vegetation are visible on the surface; in these cases, information such as salt types and quantity as well as the crust thickness can be provided based on the soil surface spectral signatures [67,101]. Yet, the direct deduction of soil salinity in arid and semi-arid regions that are characterized by dense vegetation cover would be difficult, as the vegetation will cause spectral mixing [102,103].

On the other hand, different studies such as Tashi et al. [104], Fernandez-Buces, et al. [42] have successfully used indirect indicator to map soil salinity through monitoring the vegetation condition by using NDVI. Generally, these methods assume that soil salinity is the only stressor decreasing and damaging the crop condition, whereas other factors such as inappropriate soil management and water quality are neglected. Moreover, reductions in vegetation growth and vigor could be related to a lack of necessary nutrients rather than the occurrence of salt. Additionally, different plants generally grow in different levels of salinity, so that the NDVI is considered an uncertain indicator for soil salinity monitoring and mapping. Besides, the existence of halophytic plants may confuse soil salinity detection based on the NDVI due to mixing with the spectral signature of salt, which then will lead to classification errors [98]. Hence, to overcome this issue and remove classification errors to some degree, the SAVI index and other indices and enhancement models have helped to separate soil and vegetation signals [105].

Additionally, low spectral resolution satellite images limit the direct detection and mapping of soil salinity, primarily due to their inability to detect particular absorption bands of some salt types and the frequent occurrence of problems with mixedspectral signatures that come from a variety of surface components [74,106]. Nevertheless, the advance of hyperspectral sensors has enabled spectral features associated to the characteristic absorption bands of salt minerals to be mapped with more detail [107]. The above shortcomings indicate that detecting and mapping soil salinity in arid and semi-arid regions using remote sensing is challenging. This is primarily attributed to the weakness of the spectral signals from saline soils compared to the noise caused by other factors. Obviously, there is no agreed-on best approach to this technology for monitoring and mapping soil salinity, as many researchers have used and applied different tools and techniques to map and monitor saline soils with varying degrees of success. For example, in the case of vegetation indices, some researchers like Elnaggar & Noller [51] have found that vegetation indices (NDVI, SAVI, and GVI) had a weak correlation with the EC measurements, which suggests that halophytes could not be used to identify salt-affected soils under vegetation cover. Sethi et al. [108] in a study in India, found the same, whereas Pérez González et al. [83] found the NDVI to be very useful in detecting halophytic plants and relating it to saline soils. Despite these varying results of using halophytic plants as indirect indicators for soil salinity detection and mapping, it is ineffective to monitor and map soil salinity through non-halophyte plants, as they cannot live in highly saline areas. Thus, taking into consideration the spectral reflectance of halophytic plants is necessary for soil salinity detection and mapping, particularly in highly saline areas.

## 6. CONCLUSIONS

Soil salinity, either naturally occurring or human-induced, is a serious global environmental problem, especially in arid and semi-arid regions. This is a complex dynamic process with serious consequences for the soil environment as well as, geochemical, hydrological, climatic, agricultural, and economic impacts. Being a severe environmental hazard, the frequent detection of soil salinity and assessment of its extent and severity at an early stage become very important at both local and regional scales. Traditionally, soil salinity was assessed via collecting in situ soil samples and analyzing those samples in the laboratory. Undertaking this method, especially over a large area, is expensive and time consuming. Remote sensing represents a good alternative for monitoring and mapping changes in soil salinity.

Remote sensing data have been used extensively to identify and map saline areas, and the potential of remote sensing for assessing and mapping soil salinity is enormous. Multispectral satellite sensors are the preferred method for mapping and monitoring soil salinity, largely due to the low cost of such imagery and the ability to map extreme surface expressions of salinity. However, multispectral data have limited capabilities due to their spatial and spectral resolution. Hyperspectral imagery, with its fine spatial and spectral resolutions, allows soil salinity mapping in greater detail and represents another alternative.

Surface reflectance is highly affected by soil's moisture content, salt content, color, and surface roughness. High salt concentrations can be identified through the existence of characteristic vegetation types and growth patterns or by the salt efflorescence and crust that are present on bare soils. Similar to vegetation indices, researchers have developed different salinity indices to detect and map soil salinity. As discussed, these indices have been applied with varying degree of success. Field sites differ in terms of levels of salinity and the amount of vegetation cover; hence, a single selected index may not perform best in all cases. Each site needs to be assessed regarding the strengths and weaknesses of the proposed indices before appropriate remote sensingbased indices are used for soil salinity mapping and assessing.

## REFERENCES

1. L. Richards, "Diagnosis and Improvement of Saline and alkali Soils," Soil Science, Vol. 78, No. 2, 1954, p. 154.
2. D. P. Shrestha and A. Farshad, "Chapter 13. Mapping Salinity Hazard: An Integrated Application of Remote Sensing and Modeling-Based Techniques," In: G. Metternicht and J. A. Zinck, Eds., Remote Sensing of Soil Salinization: Impact on Land Management, CRC Press, Boca Raton, 2008, p. 257.

3. D. Hillel, "Salinity Management for Sustainable Irrigation: Integrating Science, Environment, and Economics," World Bank Publications, Washington DC, 2000.

4. P. Koohafkan, "Water and Cereals in Drylands," The Food and Agriculture Organization of the United Nations and Earthscan, Rome, 2012.

5. ABS Australian Bureau of Statistics, "Salinity," Canberra, 2002. http://www.abs.gov.au/ausstats/abs@.nsf/Lookup/by%20Subject/1370.0~2010~Chapter~Salinity%20(6.2.4.4)

6. D. J. McFarlane, et al., "The Extent and Potential Area of Salt-Affected Land in Western Australia Estimated Using Remote Sensing and Digital Terrain Models," Engineering Salinity Solutions: 1st National Salinity Engineering Conference 2004, Barton, 2004, pp. 55-60.

7. J. K. Zhu, "Plant Salt Tolerance," Trends in Plant Science, Vol. 6, No. 2, 2001, pp. 66-71.

8. D. Corwin and S. Lesch, "Application of Soil Electrical Conductivity to Precision Agriculture," Agronomy Journal, Vol. 95, No. 3, 2003, pp. 455-471.

9. J. Rhoades and J. Loveday, "Salinity in Irrigated Agriculture," In: B. A. Stewart and D. R. Nielsen, Eds., American Society of Civil Engineers, Irrigation of Agricultural Crops, Vol. 30, Monograph, American Society of Agronomists, Madison, 1990, pp. 1089-1142.

10. A. Naifer, et al., "Economic Impact of Salinity: The Case of Al-Batinah in Oman," International Journal of Agricultural Research, Vol. 6, No. 2, 2011, pp. 134-142.

11. P. Brunner, et al., "Generating Soil Electrical Conductivity Maps at Regional Level by Integrating Measurements on the Ground and Remote Sensing Data," International Journal of Remote Sensing, Vol. 28, No. 15, 2007, pp. 3341-3361.

12. D. Dent and A. Young, "Soil Survey and Land Evaluation," George Allen & Unwin, Crows Nest, 1981.

13. R. Dehaan and G. Taylor, "Field-Derived Spectra of Salinized Soils and Vegetation as Indicators of IrrigationInduced Soil

Salinization," Remote Sensing of Environment, Vol. 80, No. 3, 2002, pp. 406-417.

14. M. R. D. Nanni, "Spectral Reflectance Methodology in Comparison to Traditional Soil Analysis," Soil Science Society of America Journal, Vol. 70, No. 2, 2006, pp. 393-407.

15. T. Ghabour and L. Daels, "Mapping and Monitoring of Soil Salinity of ISSN," Egyptian Journal of Soil Science, Vol. 33, No. 4, 1993, pp. 355-370.

16. P. Dale, et al., "Classification of Reflectance on Colour Infrared Aerial Photographs and Sub-Tropical Salt-Marsh Vegetation Types," International Journal of Remote Sensing, Vol. 7, No. 12, 1986, pp. 1783-1788.

17. J. Farifteh, "Imaging Spectroscopy of Salt-Affected Soils: Model-Based Integrated Method," International Institute Geo-Information Science and Earth Observation (ITC) and Utrecht University, Utrecht, 2007.

18. Y. Weng, et al., "Soil Salt Content Estimation in the Yellow River Delta with Satellite Hyperspectral Data," Canadian Journal of Remote Sensing, Vol. 34, No. 3, 2008, pp. 259-270.

19. S. Teggi, et al., "SPOT 5 Imagery for Soil Salinity Assessment in Iraq," Proceedings of SPIE—Earth Resources and Environmental Remote Sensing/GIS Applications III, Vol. 8538, 2012, pp. 85380V-85380V-12.

20. A. K. Koshal, "Spectral Characteristics of Soil Salinity Areas in Parts of South-West Punjab through Remote Sensing and GIS," International Journal of Remote Sensing and GIS, Vol. 1, No. 2, 2012, pp. 84-89.

21. A. Dehni and M. Lounis, "Remote Sensing Techniques for Salt Affected Soil Mapping: Application to the Oran Region of Algeria," Procedia Engineering, Vol. 33, 2012, pp. 188-198.

22. R. Setia, et al., "Severity of Salinity Accurately Detected and Classified on a Paddock Scale with High Resolution Multispectral Satellite Imagery," Land Degradation & Development, Vol. 24, No. 4, 2011, pp. 375-384.

23. R. Dwivedi, et al., "5 Generation of Farm-Level Information on Salt-Affected Soils Using IKONOS-II Multispectral Data," In: G. Metternicht and J. Zinck, Eds., Remote Sensing of Soil Salinization: Impact on Land Management, CRC Press, Boca Raton, 2008.

24. E. Weiss, et al., "Application of NOAA-AVHRR NDVI Time-Series Data to Assess Changes in Saudi Arabia's Rangelands," International Journal of Remote Sensing, Vol. 22, No. 6, 2001, pp. 1005-1027.

25. D. R. Tilley, et al., "Hyperspectral Reflectance Response of Freshwater Macrophytes to Salinity in a Brackish Subtropical Marsh," Journal of Environmental Quality, Vol. 36, No. 3, 2007, pp. 780-789.

26. H. R. Matinfar, et al., "Detection of Soil Salinity Changes and Mapping Land Cover Types Based upon Remotely Sensed Data," Arabian Journal of Geosciences, Vol. 6, No. 3, 2013, pp. 913-919.

27. M. S. Alhammadi and E. P. Glenn, "Detecting Date Palm Trees Health and Vegetation Greenness Change on the Eastern Coast of the United Arab Emirates Using SAVI," International Journal of Remote Sensing, Vol. 29, No. 6, 2008, pp. 1745-1765.

28. F. Iqbal, "Detection of Salt Affected Soil in Rice-Wheat Area Using Satellite Image," African Journal of Agricultural Research, Vol. 6, No. 21, 2011, pp. 4973-4982.

29. T. T. Zhang, et al., "Using Hyperspectral Vegetation Indices as a Proxy to Monitor Soil Salinity," Ecological Indicators, Vol. 11, No. 6, 2011, pp. 1552-1562.

30. Y. Y. Aldakheel, "Assessing NDVI Spatial Pattern as Related to Irrigation and Soil Salinity Management in Al-Hassa Oasis, Saudi Arabia," Journal of the Indian Society of Remote Sensing, Vol. 39, No. 2, 2011, pp. 171- 180.

31. G. Metternicht and A. Zinck, "Remote Sensing of Soil Salinization: Impact on Land Management," CRC Press, 2008.

32. J. A. M. Demattê, et al., "Visible-NIR Reflectance: A New Approach on Soil Evaluation," Geoderma, Vol. 121, No. 1-2, 2004, pp. 95-112.

33. D. J. Brown, et al., "Global Soil Characterization with VNIR Diffuse Reflectance Spectroscopy," Geoderma, Vol. 132, No. 3, 2006, pp. 273-290.

34. M. Baumgardner, et al., "Reflectance Properties of Soils," Advances in Agronomy, Vol. 38, 1985, pp. 2-44.

35. S. De Jong, "The Analysis of Spectroscopical Data to Map Soil Types and Soil Crusts of Mediterranean Eroded Soils," Soil technology, Vol. 5, No. 3, 1992, pp. 199-211.

36. D. Shrestha, D. E. Margateb, F. van der Meera and H. V. Anhc, "Analysis and Classification of Hyperspectral Data for Mapping Land Degradation: An Application in Southern Spain," International Journal of Applied Earth Observation and Geoinformation, Vol. 7, No. 2, 2005, pp. 85-96.

37. T. Schmid, M. Koch and J. Gumuzzio, "Application of Hyperspectral Imagery to Soil Salinity Mapping," In: G. Metternicht and J. Zinck, Eds., Remote Sensing of Soil Salinization: Impact on Land Management, CRC Press, Boca Raton, 2008, pp. 113-137.

38. D. S. G. Thomas, "Arid Zone Geomorphology: Process, Form and Change in Drylands," John Wiley & Sons, Chichester, 2011.

39. R. P. Singh and A. Sirohi, "Spectral Reflectance Properties of Different Types of Soil Surfaces," ISPRS Journal of Photogrammetry and Remote Sensing, Vol. 49, No. 4, 1994, pp. 34-40.

40. B. R. M. Rao, R. C. Sharma, T. Ravi Sankar, S. N. Das, R. S. Dwivedi, S. S. Thammappa and L. Venkataratnam, "Spectral Behaviour of Salt-Affected Soils," International Journal of Remote Sensing, Vol. 16, No. 12, 1995, pp. 2125-2136.

41. G. Metternicht and J. A. Zinck, "Spatial Discrimination of Saltand Sodium-Affected Soil Surfaces," International Journal of Remote Sensing, Vol. 18, No. 12, 1997, pp. 2571-2586.

42. N. Fernandez-Buces, C. Siebe, S. Cram and J. L. Palacio, "Mapping Soil Salinity Using a Combined Spectral Response Index for Bare Soil and Vegetation: A Case Study in the Former Lake Texcoco, Mexico," Journal of Arid Environments, Vol. 65, No. 4, 2006, pp. 644-667.

43. E. P. Glenn, J. Jed Brown and E. Blumwald, "Salt Tolerance and Crop Potential of Halophytes," Critical Reviews in Plant Sciences, Vol. 18, No. 2, 1999, pp. 227-255.

44. G. I. Metternicht, "Analyzing the Relationship between Ground Based Reflectance and Environment Indicators of Salinity Processes in the Cochabamba Valley (Bolivia)," International Journal of Ecology and Environmental Sciences, Vol. 24, No. 4, 1998, pp. 359-370.

45. C. Wiegand, G. Anderson, S. Lingle and D. Escobar, "Soil Salinity Effects on Crop Growth and Yield-Illustration of an Analysis and Mapping Methodology for Sugarcane," Journal of Plant Physiology, Vol. 148, No. 3-4, 1996, pp. 418-424.

46. C. Wiegand, J. D. Rhoades, D. E. Escobar and J. H. Everitt, "Photographic and Videographic Observations for Determining and Mapping the Response of Cotton to Soil Salinity," Remote Sensing of Environment, Vol. 49, No. 3, 1994, pp. 212-223.

47. D. B. Lobell, J. I. Ortiz-Monasterio, F. C. Gurrola and L. Valenzuela, "Identification of Saline Soils with Multiyear Remote Sensing of Crop Yields," Soil Science Society of America Journal, Vol. 71, No. 3, 2007, pp. 777-783.

48. D. B. Lobell, S. M. Leschb, D. L. Corwinc, M. G. Ulmerd, K. A. Andersone, D. J. Pottsf, J. A. Doolittleg, M. R. Matosh and M. J. Baltes, "Regional-Scale Assessment of Soil Salinity in the Red River Valley Using Multi-Year MODIS EVI and NDVI," Journal of Environmental Quality, Vol. 39, No. 1, 2010, pp. 35-41.

49. R. S. Dwivedi, "Soil Resources Mapping: A Remote Sensing Perspective," Remote Sensing Reviews, Vol. 20, No. 2, 2001, pp. 89-122.

50. K. Verma, R. K. Saxena, A. K. Barthwal and S. N. Deshmukh, "Remote Sensing Technique for Mapping Salt Affected Soils," International Journal of Remote Sensing, Vol. 15, No. 9, 1994, pp. 1901-1914.

51. A. A. Elnaggar and J. S. Noller, "Application of RemoteSensing Data and Decision-Tree Analysis to Mapping SaltAffected Soils over Large Areas," Remote Sensing, Vol. 2, No. 1, 2009, pp. 151-165.

52. R. Katawatin and W. Kotrapat, “Use of LANDSAT-7 ETM+ with Ancillary Data for Soil Salinity Mapping in Northeast Thailand,” Third International Conference on Experimental Mechanics and Third Conference of the Asian 2004, Sevilla, 3 June 2005, pp. 708-716.

53. R. T. Mehrjardi, Sh. Mahmoodi, M. Taze and E. Sahebjalal, “Accuracy Assessment of Soil Salinity Map in Yazd-Ardakan Plain, Central Iran, Based on Land sat ETM+ Imagery,” American-Eurasian Journal of Agricultural & Environmental Sciences, Vol. 3, No. 5, 2008, pp. 708-712.

54. R. Yu, T. X. Liu, Y. P. Xu, C. Zhu, Q’ Zhang, Z. Y. Qu, X. M. Liu and C. Y. Li, “Analysis of Salinization Dynamics by Remote Sensing in Hetao Irrigation District of North China,” Agricultural Water Management, Vol. 97, No. 12, 2010, pp. 1952-1960.

55. A. E. K. Douaoui, H. Nicolasb and C. Walter, “Detecting Salinity Hazards within a Semiarid Context by Means of Combining Soil and Remote-Sensing Data,” Geoderma, Vol. 134, No. 1-2, 2006, pp. 217-230.

56. J. Farifteh, A. Farshad and R. J. George, “Assessing SaltAffected Soils Using Remote Sensing, Solute Modelling, and Geophysics,” Geoderma, Vol. 130, No. 3-4, 2006, pp. 191-206.

57. A. Eldeiry and L. A. Garcia, “Detecting Soil Salinity in Alfalfa Fields Using Spatial Modeling and Remote Sensing,” Soil Science Society of America Journal, Vol. 72, No. 1, 2008, pp. 201-211.

58. R. Goossens, M. el Badawi, T. Ghabour and M. de Dapper, “A Simulation Model to Monitor the Soil Salinity in Irrigated Arable Land in Arid Areas Based upon Remote Sensing and GIS,” EARSeL. Advances in Remote Sensing, Vol. 2, No. 3, 1993, pp. 165-171.

59. I.Ahmed and H. H. Andrianasolo, “Comparative Assessment of Multisensor Data for Suitability in Study of the Soil Salinity Using Remote Sensing and GIS in the Fordwah Irrigation Division,” IEEE International Conference on Geoscience and Remote Sensing, Singapore, 3-8 August 1997, pp. 1627-1629.

60. S. K. Alavi Panah and R. Goossens, “Relationship between the Landsat TM, MSS Data and Soil Salinity,” Journal of Agricultural Science and Technology (JAST), Vol. 3, No. 2, 2001, pp. 21-31.

61. R. Goossens, A. P. S. Kazem, M. De Dapper and O. Kissyar, "The Use of Thermal Band of Landsat TM for the Study of Soil Salinity in Iran (Ardakan Area) and Egypt (Ismailia Province)," International conference on Geoinformatics for Natural Resource Assessment, Monitoring and Management, Indian Institute of Remote Sensing, Dehradun, 3-11 March 1999, pp. 454-459.

62. S. Y. Huang, Q. H. Liu, X. W. Li and Q. H. Liu, "Spectral Model of Soil Salinity in Xinjiang of China," IEEE International Geoscience and Remote Sensing Symposium, Seoul, 25-29 July 2005, pp. 4458-4460.

63. K. Navulur, "Multispectral Image Analysis Using the Object-Oriented Paradigm," CRC Press, Boca Raton, 2006.

64. A. Elhaddad and L. Garcia, "Detecting Soil Salinity Levels in Agricultural Lands Using Satellite Imagery," Proceedings of the American Society for Photogrammetry and Remote Sensing Annual Conference, Reno, 2006.

65. B. Spies and P. Woodgate, "Salinity Mapping Methods in the Australian Context," Department of the Environment and Heritage, Canberra, 2005.

66. S. Furby, et al., "Detecting and Monitoring Salt-Affected Land: A Report from the LWRRDC Project Detecting and Monitoring Changes in Land Condition through Time Using Remotely Sensed Data," CSIRO, Division of Mathematics & Statistics, 1995.

67. F. Howari, "The Use of Remote Sensing Data to Extract Information from Agricultural Land with Emphasis on Soil Salinity," Australian Journal of Soil Research, Vol. 41, No. 7, 2003, pp. 1243-1253.

68. D. Fraser and S. Joseph, "Mapping Soil Salinity in the Murray Valley (NSW) Using Satellite Imagery," Proceeding of the 9th Australaian Remote Sensing and Photogrammetry Conference, Australia Causal Productions Sydney, Sydney, 1998.

69. P. Hick and W. Russell, "Remote Sensing of Agricultural Salinity," CSIRO, Division of Exploration Geoscience, Perth, 1987.

70. R. P. Gupta, "Remote Sensing Geology," Springer Verlag, New York, 2003.

71. A. Dutkiewicz, "Evaluating Hyperspectral Imagery for Mapping the Surface Symptoms of Dryland Salinity," The University of Adelaide, Adelaide, 2006.

72. G. R. Taylor, B. A. Bennett, A. H. Mah and R. D. Hewson, "Spectral Properties of Salinised Land and Implications for Interpretation of 24 Channel Imaging Spectrometry," Proceedings of the International Airborne Remote Sensing Conference and Exhibition, Taylor & Francis Strasbourg, France, 1994, pp. 504-513.

73. R. L. Dehaan and G. R. Taylor, "Field-Derived Spectra of Salinized Soils and Vegetation as Indicators of Irrigation-Induced Soil Salinization," Remote Sensing of Environment, Vol. 80, No. 3, 2002, pp. 406-417.

74. R. Dehaan and G. R. Taylor, "Image-Derived Spectral Endmembers as Indicators of Salinisation," International Journal of Remote Sensing, Vol. 24, No. 4, 2003, pp. 775- 794.

75. J. Farifteh, F. Van der Meera, C. Atzbergerb and E. J. M. Carranzaa, "Quantitative Analysis of Salt-Affected Soil Reflectance Spectra: A Comparison of Two Adaptive Methods (PLSR and ANN)," Remote Sensing of Environment, Vol. 110, No. 1, 2007, pp. 59-78.

76. G. Metternicht and A. Zinck, "Remote Sensing of Soil Salinization: Impact on Land Management," CRC Press, Taylor and Francis Publisher, Boca Raton, 2008.

77. O. Satir, S. Berberoglu, S. Kapur, T. Nagano, E. Akça, M. A. Erdogan, C. Donmez, N. Yonca Satir and K. Tanaka, "Soil Salinity Mapping Using Chris-Proba Hyperspectral Data," Proceedings of Hyperspectral Workshop 2010, Frascati, 17-19 March 2010, 8p.

78. S. Hamzeh, A. A. Naseri, S. K. Alavi Panah, B. Mojaradi, H. M. Bartholomeus and M. Herold, "Mapping Salinity Stress in Sugarcane Fields with Hyperspecteral Satellite Imagery," Proceedings SPIE 8531, Remote Sensing for Agriculture, Ecosystems, and Hydrology XIV, Edinburgh, 24 September 2012, pp. 85312B-1.

79. M. T. Jabbar and X. Chen, "Land Degradation Due to Salinization in Arid and Semi-Arid Regions with the Aid of Geo-Information Techniques," Geo-Spatial Information Science, Vol. 11, No. 2, 2008, pp. 112-120.

80. D. Wang, J. A. Poss, T. J. Donovan, M. C. Shannon and S. M. Lesch, "Biophysical Properties and Biomass Production of Elephant Grass under Saline Conditions," Journal of Arid Environments, Vol. 52, No. 4, 2002, pp. 447-456.

81. D. Lobell, S. M. Lesch, D. L. Corwin, M. G. Ulmer, K. A. Anderson, D. J. Potts, J. A. Doolittle, M. R. Matos and M. J. Baltes "Regional-Scale Assessment of Soil Salinity in the Red River Valley Using Multi-Year MODIS EVI and NDVI," Journal of Environmental Quality, Vol. 39, No. 1, 2010, pp. 35-41.

82. D. Deering and J. Rouse, "Measuring 'Forage Production' of Grazing Units from Landsat MSS Data," 10th International Symposium on Remote Sensing of Environment, ERIM, Ann Arbor, 1975, pp. 1169-1178.

83. M. E. Pérez González, M. del Pilar García Rodríguez, V. González-Quiñones and R. Jiménez Ballesta, "Spatial Variability of Soil Quality in the Surroundings of a Saline Lake Environment," Environmental Geology, Vol. 51, No. 1, 2006, pp. 143-149.

84. A. Bannari, A. M. Guedona, A. El-Hartib, F. Z. Cherkaouic and A. El-Ghmari, "Characterization of Slightly and Moderately Saline and Sodic Soils in Irrigated Agricultural Land using Simulated Data of Advanced Land Imaging (EO-1) Sensor," Communications in Soil Science and Plant Analysis, Vol. 39, No. 19-20, 2008, pp. 2795-2811.

85. H. Q. Liu and A. Huete, "A Feedback Based Modification of the NDVI to Minimize Canopy Background and Atmospheric Noise," IEEE Transactions on Geoscience and Remote Sensing, Vol. 33, No. 2, 1995, pp. 457-465.

86. D. Wang, C. Wilson and M. C. Shannon, "Interpretation of Salinity and Irrigation Effects on Soybean Canopy Reflectance in Visible and Near-Infrared Spectrum Domain," International Journal of Remote Sensing, Vol. 23, No. 5, 2002, pp. 811-824.

87. A. R. Huete, "A Soil-Adjusted Vegetation Index (SAVI)," Remote Sensing of Environment, Vol. 25, No. 3, 1988, pp. 295-309.

88. M. S. Alhammadi, “Using QuickBird Satellite Images to Study the Salinity Effect on Date Palm Field,” International Congress Geotunis 2010, Tunis, 29 November-3 December 2010, pp. 1-6.

89. A. K. Koshal, “Indices Based Salinity Areas Detection through Remote Sensing and GIS In Parts of South Waste Punjab,” 13th Annual International Conference and Exibition on Geospatial Information Technology and Applications, Gurgaon, 2010.

90. A. Masoud and K. Koike, “Arid Land Salinization Detected by Remotely-Sensed Landcover Changes: A Case Study in the Siwa Region, NW Egypt,” Journal of Arid Environments, Vol. 66, No. 1, 2006, pp. 151-167.

91. D. Major, F. Baret and G. Guyot, “A Ratio Vegetation Index Adjusted for Soil Brightness,” International Journal of Remote Sensing, Vol. 11, No. 5, 1990, pp. 727-740.

92. N. M. Khan, V. V. Rastoskuev, Y. Sato and S. Shiozawa, “Assessment of Hydrosaline Land Degradation by Using a Simple Approach of Remote Sensing Indicators,” Agricultural Water Management, Vol. 77, No. 1, 2005, pp. 96-109.

93. A. Abbas and S. Khan, “Using Remote Sensing Techniques for Appraisal of Irrigated Soil Salinity,” In: L. Oxley and D. Kulasiri, Eds., International Congress on Modelling and Simulation (MODSIM), Modelling and Simulation Society of Australia and New Zealand, Brighton, 2007, pp. 2632-2638.

94. A. Vidal, P. Maure, H. Durand and P. Strosser, “Remote Sensing Applied to Irrigation System Management: Example of Pakistan,” EURISY Colloquium: Satellite Observation for Sustainable Development in the Mediterranean Area, Paris, 1996, pp. 132-142.

95. B. Vincent, A. Vidal, D. Tabbet, A. Baqri and M. Kuper, “Use of Satellite Remote Sensing for the Assessment of Water Logging or Salinity as an Indication of the Performance of Drained Systems,” 16th Congress on Irrigation and Drainage, Cairo, 15-22 September 1996, pp. 203-216.

96. M. Bouaziz, J. Matschullat and R. Gloaguen, “Improved Remote Sensing Detection of Soil Salinity from a SemiArid Climate in Northeast Brazil,” Comptes Rendus Geoscience, Vol. 343, No. 11-12, 2011, pp. 795-803.

97. X. Fan, B. Pedroli, G. Liu, Q. Liu, H. Liu and L. Shu, "Soil Salinity Development in the Yellow River Delta in Relation to Groundwater Dynamics," Land Degradation & Development, Vol. 23, No. 2, 2012, pp.175-189.

98. M. Sethi, D. S. Bundela, M. Sethi, K. Lal and S. K. Kamra, "Remote Sensing and Geographic Information System for Appraisal of Salt-Affected Soils in India," Journal of Environmental Quality, Vol. 39, No. 1, 2010, pp. 5-15.

99. T. J. Schmugge, W. P. Kustas, J. C. Ritchie, T. J. Jackson and A. Rango, "Remote Sensing in Hydrology," Advances in Water Resources, Vol. 25, No. 8-12, 2002, pp. 1367- 1385.

100. O. Dewitte, A. Jones, H. Elbelrhiti, S. Horion and L. Montanarella, "Satellite Remote Sensing for Soil Mapping in Africa An Overview," Progress in Physical Geography, Vol. 36, No. 4, 2012, pp. 514-538.

101. H. A. Ghrefat, P. C. Goodell, B. E. Hubbard, R. P. Langford and R. E. Aldouri, "Modeling Grain Size Variations of Aeolian Gypsum Deposits at White Sands, New Mexico, Using AVIRIS Imagery," Geomorphology, Vol. 88, No. 1-2, 2007, pp. 57-68.

102. A. L. Kaleita, L. F. Tian and M. C. Hirschi, "Relationship between Soil Moisture Content and Soil Surface Reflectance," Transactions of the ASAE, Vol. 48, No. 5, 2005, pp. 1979-1986.

103. J. Ding, M.-C. WU and T. Tiyip, "Study on Soil Salinization Information in Arid Region Using Remote Sensing Technique," Agricultural Sciences in China, Vol. 10, No. 3, 2011, pp. 404-411.

104. Y. Tashi, P. C. Chamard, M.-F. Courel, T. Tiyip, Y. Tuerxun and S. Drake, "The Recent Evolution of the Oasis Environment in the Taklimakan Desert, China," In: Water and Sustainability in Arid Regions, Springer, Netherlands, 2010, pp. 51-74.

105. V. Mulder, S. de Bruin, M. E. Schaepman and T. R. Mayr, "The Use of Remote Sensing in Soil and Terrain Mapping—A Review," Geoderma, Vol. 162, No. 1-2, 2011, pp. 1-19.

106. B. Mougenot, M. Pouget and G. F. Epema, "Remote Sensing of Salt Affected Soils," Remote Sensing Reviews, Vol. 7, No. 3-4, 1993, pp. 241-259.

107. E. Cloutis, "Review Article Hyperspectral Geological Remote Sensing: Evaluation of Analytical Techniques," International Journal of Remote Sensing, Vol. 17, No. 12, 1996, pp. 2215-2242.

108. M. Sethi, G. S. Dasog, A. Van Lieshoutc and S. B. Salimathd, "Salinity Appraisal Using IRS Images in Shorapur Taluka, Upper Krishna Irrigation Project, Phase I, Gulbarga District, Karnataka, India," International Journal of Remote Sensing, Vol. 27, No. 14, 2006, pp. 2917-2926.

# Chapter 5

# Vegetation Cover Density and Land Surface Temperature Interrelationship Using Satellite Data, Case Study of Wadi Bisha, South KSA

**Amalyahya Alshaikh**

*King Abdulaziz University, Jeddah, Saudi Arabia*

## ABSTRACT

The aim of this study is to identify the relationship between Vegetation Cover (VC) and the land Surface Temperature (LST), using satellite data of Wadi Bisha, south the Kingdome of Saudi Arabia (KSA). The Landsat 7 Thematic Mapper (ETM) thermal band (band 6) was used for calculating the (LST) values. The near-infrared (NIR) and red band (bands 3 and 4 respectively) were used for estimating the vegetation cover. ERDAS Imagine 9.3 and ArcGIS 10.2 were used in the current study. The results of the study show that the increase of vegetation cover (VC) coincides with decrease of (LST), while the decrease in vegetation cover is linked with increase of (LST). It was found that there was no vegetation observed in areas practiced the highest temperature of 49˚C, while areas of lowest temperature of 28˚C were characterized by dense vegetation cover. Thus, a quite significant correlation is approved between the (VC) and the (LST), based on the validation of (50) locations. It was concluded that availability and continuity of Satellite remote sensing data was required for elaborating a continuous monitoring of vegetation cover conditions and mapping was recommended in Wadi Bisha. Operational monitoring is recommended to ensure the adoption of flexible land cover validation protocols.

**Keywords:** Relationship, Vegetation Cover (VC), Land Surface Temperature (LST), Satellite Data, Wadi Bisha (South KSA)

## 1. INTRODUCTION

Advances in space technology have increasingly allowed the use of satellite data to study complex physical processes on the Earth's surface [1]. The importance of this study is to evaluate the use of remote sensed information, especially thermal bands to gain land surface temperature (LST), according to [2]. The LST values will be compared with vegetation cove density, extracted from the normalized difference vegetation index (NDVI). Land-surface temperature (LST) can be defined as the thermal emission from the landscape "surface", including the top of the canopy, for vegetated surfaces, as well as other surfaces (e.g. bare soils). Jinn and Dickinson [3] referred that LST controlled the surface heat and water exchange with atmosphere. Estimation of LST, from satellites infrared radiometers, has been proven useful. Most studies have focused on the use of polar orbiting satellite systems because of their high spatial resolution [4]. In estimation of LST from satellite thermal data, the digital number (DN) of image pixels needs to be converted into spectral radiance, using the sensor calibration data [5]. Landsat-7 Enhanced Thematic Mapper (ETM) scenes include a thermo band (band 6), which can detect thermal radiation released from objects on the earth surface, in addition to the red and near-infrared bands (band 3 and band 4). A vegetative index is a value derived from sets of remotely-sensed data that are used to quantify the vegetative cover on the Earth's surface. The NDVI is calculated as a ratio between measured reflectivity in the red and near infrared portions of the electromagnetic spectrum. These two spectral bands are chosen because they are most affected by the absorption of chlorophyll in leafy green vegetation and by the density of green vegetation on the surface. Also, in red and near-infrared bands, the contrast between vegetation and soil is at a maximum [6]. In this study, Landsat 7 ETM was used in order to estimate LST over Wadi Bisha (South KSA) and compared it with vegetation cover.

## 2. OBJECTIVES

The main objective of the current study is to investigate the generality of the LST-NDVI relationship over a wide range of moisture and climatic/radiation conditions, through the following steps:

1) Estimate NDVI Value from Satellite Images, using red and near-infrared bands (band 3 and band 4 of Landsat ETM 7).
2) Calculation of Land Surface Temperature (LST) from Satellite remote sensing data using the thermal band (band 6).
3) Finding out the relationship between normalized difference vegetation index (NDVI) and the land Surface Temperature (LST) in WadiBisha.
4)

## 3. STUDY AREA

The study area is located at the South of the Kingdom of Saudi Arabia (KSA) as shown in (Figure 1). The study area is characterized by diversity of its natural terrain, including plains, mountains and valleys where graded region height of (750 - 2850 m) above sea level (Figure 2).

## 4. DATA

The ETM data used in this study were obtained from a remote sensing satellite (Landsat 7) launched by National Aeronautics and Space Administration (NASA) of USA. A number of two images were used (Figure 3), images characteristics [7] are shown in Table 1.

## 5. METHODOLOGY AND ANALYSES

In order to achieve the objectives of this study, the activities were elaborated in six main stages, while conducting subdividing the study area “WADI BISHA” into 4 parts to facilitate the application and analyses.

## 5.1. Data Collection, Processing and Calibration of Raw Data and Geometric Correction

The image processing software system (Erdas Imagine 9.3) was used for the geometric correction; calibration and processing of raw satellite data [8].

## 5.2. Conversion of Thermo Band Digital Number (DN) to Spectral Radiance (L)

For Landsat images that are not in the original United States Geological Survey (USGS) "GeoTIFF with Metatdata" format, it was needed to manually convert these data to radiance values. There are two formulas that can be used to convert DN's to radiance; the method depends on the scene calibration data available in the header file ([9] and [10] . One method uses the Gain and Bias (or Offset) values from the header file. The longer method uses the LMin and LMax spectral radiance scaling factors. The following formulas have been adapted from the USGS Landsat Users handbook [11].

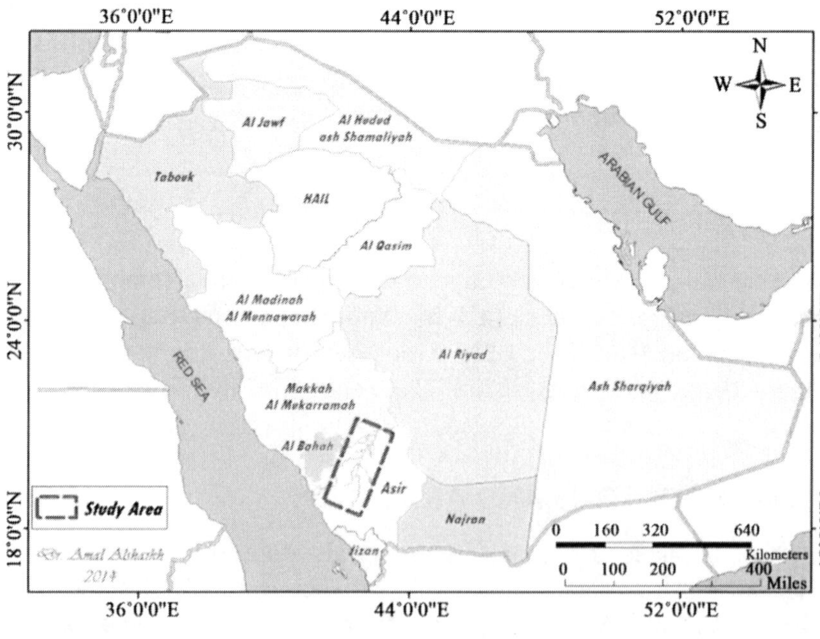

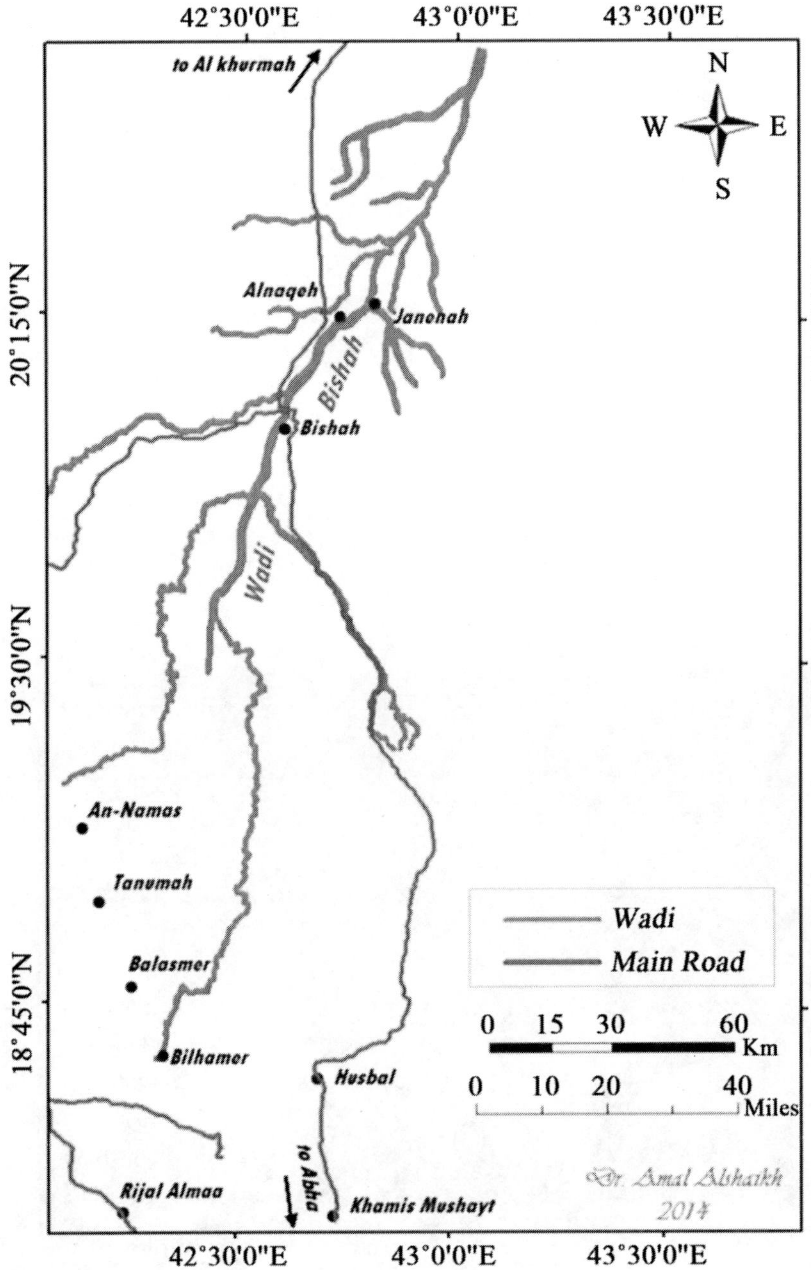


**Figure 1**. Location map of study area (between (18° - 21°) N longitude and (42° - 43°) E latitude.

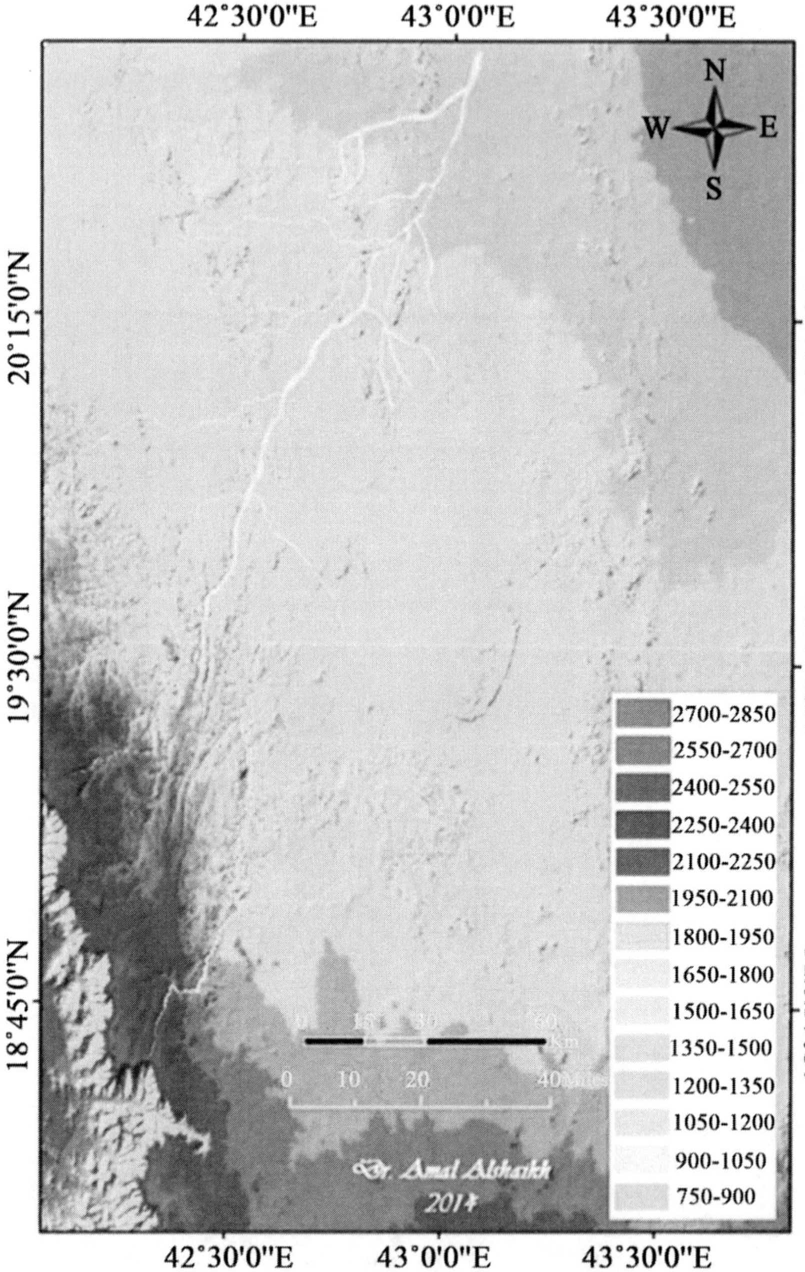


**Figure 2**. Topography of study area (altitude over sea level).

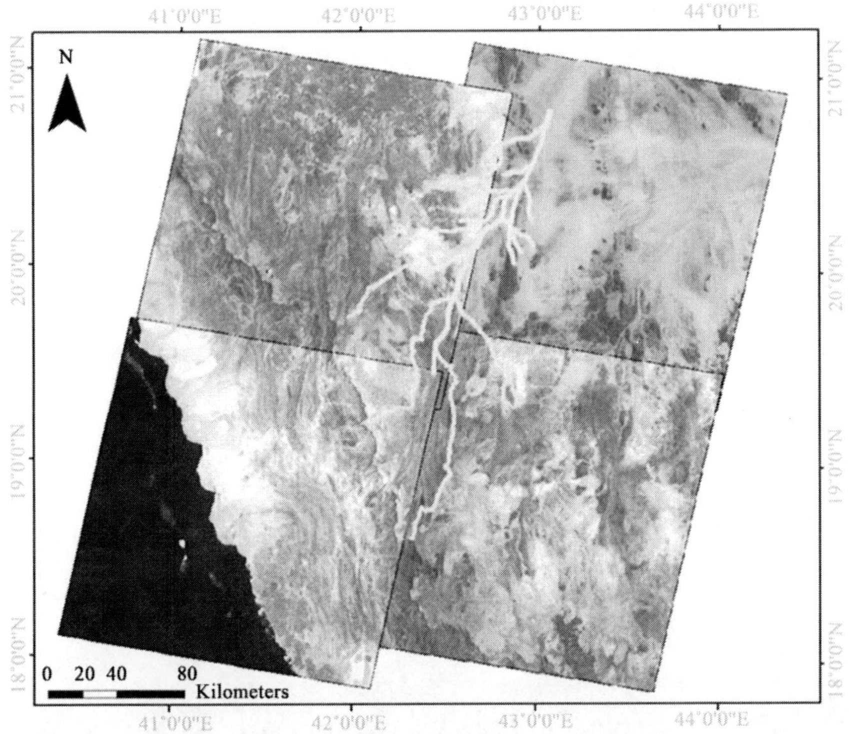


**Figure 3**. Satellite images covering study area.

**Table 1**. Spectral characteristics of Landsat 7 ETM images.

| Band | Wave length (Micrometers) | Spatial resolution (Meters) |
|---|---|---|
| Band 1 | 0.45 - 0.52 | 30 |
| Band 2 | 0.52 - 0.60 | 30 |
| Band 3 | 0.63 - 0.69 | 30 |
| Band 4 | 0.77 - 0.90 | 30 |
| Band 5 | 1.55 - 1.75 | 30 |
| Band 6 | 10.40 - 12.50 | 60 |
| Band 7 | 2.9 - 2.35 | 30 |

METADATA; SPACECRAFT_ID = "Landsat7"; SENSOR_ID = "ETM+"; ACQUISITION_DATE = 2000-08-24 and 2000-06-21; WRS_PATH = 167; WRS_ROW = 046 and 047; REFERENCE_DATUM = "WGS8"; REFERENCE_ELLIPSOID = "WGS84"; MAP_PROJECTION = "UTM (Universal Transverse Mercator)"; ZONE_NUMBER = +38.

A. Gain and Bias Method:

The formula to convert DN to radiance using gain and bias values is:

$$CV_R = G(CV_{DN}) + B \tag{1a}$$

where:

$CV_R$ is the cell value as radiance,

$CV_{DN}$ is the cell value digital number,

G is the gain value for a specific band,

B is the bias value for a specific band,

B. $L_{Min}$ and $L_{Max}$ Method:

The formula to convert DN to radiance using $L_{min}$ and $L_{max}$ values is:

$$L = LMIN + (LMAX - LMIN) * DN/255 \tag{1b}$$

where:

L = Spectral radiance,

LMIN = (Spectral radiance of DN value 1),

LMAX = (Spectral radiance of DN value 255),

DN = Digital Number.

Table 2 lists these values for ETM band 6 for the two images

Using these values in Table 2 absolute radiance values can calculate for each pixel in the two images. It should be noticed that the radiance values are real numbers, while the DN values in Band 6 are integers.

## 5.3. Conversion of Spectral Radiance to Temperature in Kelvin

Once the DN's values for the thermal bands have been converted to radiance values, it is simply a matter of applying the inverse of the Planck function to derive temperature values.

The formula to convert radiance to temperature is:

$$T = K_2 / Ln\{(K_1 * \varepsilon)/CV_{R1} + 1\} \quad (2)$$

where:

T = Effective at-satellite temperature in Kelvin,

$K_2$ = Calibration constant 2 from Table 3,

$K_1$ = Calibration constant 1 from Table 3,

$CV_{R1}$ = is the cell value as radiance,

ε = is emissivity (typically 0.95).

## 5.4. Conversion of Temperatures

The temperature values are estimated in degrees Kelvin, and are then converted to degree Celsius.

## 5.5. Calculation (NDVI) Values

A vegetative index is a value that is derived from sets of remotely-sensed data that is used to quantify the vegetative cover on the Earth's surface. The NDVI is calculated as a ratio between measured reflectivity in the red and near infrared portions of the electromagnetic spectrum. These two spectral bands are chosen because they are most affected by the absorption of chlorophyll in leafy green vegetation and by the density of green vegetation on the surface. Also, in red and near-infrared bands, the contrast between vegetation and soil is at a maximum.

data sets with the following formula.

**Table 2**. Values of LMIN and LMAX for the Landsat 7 ETM.

| Parameters | LMIN | LMAX |
|---|---|---|
| Landsat-7 ETM P/R-167/46 (2000-08-24) | 3.200 | 12.65 |
| Landsat-7 ETM P/R-167/47 (2000-06-21) | 3.200 | 12.65 |

**Table 3**. Landsat 7 ETM thermal band calibration constants.

| Satellite | Constant 1-K1 watts/(meter squared/ster/µm) | Constant 2-K2 Kelvin |
|---|---|---|
| Landsat 7 | 666.09 | 1282.71 |

The Thematic Mapper bands 3 and 4 provide red and near-infrared measurements respectively and therefore can be used to generate NDVI

$$NDVI = (NIR - RED)/(NIR + RED)$$

$$NDVI = (Band4 - Band3)/(Band4 + Band3) \quad (3)$$

## 5.6. Relationship between Vegetation Cover and (LST)

Software (ArcGIS-ArcInfo 10.0) were used to prepare maps that represent the relationship between vegetation cover and (LST) in addition to the preparation of all study result maps.

# 6. RESULTS

## 6.1. VC and LST Pattern and Interrelationship in Wadi Pisha

The relationship, between Vegetation Cover (VC) and (LST) of the Four Parts of Wadi Bisha, are presented in the following Figures, Tables, and graphs. Regarding the first part, Figure 4 & Figure 5and Table 4 demonstrate the distribution of LST and VC respectively. Table 4 demonstrates the relationship between the two parameters. It is found that LST values range between 30˚C to 48˚C while NDVI values between 0.19 and 0.34. Figure 6 shows a negative correlation coefficient of −0.96 between the two parameters.

The first part is characterized as follow: LST (Kelvin) = (303.1976768 - 320.5987138). LST (Celsius) = (30˚C - 48˚C). NDVI Value (0.19 - 0.34). R = −0.98322. $R^2$ = 0.966 (Figure 6).

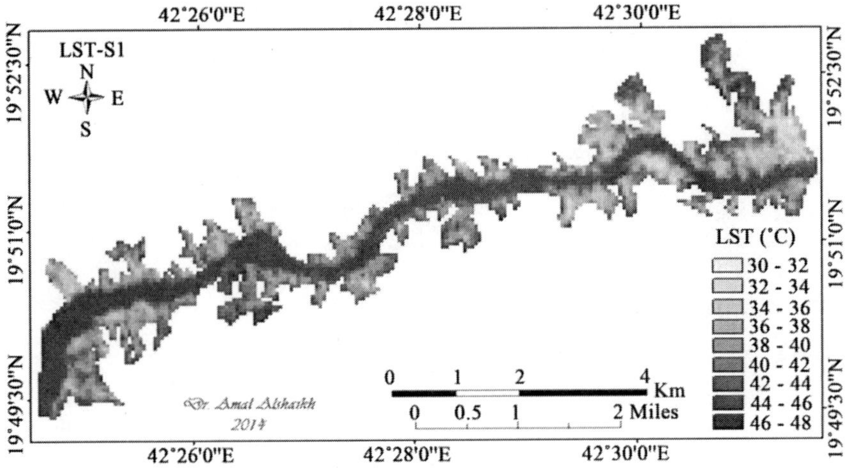


**Figure 4**. LST (Celsius) in the first part.

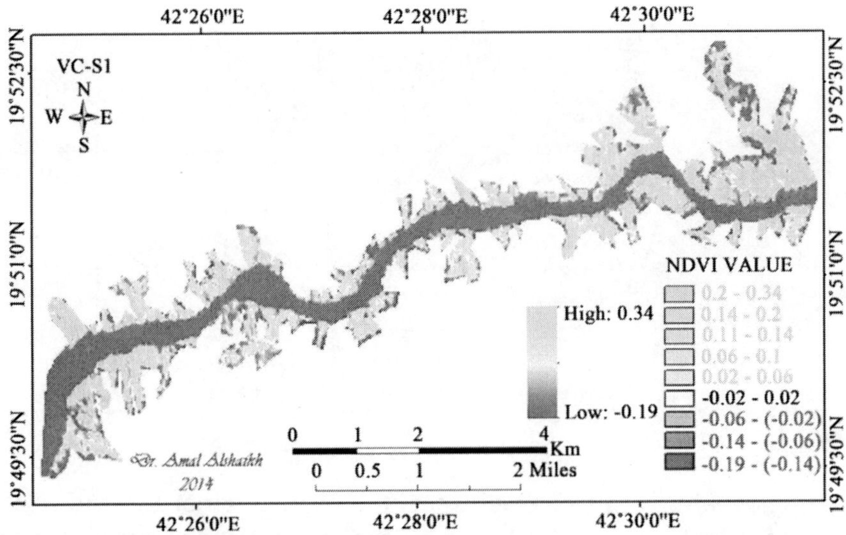


**Figure 5**. NDVI value in the first part.

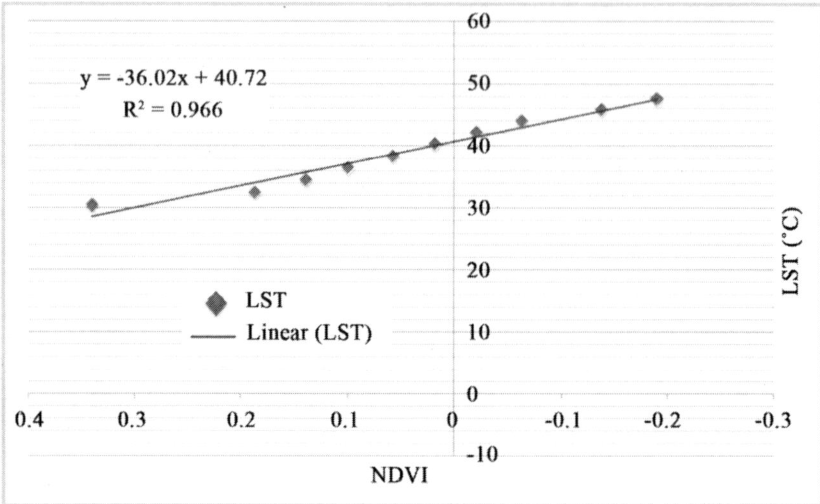


**Figure 6**. Relationship between LST and NDVI at part 1.

**Table 4**. LST and NDVI measurement at part 1.

| Pixel Value | Radiance | LST (Kelvin) | LST (Degree) | NDVI_VALUE |
|---|---|---|---|---|
| 166 | 9.338779528 | 303.1976768 | 30.1976768 | 0.3405 |
| 174 | 9.636417323 | 305.4292583 | 32.4292583 | 0.1868 |
| 181 | 9.896850394 | 307.3512105 | 34.3512105 | 0.139 |
| 188 | 10.15728346 | 309.2458629 | 36.2458629 | 0.09957 |
| 195 | 10.41771654 | 311.1143996 | 38.1143996 | 0.05802 |
| 203 | 10.71535433 | 313.2193053 | 40.2193053 | 0.01855 |
| 210 | 10.9757874 | 315.0355067 | 42.0355067 | −0.02092 |
| 217 | 11.23622047 | 316.828828 | 43.828828 | −0.06247 |
| 224 | 11.49665354 | 318.6001664 | 45.6001664 | −0.1372 |
| 232 | 11.79429134 | 320.5987138 | 47.5987138 | −0.1892 |

In regard with part 2 of Wadi Bisha, the distribution of LST and NDVI values are shown in Figure 7and Figure 8 respectively. Table 5 demonstrates the values of the two parameters in different locations. It is found that LST ranges between 30˚C to 49˚C, while NDVI between 0.20 and 0.48.Table 5 and Figure 9 show that a negative correlation coefficient of 0.98 cauterizes the relation between the two parameters.

The second part is characterized as follow: (Figure 7 and Figure 8) (Table 5). LST (Kelvin) = (302.633713 - 322.3257189). LST (Celsius) = (30˚C - 49˚C). NDVI Value (0.2 - 0.48). R = −0.96896. $R^2$ = 0.938 (Figure 9).

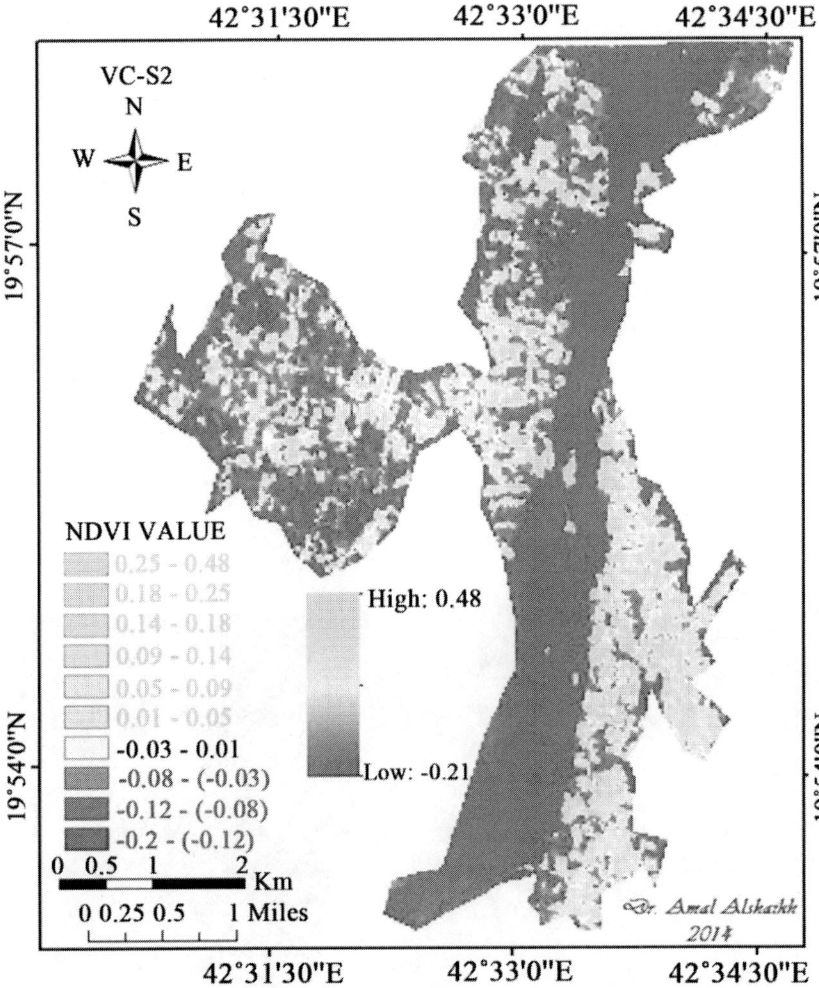


**Figure 7**. LST (Celsius) in the second part.

Distribution of LST and NDVI, at the third part 3 of Wadi Bisha, is shown in Figure 10 and Figure 11respectively. Table 6 demonstrates the values of the two parameters in different locations. It is found that LST ranges between 31˚C to 46˚C, while NDVI ranged between 0.21 and 0.29. A negative correlation coefficient of 0.98 characterizes the relation between the two parameters.

The third part is characterized as follow: LST (Kelvin) = (304.3182571 - 319.3527926). LST (Celsius) = (31˚C - 46˚C). NDVI Value (0.21 - 0.29). R = −0.97987. $R^2$ = 0.960 (Figure 12).

Regarding the Forth part of Wadi Bisha, the distribution of LST and NDVI is shown in Figure 13 andFigure 14 respectively. It is found that LST ranges between 28˚C to 47˚C, while NDVI lies between −0.30 and 0.42. A negative correlation coefficient 0.98 (Table 7 and Figure 15) is found between the two parameters.

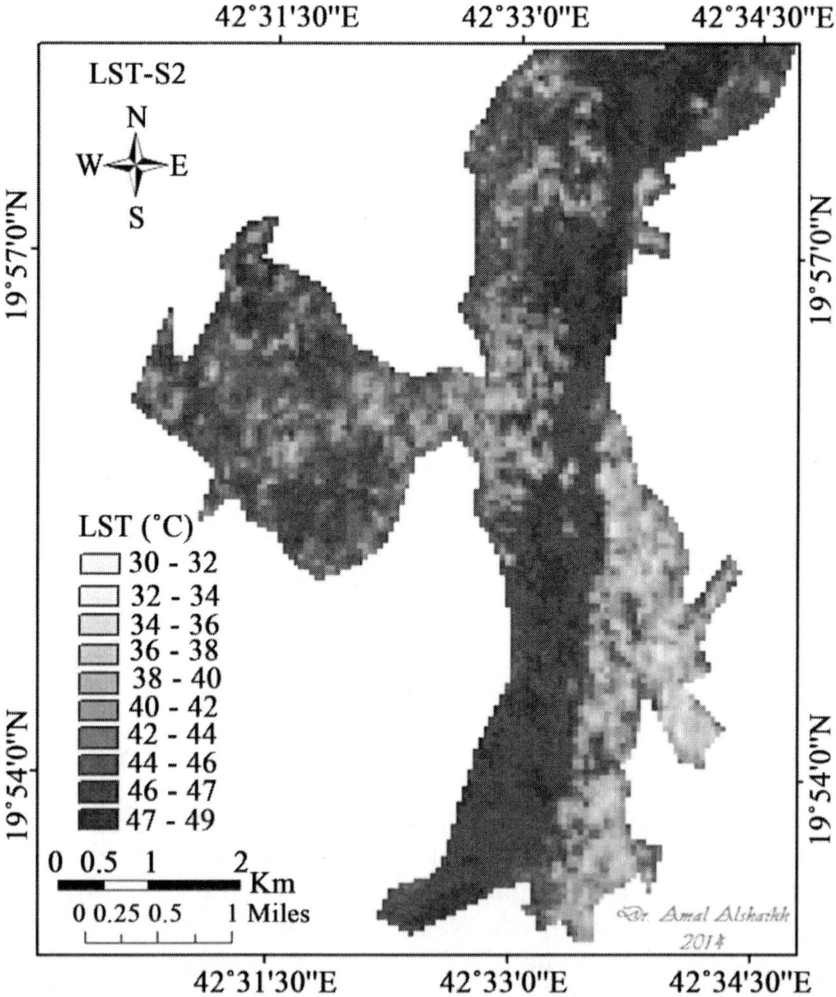


**Figure 8**. NDVI Value in the second part.

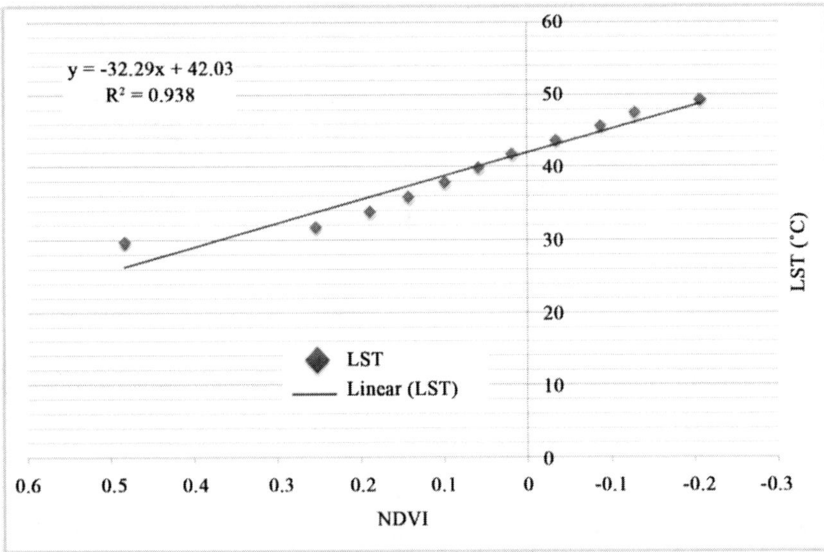


**Figure 9**. Relationship between LST and NDVI at part 2.

**Table 5**. LST and NDVI measurement at part 2.

| Pixel Value | Radiance | LST (Kelvin) | LST (Degree) | NDVI_VALUE |
|---|---|---|---|---|
| 164 | 9.264370079 | 302.633713 | 29.633713 | 0.4839 |
| 171 | 9.52480315 | 304.5968955 | 31.5968955 | 0.25346 |
| 179 | 9.822440945 | 306.804921 | 33.804921 | 0.1884 |
| 186 | 10.08287402 | 308.7072487 | 35.7072487 | 0.14231 |
| 194 | 10.38051181 | 310.849022 | 37.849022 | 0.098943 |
| 201 | 10.64094488 | 312.6960558 | 39.6960558 | 0.058281 |
| 209 | 10.93858268 | 314.7774719 | 41.7774719 | 0.01762 |
| 216 | 11.19901575 | 316.5740054 | 43.5740054 | −0.033885 |
| 224 | 11.49665354 | 318.6001664 | 45.6001664 | −0.0881 |
| 231 | 11.75708661 | 320.3503644 | 47.3503644 | −0.12876 |
| 239 | 12.05472441 | 322.3257189 | 49.3257189 | −0.20737 |

LST and NDVI measurement at several locations). LST (Kelvin) = (306.2563744 - 321.5880036). LST (Celsius) = (33°C - 49°C). NDVI value (0.169 - 0.229). R = −0.942713. $R^2$ = 0.888.

The forth part is characterized as follow: (Figure 13 and Figure 14) (Table 7). LST (Kelvin) = (300.6399705 - 319.8524179). LST (Celsius) =

(28°C - 47°C). NDVI value (0.303 - 0.42). R = - 0.99245. $R^2$ = 0.985 (Figure 15).

## 6.2. Validation of VC-LST Relationship

Finally, a number of 10 geographic locations were chosen randomly to represent the study area, for measuring both LST and NDVI. The measurements show that LST ranges between 33°C to 49°C, while NDVI values range is 0.16 to 0.23. Table 8 and Figure 16 show that a negative correlation coefficient of 0.94 characterizes the relation between the two parameters.

Table 9 and Table 10 summaries the overall obtained results related to SLT and NDVI values found in different study portions. It can be outlined that the abundance of vegetation cover is one of the most influential in controlling (LST) in Wadi Bisha. As per the (NDVI values) the data indicates that areas having lowest surface temperatures were rich in vegetation (dense vegetation). A clear example may be highlighted at the part 4 of Wadi Pisha, where the Lowest LST was (28°C) and highest NDVI (0.41) are recorded. Also, part 2 includes the bar land area of highest LST (49°C) characterized by the lowest NDVI value (−0.20).

The statistical test of the obtained data identified the relationship between vegetation cover and the land surface temperature. The correlation between the (VC) and the (LST) over (50) locations in the study area is quite significant. This means that the increased vegetation cover (VC) indicates decrease in (LST), and the decreased vegetation cover (VC) indicates increased (LST).

**Table 6**. LST and NDVI measurement at part 3.

| Pixel Value | Radiance | LST (Kelvin) | LST (Degree) | NDVI_VALUE |
|---|---|---|---|---|
| 170 | 9.487598425 | 304.3182571 | 31.3182571 | 0.2874 |
| 177 | 9.748031496 | 306.2563744 | 33.2563744 | 0.1675 |
| 184 | 10.00846457 | 308.1664763 | 35.1664763 | 0.1224 |
| 191 | 10.26889764 | 310.049796 | 37.049796 | 0.08309 |
| 198 | 10.52933071 | 311.9074828 | 49.000000 | 0.0497 |
| 205 | 10.78976378 | 313.7406093 | 40.7406093 | 0.01631 |
| 212 | 11.05019685 | 315.5501786 | 42.5501786 | −0.03082 |
| 219 | 11.31062992 | 317.33713 | 44.33713 | −0.1506 |
| 227 | 11.60826772 | 319.3527926 | 46.3527926 | −0.2135 |

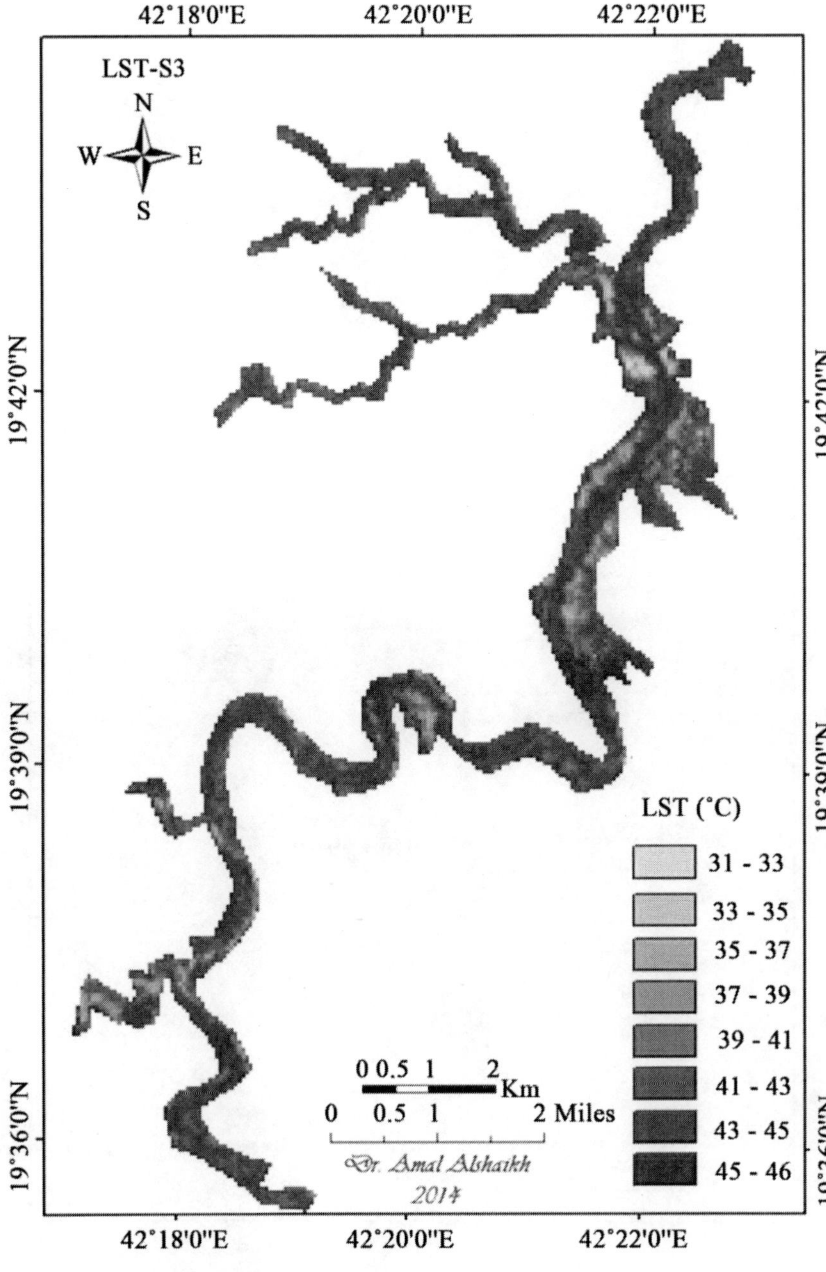


**Figure 10**. LST (Celsius) in the third part.

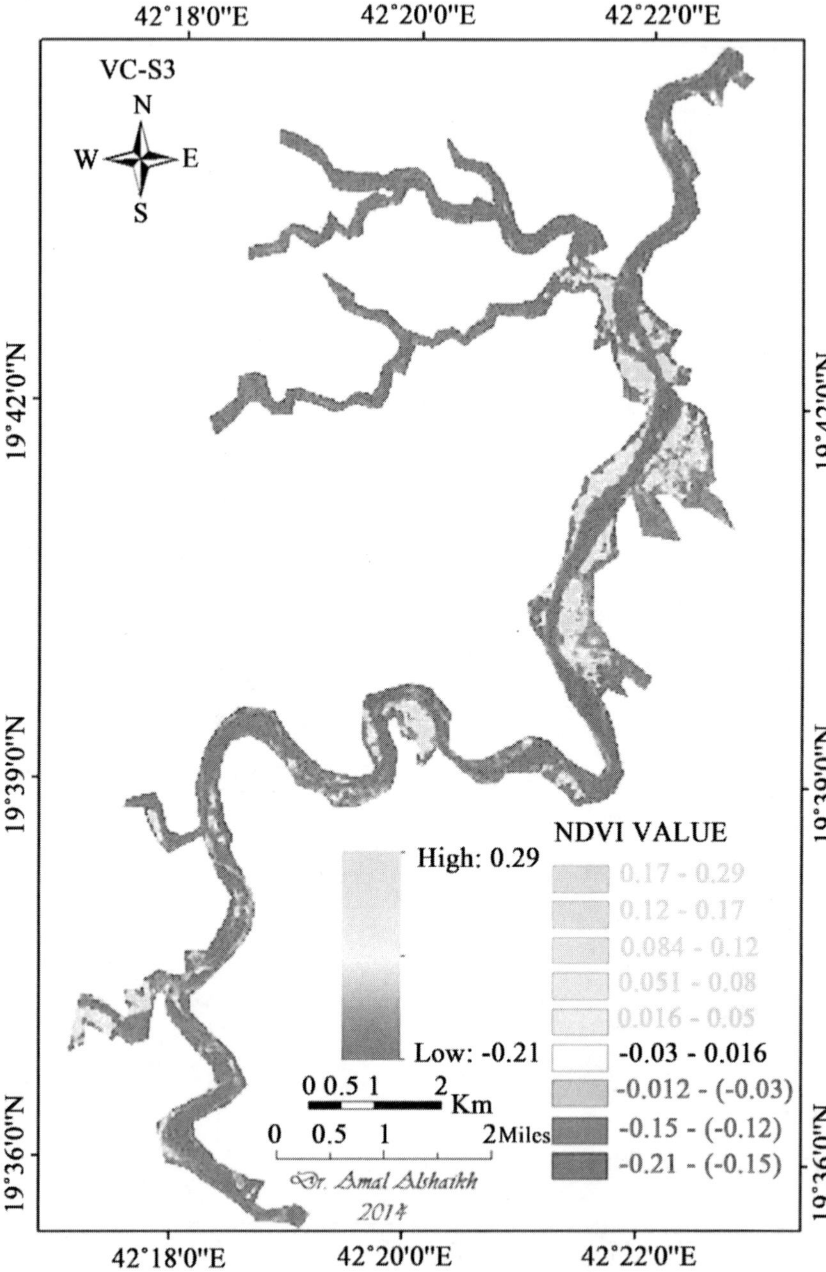


**Figure 11**. NDVI Value in the third part.

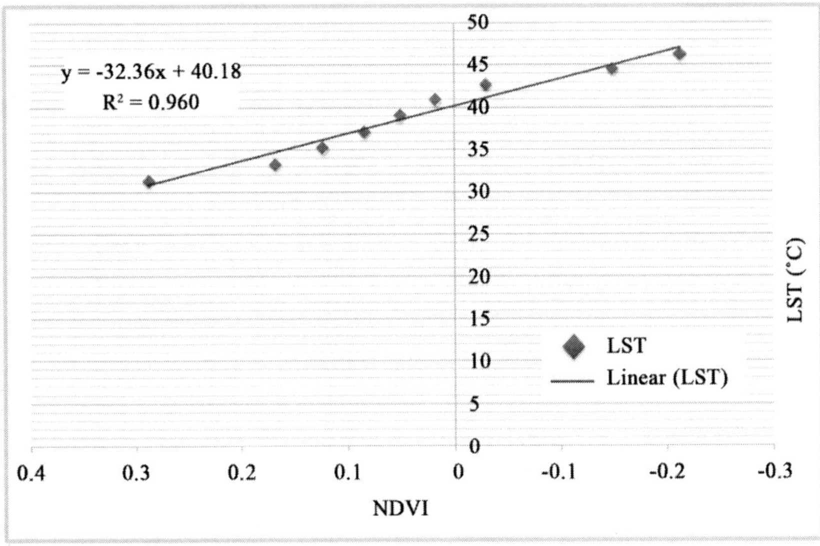


**Figure 12**. Relationship between LST and NDVI at part 3.

**Table 7**. LST and NDVI measurement at part 4.

| Pixel Value | Radiance | LST (Kelvin) | LST (Degree) | NDVI_VALUE |
|---|---|---|---|---|
| 157 | 9.003937008 | 300.6399705 | 27.6399705 | 0.4170 |
| 164 | 9.264370079 | 302.633713 | 29.633713 | 0.3097 |
| 171 | 9.52480315 | 304.5968955 | 31.5968955 | 0.2306 |
| 178 | 9.78523622 | 306.5309316 | 33.5309316 | 0.1543 |
| 186 | 10.08287402 | 308.7072487 | 35.7072487 | 0.0893 |
| 193 | 10.34330709 | 310.583131 | 37.583131 | 0.0271 |
| 200 | 10.60374016 | 312.4336943 | 39.4336943 | −0.0293 |
| 207 | 10.86417323 | 314.2599907 | 41.2599907 | −0.0773 |
| 214 | 11.1246063 | 316.0630045 | 43.0630045 | −0.1197 |
| 221 | 11.38503937 | 317.8436579 | 44.8436579 | −0.1592 |
| 229 | 11.68267717 | 319.8524179 | 46.8524179 | −0.3033 |

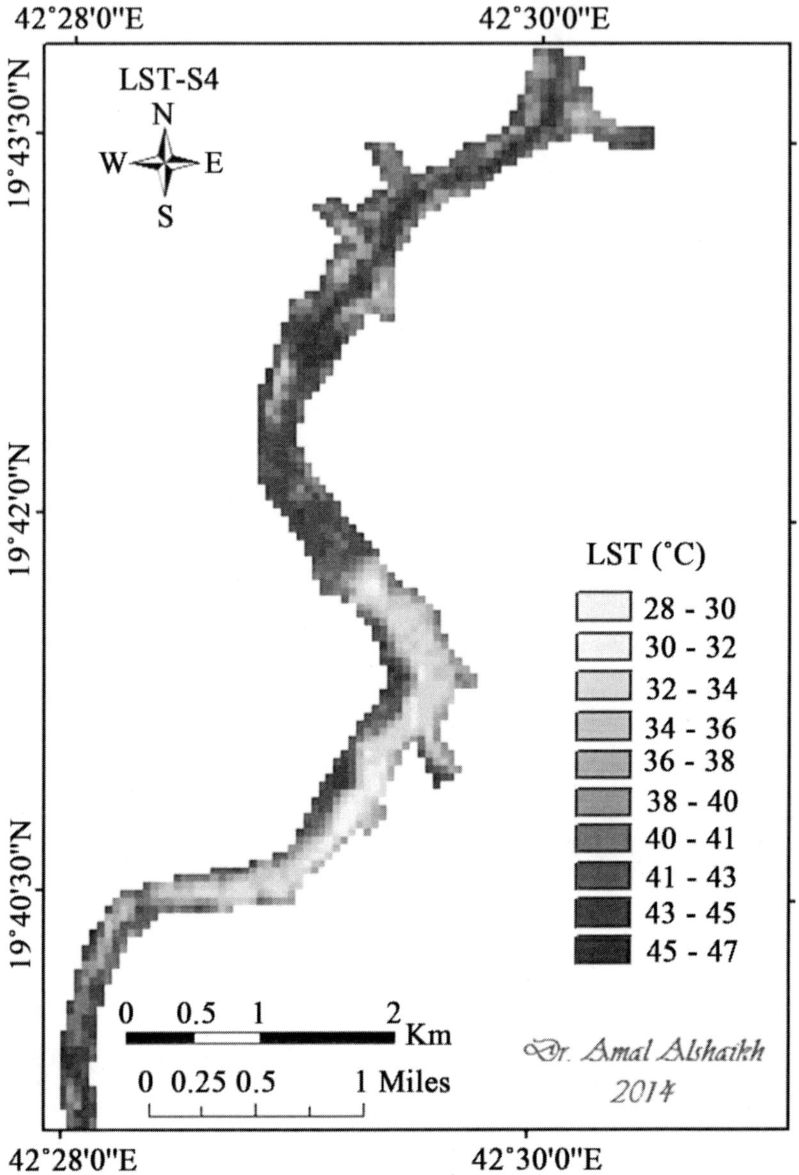


**Figure 13**. LST (Celsius) in the fourth part.

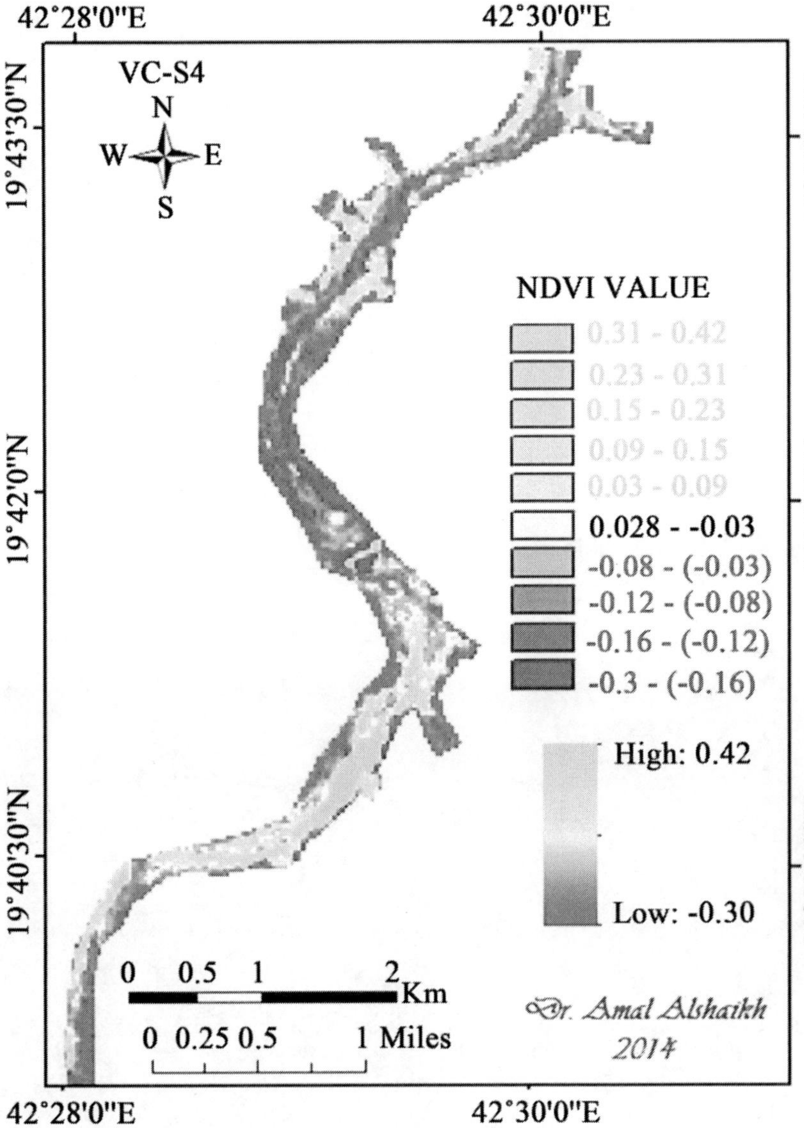


**Figure 14**. NDVI value in the fourth part.

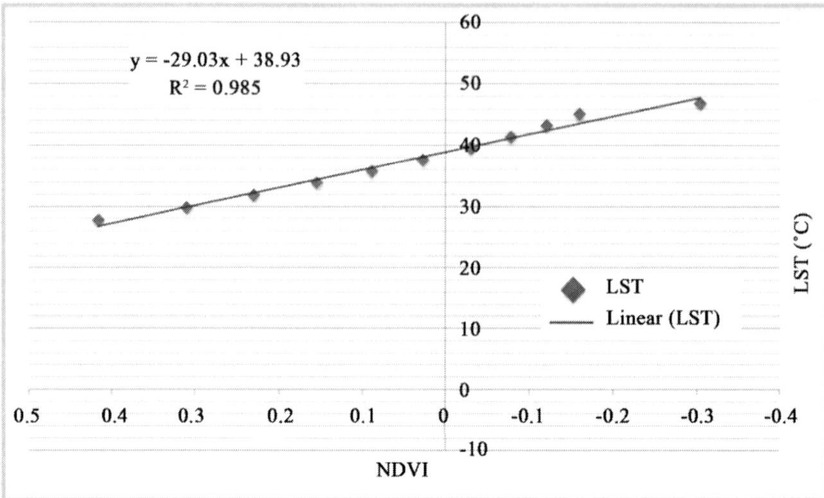


**Figure 15**. Relationship between LST and NDVI at part 4.

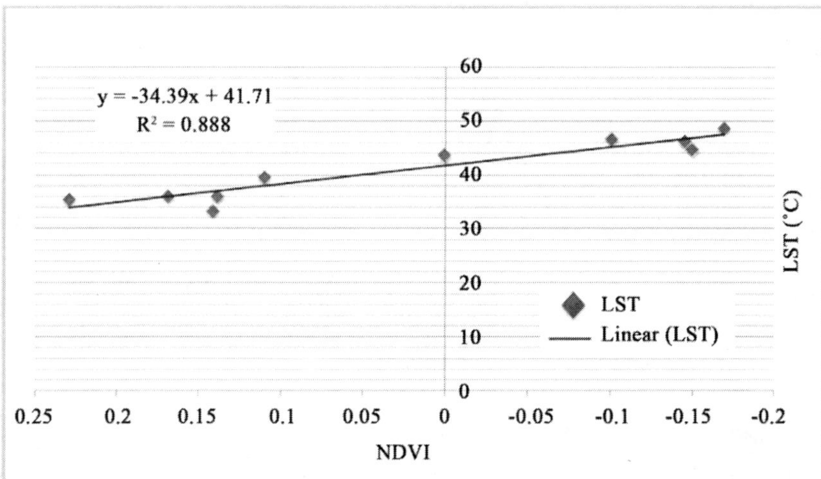


**Figure 16**. Relationship between LST and NDVI at several locations.

**Table 8**. LST and NDVI measurement at several locations.

| No. | Lat | Long | LST (Kelvin) | LST (Degree) | NDVI_VALUE |
|---|---|---|---|---|---|
| 1 | 19 50 30 | 42 24 53 | 306.2563744 | 33.2563744 | 0.141 |
| 2 | 19 50 17 | 42 25 02 | 317.5906145 | 44.5906145 | −0.15 |
| 3 | 19 50 54 | 42 27 35 | 319.1023446 | 46.1023446 | −0.145 |
| 4 | 19 50 37 | 42 24 49 | 308.9768239 | 35.9768239 | 0.138 |
| 5 | 20 23 21 | 42 49 04 | 308.437134 | 35.437134 | 0.229 |
| 6 | 20 28 07 | 42 49 29 | 319.6028162 | 46.6028162 | −0.101 |
| 7 | 20 35 47 | 42 59 22 | 316.5740054 | 43.5740054 | 0 |
| 8 | 20 35 35 | 42 59 02 | 312.4336943 | 39.4336943 | 0.11 |
| 9 | 19 50 33 | 42 27 32 | 308.9768239 | 35.9768239 | 0.168 |
| 10 | 20 38 16 | 42 56 13 | 321.5880036 | 48.5880036 | −0.169 |

**Table 9**. Conclusions from analyzing the (VC) and (LST) values.

| Wadi Bisha | Part-1 | Part-2 | Part-3 | Part-4 | Several Locations |
|---|---|---|---|---|---|
| LST (Kelvin) | 303 - 320 | 302 - 322 | 304 - 319 | 300 - 319 | 306 - 321 |
| LST (Celsius) | 30°C - 48°C | 30°C - 49°C | 31°C - 46°C | 28°C - 47°C | 33°C - 49°C |
| NDVI (Values) | 0.19 - 0.34 | 0.2 - 0.48 | 0.21 - 0.29 | 0.303 - 0.42 | 0.169 - 0.229 |

**Table 10**. Correlation coefficient between (VC) and (LST).

| Wadi Bisha | Part-1 | Part-2 | Part-3 | Part-4 | Several Locations |
|---|---|---|---|---|---|
| R | 0.98322 | 0.96896 | 0.97987 | 0.99245 | 0.942713 |
| $R^2$ | 0.966 | 0.938 | 0.960 | 0.985 | 0.888 |

## 7. CONCLUSIONS

As Land surface temperature (LST) controls the surface heat and water exchange with atmosphere, Land use/ Cover is very important factor having a significant impact on Erath Ecosystem. As land surface temperature is considered the thermal landscape emission, influence of landscape on controlling surface heat and water exchange with atmosphere can be predicted. The concrete conclusion of the current research can be outlined as follows:

- Remote-sensing data is the most important data sources in land use modeling.
- Land's surface temperature is affected by many factors; the most important are richness of water and vegetation.
- There is a strong negative relationship between LST and vegetation cover.
- The temperature was increased in the urban areas.

## 8. RECOMMENDATIONS

It is possible to recommend the following:

1. Availability and continuity of Satellite remote sensing data is required to support controlling the ecosystem.
2. Continuous monitoring of vegetation cover conditions and mapping is recommended in WadiBisha for its ecological tourism potentialities.
3. Encouraging research and studies related to the drought in WadiBisha to support its environmental conservation and sustainable development.
4. Presidency of Meteorology and Environment (PME) should continue their development and ensure the adoption of flexible land cover validation protocols.

## CITE THIS PAPER

AmalyahyaAlshaikh, (2015) Vegetation Cover Density and Land Surface Temperature Interrelationship Using Satellite Data, Case Study of Wadi Bisha, South KSA. *Advances in Remote Sensing*,**04**,248-262.

## REFERENCES

1. Ahmed, A., Noorazuan, H. and Zolkepli, B. (2006) Estimation of Land Surface Temperature Using Landsat Tm Thermal Infrared in Selangor-Negeri Sembilan. Proceedings of the National Seminar on Science and Its Applications in Industry (SSASI 2006), Malacca, 14-15 February 2006.

2. Hassan, Q.K. and Rahman, K.M. (2012) Applicability of Remote Sensing-Based Surface Temperature Regimes in Determining Deciduous Phenology over Boreal Forest. Journal of Plant Ecology, 6, 84-91.

3. Dickinson, R.E. (2010) Land Surface Skin Temperature Climatology: Benefitting from the Strengths of Satellite Observations. Environmental Research Letters, 5, Article ID: 044004.

4. Sun, D. and Pinker, T. (2004) Case Study of Soil Moisture Effect on Land Surface Temperature Retrieval. IEEE Geoscience and Remote Sensing Letters, 1, 127.

5. Markham, B.L. and Barker, J.L. (1986) Landsat MSS and TM Post-Calibration Dynamic Rangers, Exoatmospheric Reflectance and At-Satellite Temperatures. EOSAT Landsat Tech. Notes, 3-8.

6. Schott, J.R. and Volchok, W.J. (1985) Thematic Mapper Thermal Infrared Calibration. Photogrammetric Engineering and Remote Sensing, 51, 1351-1357.

7. Townshend, J.R.G. and Justice, C.O. (1986) Analysis of the Dynamics of African Vegetation Using the Normalized Difference Vegetation Index. International Journal of Remote Sensing, 8, 1189-1207.

8. Barsi, J.A., Barker, J.L. and Schott, J.R. (2003) An Atmospheric Correction Parameter Calculator for a Single Thermal Band Earth-Sensing Instrument. IGARSS 03, Toulouse, 21-25 July 2003.

9. Govindha Raj, B. and Fleming, K. (2008) Surface Temperature Estimation from Landsat ETM+ Data for a Part of the Baspa Basin, NW Himalaya, India. Bulletin of Glaciological Research, 25, 19-26.

**10.** Sobrino, J.A., Jiménez-Mu&ntildeoz, J.C., Zarco-Tejada, P.J., Sepulcre-Cantó, G. and de Miguel, E. (2006) Land Surface Temperature Derived from Airborne Hyperspectral Scanner Thermal Infrared Data. Remote Sensing of Environment, 102, 99-115.

# Chapter 6

# Investigating the Use of Remote Sensing and GIS Techniques to Detect Land Use and Land Cover Change: A Review

**Abdullah F. Alqurashi[1,2], Lalit Kumar[1]**

[1]*School of Environmental and Rural Science, University of New England, Armidale, Australia*

[2]*Department of Geography, Umm Al-Qura University, Makkah, Saudi Arabia*

## ABSTRACT

The accuracy of change detection on the earth's surface is important for understanding the relationships and interactions between human and natural phenomena. Remote Sensing and Geographic Information Systems (GIS) have the potential to provide accurate information regarding land use and land cover changes. In this paper, we investigate the major techniques that are utilized to detect land use and land cover changes. Eleven change detection techniques are reviewed. An analysis of the related literature shows that the most used techniques are post-classification comparison and principle component analysis. Post-classification comparison can minimize the impacts of atmospheric and sensor differences between two dates. Image differencing and image ratioing are easy to implement, but at times they do not provide accurate results. Hybrid change detection is a useful technique that makes full use of the benefits of many techniques, but it is complex and depends on the characteristics of the other techniques such as supervised and unsupervised classifications. Change vector analysis is complicated to implement, but it is useful for providing the direction and magnitude of change. Recently, artificial neural networks, chi-square, decision tree and image fusion have been frequently used in change detection. Research on integrating remote sensing data and GIS into change detection has also increased.

**Keywords:** Change Detection Techniques; Remote Sensing; GIS; Land Use and Land Cover Change

## 1. INTRODUCTION

Change detection is the process of identifying differences in the state of an object or phenomenon by observing it at different times [1]. The time and accuracy of change detection on the earth's surface can provide a better understanding of the relationships and interactions between human and natural phenomena and can provide guidance in the management of the use of resources. In change detection applications, it is necessary to use multi-temporal datasets to analyze the temporal effects of the object or phenomenon [2]. Currently, with increased computer capability and data availability, Remote Sensing and Geographic Information Systems (GIS) have become effective tools for detecting objects and phenomena change.

Remote Sensing means the ability to detect change on the earth's surface through space-borne sensors [3]. The repetitive coverage of satellite images and the improvement of image quality can provide valuable assistance in the identification of changes [4]. Temporal and spatial resolutions allow scientists to monitor and detect changes over a broad scale and help planners to obtain or maintain information on various phenomena, such as shifting agricultural patterns, crop stress, disaster monitoring, land use and land cover changes [5].

Geographic Information System is a useful tool for measuring the change between two or more time periods. It has the ability to incorporate multi-sources of data into a change detection platform [2]. For example, the use of multiple layers, such as classified images, topographical maps, soil maps and hydrological maps, provides a greater ability to extract useful information about the changes over a particular area. Moreover, GIS can measure the trends in these changes by modeling the available data and using statistical and analytical functions. The benefit of GIS is the provision of different outputs in different formats (e.g. maps or tables), which allows users to select the appropriate output for extracting the desired information.

Remotely sensed data and GIS are widely used for detecting land use and land cover changes. Many studies have attempted to use remotely sensed data and GIS to address land use change detection e.g. [6-9]. A variety of procedures or methods of Remote Sensing technologies are used to detect land use and land cover changes. Some studies have actually utilized Remote Sensing techniques; others have integrated remotely sensed data with GIS data e.g. [10-12]. In addition, many studies have reviewed and summarized the various change detection techniques [1,2,5,13-15]. The articles reviewed here consider these change detection techniques in many applications, including land use and land cover changes, vegetation and forest changes, urban changes, environmental changes, crop monitoring, forest fires, deforestation and other applications. **Table 1**shows examples of applications that are investigated using change detection techniques.

In this paper, we firstly review the major land use and land cover change detection techniques, including image differencing, image ratioing, change vector analysis (CVA), principal component analysis (PCA), chi-square, post-classification comparison, decision trees, image fusion, hybrid change detection, artificial neural networks (ANN) and Geographic Information Systems (GIS), by giving overview about the characteristics, strengths and weaknesses of each technique. Second, we discuss the importance of accuracy assessment in change detection and the use of different matrices in previous studies. Finally, we evaluate the change detection techniques according to the analysis of related literature.

## 2. LAND USE AND LAND COVER CHANGE TECHNIQUES

### 2.1. Image Differencing

Image differencing, also referred to as image delta [16], is a simple, easy-to-use technique for implementing and interpreting change detection [17]. It divides the image pixels into two results: change or no change. The process by which these results are obtained involves subtracting a pixel's digital number on the image for date one from the corresponding pixel's digital number on the image for date two. The general process for detecting the change in two dates in image differencing is extracting the change of the image of date 2 from the image of date 1 (e.g. image of date 1—image of date 2) [18]. In this technique, it is necessary to select thresholds for determining the changed area [16,19]. However, the image differencing technique cannot provide sufficient information about the change itself. Atmospheric and other non-surface radiance characteristics can affect the results of image differencing [20].

Image differencing is widely utilized for change detection in the geographical environment [16,21-23]. It has been used either as a single-band difference [24] or as a color composite of three bands [25]. Sohl [22] examined five change detection techniques, including univariate image differencing, "enhanced" image differencing, vegetation index differencing, post-classification differencing and change vector analysis, to detect the landscape change in the Abu Dhabi Emirate using Landsat Thematic Mapper (TM) data. He found that the "enhanced" image differencing technique provided the most accurate values of change when compared to other techniques, while change vector analysis was a useful technique for providing rich qualitative detail about the nature of the change. Ridd and Liu [26] compared four techniques for change detection in an urban environment; image differencing, image regression, tasseled cap transformation and chi-square were used. The results showed that the most accurate technique for detecting change was the regression of TM band 3, whereas the image differencing of TM 4 was found to be the least accurate.

In arid and semi-arid environments, red band image differencing is more effective than using the Normalized Difference Vegetation Index (NDVI) for vegetation change detection [27]. Also, Pilon, Howarth [28] concluded that visible red band data provided the most accurate identification of spectral change for their semi-arid study area in north-western Nigeria in Sub-Sahelian Africa.

**Table 1**. Examples of the applications that can be investigated using change detection techniques.

| Application | Most Commonly Used Techniques | Examples |
|---|---|---|
| Land use/Land cover change | Image differencing, image ratioing, NDVI, CVA, PCA, chi-square, post-classification, hybrid change detection, ANN, decision tree, GIS | [21,22,26,31,33-41] |
| Urban change | Image differencing, post-classification, hybrid change detection, PCA, GIS, chi-square, image fusion | [26,42-44] |
| Environmental change | NDVI, ANN, CVA, post-classification, image differencing | [30,38,45] |
| Vegetation change | NDVI, CVA, image differencing, post-classification | [46-48] |
| Landscape change | Post-classification, GIS | [49-51] |
| Deforestation | Post-classification, NDVI, image differencing, PCA | [52-54] |
| Wetland change | Post-classification, GIS | [55-57] |

## 2.2. Image Ratioing

Image ratioing is extracting the information between two or more different images by using the same bands of two or more images. For example, to compute changes between two times using band 2, the simple process of image ratioing could be: (Band 2 of $t_1$ divided by Band 2 of $t_2$). It is used to highlight subtle variations between the pixels of the various land covers. In image ratioing, the unchanged pixel takes the same number for both dates with a grey level. The changed pixel takes a different value and is displayed at a lighter or darker level. To display the different changes between two or more images, selecting the appropriate threshold value is important in image ratioing. The image ratioing technique is useful for extracting vegetation cover information. The advantage of this technique is that the effect of shadows, the radiation change, image noise and the sun angle can be reduced [29]. However, there are two major disadvantages: it is difficult to select the threshold value, and the types of changes cannot be analyzed [2].

Image ratioing was used by Howarth and Wickware [30] to generate three different color enhancements—an overlay of band 5, ratios of bands 5 and 7 and a vegetation index. The results showed that the major change was emphasized by using band ratioing and band 5 overlay, which gave additional information about the change, while the vegetation index enhancement results showed less information about the change. Nelson [31] tested three change detection techniques, including image differencing, image ratioing and vegetation index differencing, to determine the most appropriate technique for detecting changes due to gypsy moth defoliation. The results indicated that the vegetation index difference and the band 5 ratio were more accurate for delineation of forest canopy changes than other single bands or band combinations. Prakash and Gupta [32] used image ratioing and the normalized difference vegetation index image and they found that image ratioing provided very useful land use mapping information.

## 2.3. Change Vector Analysis (CVA)

Change vector analysis can represent both the direction and magnitude of a change. The total change magnitude per pixel $(CM_{\text{pixel}})$ can be computed by determining the Euclidean distance between end points through n-dimensional change space [58]:

$$CM_{\text{pixel}} = \sum_{i=1}^{n} (X_2 - X_1)^2 \tag{1}$$

where $X_1$ and $X_2$ are date one and date two pixel values in i band.

CVA is complicated to implement, but it is useful for providing information about the change. Change vector analysis was used by Schoppmann and Tyler [46] to monitor changes in and around the Chernobyl nuclear power plant. They concluded that CVA was able to isolate a number of phenomena, including construction and excavation, water level variations, changes in agricultural cover type and riparian vegetation. In addition, change vectors were consistent across both change images. Allen and Kupfer [36] used change vector analysis to examine spruce-fir ecosystems. The results showed the ability of change vectors in multiple biophysical dimensions to differentiate forest disturbance and regeneration trends as an alternative to nominal forest or land cover classifications. Berberoglu and Akin [59] used four change detection techniques, including image differencing, image ratioing, image regression and CVA to detect land use and land cover change in a Mediterranean environment. The results showed that CVA was the most accurate technique for handling the variability present in Mediterranean land use and land cover change.

Combining CVA with other techniques provides more information than either technique alone. For example, Silapaswan, Verbyla [37] used CVA, unsupervised classification and visual interpretation of aerial photographs to detect land cover changes on the Seward Peninsula. They found that, in a transitional region between tundra and boreal forest, the use of CVA and unsupervised classification together provided a more powerful interpretation of change than either method alone.

### 2.4. Principle Component Analysis (PCA)

Principle component analysis, also referred to as eigenvector transformation, Hotelling transformation and Karhunen Loeve transformation in remote sensing, is a multivariate technique [60] that is used to reduce dataset dimensionality. In this technique, the original remote sensing dataset, which is a correlated variable, is transformed into a simpler dataset for interpretation. This allows the dataset to be uncorrelated variables representing the most important information from the original [15]. The calculation of the variance-covariance matrix (C) of multiband images is expressed as:

$$\boldsymbol{C} = \frac{\sum_{j=1}^{n}\left(\boldsymbol{X}_j - \boldsymbol{M}\right)\left(\boldsymbol{X}_j - \boldsymbol{M}\right)^{\mathrm{T}}}{n-1} \qquad (2)$$

where M and X are the multiband image mean and individual pixel value vectors respectively, and n is the number of pixels.

In change detection, there are two ways to apply PCA. The first method is adding two image dates to a single file, and the second methods is subtracting the second image date from the corresponding image of the first date after performing PCA separately. The disadvantages of PCA can be grouped into three issues: 1) it requires the selection of thresholds for identifying the change; 2) PCA results are difficult to interpret and label; and 3) it does not provide a complete matrix of change class [2].

Principal components analysis has been used in the case of land use and land cover change detection using standardized methods. For example, Baronti, Carla [33] applied PCA to investigate the changes occurring in multi-temporal polarimetric synthetic aperture radar (SAR) images. They used correlation instead of a covariance matrix in the transformation to reduce gain variations that are introduced by the imaging system and that give weight to each polarization. In another example, Liu, Nishiyama [43] compared four techniques, including image differencing, image ratioing, image regression and PCA, from a mathematical perspective. They recognized that standardized PCA achieved the best performance for change detection. Standardized PCA is better than unstandardized PCA for change detection because, if the images subjected to PCA are not measured in the same scale, the correlation matrix normalizes the data onto the same scale [61].

## 2.5. Chi-Square Transformation

Chi-square transformation uses multi-bands together to produce a single change image. The original chi-square test, also referred to as Pearson's chi-square, was presented by Karl Pearson in early 1990s [62]. Ridd and Liu [26] introduced the chi-square formula that is currently used in change detection:

$$Y = (X - M)^{\mathrm{T}} \Sigma^{-1} (X - M) \tag{3}$$

where Y is the digital value of the changed image; X is the vector of the difference of the six digital values between the two dates; M is the vector of the mean residuals of each band; T is the transverse of the matrix; and $\Sigma^{-1}$ is the inverse covariance matrix of the six bands [2]. Y is distributed as a chi-square random variable with P degrees of freedom, where P is the number of bands when $Y = 0$, meaning no change. Chi-square can produce a single change image from multiple bands. However, it is not true that $Y = 0$ represents an unchanged pixel when a large portion of the image is changed. In addition, a change related to a specific spectral direction is difficult to identify [2].

The chi-square transformation change detection technique was introduced by Ridd and Liu [26]. They also used other digital change detection techniques, including image differencing, image regression, and tasseled cap transformation. Chi-square was used with six reflective bands to create a single band change image.

## 2.6. Post-Classification Comparison

Post-classification comparison is a useful technique for extracting land use and land cover information. It is a common method in change detection. Post-classification requires the comparison of independently produced classified changes. It could be that the produced map shows the complete matrix of changes by coding the classification results for time 1 and time 2. Moreover, the subset of changes is observed by selective grouping of the classification results. In post-classification comparison, the images are classified separately, pixel by pixel, on two dates, minimizing the atmospheric and sensor differences between the two dates. In addition, the accurate registration problem of multi-date images is reduced in this technique [1]. However, a great deal of time and expertise is required for the performance of post-classification comparison and the quality of the classified image for each date affects the final accuracy.

Post-classification comparison includes two scheme classification and unsupervised classification algorithms. Supervised classification is a process when the analyst selects a number of areas for an image and then identifies the type of each phenomenon on the computer screen. In accordance with the phenomenon identified by the analyst, the computer identifies the characteristics of the data that comprise each type, and classifies the most similar remaining image pixels [14]. Supervised classification usually requires training data and prior knowledge of the objects that are selected for classification.

Unsupervised classification is a process by which the computer partitions the data without prior knowledge and then applies thematic labels [14]. Unsupervised classification does not normally require training data nor further initial input from the analyst [15].

Post-classification comparison is the most commonly used technique in change detection. [6,21,41,63-69]. The advantage of post-classification was recognized by Mas [70], who compared six change detection techniques to monitor land cover change. These techniques included image differencing, vegetative index differencing, selective PCA, direct multi-date unsupervised classification, post-classification differencing and a combination of image enhancement and post-classification comparison. He concluded that post-classification comparison was the most accurate procedure and presented the advantage of indicating the nature of the changes. Schulz, Cayuela [71] utilized post-classification for investigating land cover changes in dry-land forest landscapes in Central Chile. The maximum likelihood algorithm was used to provide a consistent classifier for multi-date classification. They indicated that the classification accuracy increased, especially after applying the post-classification procedure.

Combining classification with other techniques, such as classification with NDVI, or classification with PCA and CVA, has also been used in some studies to increase classification accuracy as well as to avoid the leak of training sample sets for image classification. For example, Li and Yeh [72] combined interactive supervised maximum likelihood classification with combined PCA to prevent over-estimation of the amount of change. Abd El-Kawy, Rød [73] also used map interpretation to improve classification accuracy and to identify areas with effective use of water for irrigation and areas of private land reclamation.

Unsupervised classification was performed to determine the change in eelgrass meadows with Landsat TM data by Macleod and Congalton [74]. They used three techniques including post-classification, image differencing and PCA. Classification was separately performed for each image using both supervised and unsupervised methods. Brink and Eva [75] also used an unsupervised classification algorithm to monitor 25 years of land cover change dynamics in Africa.

## 2.7. Artificial Neural Networks (ANNs)

A neural network is a computing process that provides a solution through an unstructured approach founded on the adjustment of the weights connecting the neurons in a network. The concept of a neuron was inspired by the study of biological neural processing [76].

There are three types of layers-an input layer, a hidden layer and an output layer-in which an artificial neural network (ANN) is typically contained. In the input layer, the neurons can be the multispectral reflectance values for each pixel with their texture, e.g. elevation, slope and aspect. In the hidden layers, the use of neurons enables the simulation of nonlinear patterns in the input data. And lastly, in the output layer, the neuron represents a single thematic map for each individual land cover class e.g. agricultural or urban. Each one of these layers consists of interconnected nodes. That interconnection allows the different information to flow in multiple directions [15]. The back-propagation algorithm is used to train the multi-layer perceptron neural model [2].

To extract useful information from the remotely sensed data and ancillary data, the ANN requires training and testing classifications [15]. However, an ANN is sensitive to the amount of training data that is used and requires a long training time. Moreover, ANNs are not common in remote sensing software [2].

Using ANNs in remote sensing has been widely recognized by many researchers [34,35,77-80]. They have also been used to detect changes in many applications. For example, Foody, McCULLOCH [34] utilized a feedforward ANN using a variant of the back-propagation algorithm to classify agricultural crops from synthetic aperture radar data and found that the ANN was able to characterize classes better than discriminant analysis. The ANN was used to classify image data into from-to classes for detecting urban change. Principle component analysis was also applied to extract the salient features and reduce the dimensionality of the input data prior to the ANN-based change detection. By using an ANN, the accuracy increased more than with post-classification comparison. Principle component analysis also improved change detection accuracy [38]. Artificial neural networks have been used to map land use and land cover change with a combination of satellite sensor data and GIS. The ANN approach was used to generate land use and land cover classes from ancillary and spectral data [81]. This combination increased the accuracy of classification compared with spectral classification alone and established relationships between land use and land cover and environmental variables specific to the mapped area.

## 2.8. Hybrid Change Detection

A hybrid technique uses two or more techniques (e.g. supervised and unsupervised classification, PCA and CVA) to detect a change. There are two types of hybrid techniques. The first type is procedure-based hybrid analysis, which involves the use of different detection techniques in different detection procedures. The second type is result-based hybrid analysis, which means using different change detection techniques successively and then analyzing the results. This technique makes full use of the benefits of many techniques to obtain significant change detection results. On the other hand, the hybrid change detection technique is complex, difficult to perform and not very effective because it depends on the characteristics of other techniques [82].

Hybrid change detection, combining the advantages of supervised and unsupervised classification, was used to derive a land cover map from three Landsat Thematic Mapper Plus (ETM+) images from the year 2000 in three Eastern Europe countries (Slovakia, Poland and the Ukraine). The results were more accurate than using individual techniques [7]. The same results were obtained by Zhang, Ma [83], who used hybrid change detection and a decision tree classifier based on a data mining algorithm. The results showed better accuracy than using individual change detection techniques.

## 2.9. Image Fusion

Image fusion is a technology that merges two or more images from the same area in different sensors and wavelengths [61,84]. In general, different spectral and spatial resolutions are helpful when high spectral resolution is used in discriminating land use and land cover type, whereas high spatial resolution holds advantages in identifying terrain features and the earth's structure. The purpose of using image fusion is to provide additional information to help users detect and identify targets on the earth's surface. Image fusion can be implemented along with PCA, the intensity hue-saturation (HIS) transform, Brovey's methods and wavelet transformation to provide more accurate information and at high resolution [61].

The advantage of using image fusion in change detection is that fusion techniques can provide high spatial and spectral resolution, which can easily and effectively allow users to extract land use and land cover information. However, it is obvious that resized and registered images, which are obtained from different sensors, may be difficult to implement. For example, in a fusion image from Landsat TM and a SPOT-5 image, the TM image is resampled at the SPOT size, and then the two images have the same size and are registered before image fusion [61]. Therefore, users should take this point into account before performing fusion technique activities.

In a change detection context, image fusion has been used to detect land use and land cover change over urban areas. For example, Zhou, Huang [40] compared three methods for land cover classification of shaded areas from high spatial resolution imagery in an urban environment, including combined spectral information, linear-correlation correction and multisource data fusion. The results indicated that data fusion achieved better accuracy when compared with combined spectral information and linear-correlation correction. Zeng, Zhang [85] presented the results of different temporal SAR and optical image fusion algorithms for land cover change detection. The data used in their research were from SPOT-5 imagery and RADARSAT-1. Peijun, Sicong [44] used feature and decision level fusion to combine simple change detectors and to build an automatic change detection procedure. This method was tested with multi-temporal CBERS and HJ-1 images. The results were satisfactory and more effective than other methods.

## 2.10. Decision Tree

A decision tree classifier is an algorithm for the labeling of an unknown pattern using a sequence of decisions [61]. The structure of a decision tree is organized with rules, conditions and hypotheses, where rules and conditions are evaluated in order to test the hypotheses. The concept of a hierarchical (or top-down) decision tree is referred to when each hypothesis appears as the trunk of the tree, each rule is a limb of the tree and each condition is a leaf [15]. A decision tree splits a complex decision into many simpler decisions to obtain a solution that is easier to interpret [86].

There are two approaches in the design of a decision tree. The first one is called the manual design approach; it is based on the user's knowledge and relies solely on user interaction. The second approach is based on an automatic procedure. The manual design approach is time-consuming and it does not provide satisfactory results, especially when the number of classes is large and there is spectral overlap between classes [61].

Hansen, Defries [87] used decision tree to classify land cover using data for 1992-1993 from the Advanced Very High Resolution Radiometer (AVHRR). Yang, Ren [88] used a comparison of post-classification and knowledgebased decision tree (K-DT) classification to quantify land use and land cover changes in the watershed of the Inner Mongolia Autonomic Region of China using Landsat TM and ETM+ images from 1987 to 2007. The results showed significant modification and conversion of land use and land cover of the watershed over the course of 20 years.

### 2.11. Geographic Information Systems (GIS)

All the techniques discussed above are based on pixel by pixel change detection analysis of satellite images as well as being based on images only. It is important to use the benefit of collateral information, such as digital elevation models, hydrology and soil maps [15], which can be provided with the extracted information from remotely sensed images into GIS platforms. Thus, the integration of remotely sensed data with GIS data has the potential to improve the accuracy of results. The main advantage of GIS is that changes can be detected more clearly than with other techniques using multi-source data. However, using different source data with different formats and accuracies may affect the change detection results [2].

In recent years, remote sensing and GIS have been commonly integrated for analyzing and mapping land use and land cover changes [9,11,12,89]. Driving land use and land cover change maps into GIS applications has been done using supervised classification algorithms through remotely sensed software [90,91]. Utilizing ancillary data with satellite imagery, (e.g. digital elevation models and soil maps) provides more accuracy in the detection of change [92]. The use of GIS has also been recognized for detecting and mapping land use and land cover changes. Spatial statistical analysis and advanced functions (e.g. hotspots) have been used for change detection [93]. In addition, screen digitizing of satellite images and previous land use and land cover maps have been used to detect land use and land cover change [94, 95].

## 3. CHANGE DETECTION ACCURACY ASSESSMENT

Accuracy assessment is very important for understanding the obtained results and making good decisions. Accuracy is determined by the quality of the thematic information and maps that are obtained from remotely sensed data. Accuracy assessment is required to minimize the common sources of error in remotely sensed data (e.g. sensor systems errors such as detectors and cameras, the movement of spacecraft platforms, ground control errors and classification errors). The statistical methods used to assess the accuracy of land cover maps include overall accuracy, user's accuracy, producer's accuracy and the Kappa coefficient.

The consideration of classification errors is common in previous studies and reviews [96-102]. Image classification is subject to a variety of errors that occur during the classification process. The difference between the class label assigned by the classification and objects in the real field is significant error. In this case, an error matrix provides an appropriate summary of the omission and commission errors [98]. An error matrix or a confusion matrix and its measures, such as user's accuracy and producer's accuracy, is the most common approach for determining the accuracy of remote sensing image classification [96,103].

Change detection accuracy assessment is particularly difficult and requires a large amount of work [104]. The matrices that are most commonly used are single date error matrices, binary change/no change error matrices and the transition error matrix. However, the transition error matrix is less commonly reported than single date error matrices and binary change/no change error matrices [105]. The single date classification error matrix needs to be adapted to a change detection error matrix. The characteristics of the new matrix are the same as the single date classification error matrix and assesses errors in changes between two time periods (time 1 and time 2) [104].

Many previous studies on change detection failed to provide the required information to produce a stratified estimator of change area and construct confidence intervals [106]. It is not possible to check whole mapped areas; sampling is the means from which accuracy from land cover maps can be derived [107]. Thematic map accuracy assessment begins with the generation of an error matrix. The factors that should be considered in the generation of an error matrix include ground data collection, classification and sample scheme, sample size and sample unit, and spatial autocorrelation [108]. It is sometimes possible to reduce or remove ground data errors or to derive accurate estimates of change detection without ground data [109].

## 4. EVALUATION OF CHANGE DETECTION TECHNIQUES

Providing details and determining the direction of change is important. Some change detection techniques, such as image differencing, image ratioing, and PCA, do not provide sufficient change trend information. These techniques only provide change or no change results, therefore, the trend and direction of the change is difficult to determine. Other techniques, such as post-classification comparison, provide more detail about the objects that are selected and provide more information about change trends (**Table 2**).

**Table 2**. Characteristics of change detection techniques.

| Technique | Providing Change Direction | Selecting Threshold | Selecting Training | Providing Change Matrix | Data Used to Preform |
|---|---|---|---|---|---|
| 1. Image differencing | × | √ | × | × | Satellite imagery |
| 2. Image ratioing | × | √ | × | × | Satellite imagery |
| 3. Change vector analysis | √ | √ | × | × | Satellite imagery |
| 4. Principle component analysis | × | √ | × | × | Satellite imagery |
| 5. Chi-square | × | √ | × | × | Satellite imagery |
| 6. Post-classification comparison | √ | × | √ | √ | Satellite imagery |
| 7. Hybrid change detection | × | √ | × | × | Satellite imagery |
| 8. Artificial neural networks | √ | × | √ | √ | Satellite imagery & ancillary data |
| 9. Decision tree | × | × | × | × | Satellite imagery |
| 10. Image fusion | × | √ | × | × | Satellite imagery |
| 11. GIS | √ | × | × | √ | Satellite imagery & ancillary data |

In addition, some change detection techniques, such as image differencing, image ratioing, PCA, chi-square, hybrid change detection and CVA, require the selection of a threshold (**Table 2**) [19]. In this case, selecting an appropriate threshold is important for obtaining good change detection results; however, selecting a suitable threshold is difficult and time-consuming. On the other hand, some techniques are affected by errors and require more time and effort to reduce such errors in order to produce high-quality thematic change detection maps. For example, post-classification comparison always contains omission and commission errors and needs the selection of a confusion matrix and its measures (e.g. user's and producer's accuracy) to minimize these errors.

Furthermore, the strengths and limitations of individual change detection techniques should be taken into account. For example, post-classification comparison minimizes the impacts of atmospheric, sensor and environmental differences between multi-temporal images; image differencing is affected by these elements. Also, image ratioing can reduce the impact of sun angle, shadow and topography. Image differencing and image ratioing allow for easy interpretation of change detection results, whereas the results of principle component analysis are difficult to interpret. The use of additional data, such as digital elevation models, soil and hydrology maps, with satellite imagery may provide additional information about a change. However, the accuracy of different data may affect the quality of land use and land cover change results. It is also important to consider the level of difficulty in dealing with the various functions and software. For example, the limitation of ANNs could be in their functions, which are not common in image processing software [2].

## 5. SUMMARY

By analyzing the related literature, it seems that no one technique is suitable for all cases. Selection of an appropriate method for detecting change in an object or a phenomenon on the earth's surface depends on a number of elements, including the characteristics of the study area, the spatial resolution of the sensor, atmospheric effects and sun angle, which should be taken into account before applying a suitable technique for the detector. Furthermore, the characteristics objects themselves should be considered before selecting a method because some of these techniques showed more powerful in some cases than others. For instance, for detecting vegetation cover in arid environments, image differencing is more effective than Normalized Difference Vegetation Index. Also, post-classification comparison is useful in two ways. It provides details of changes, and avoids selecting appropriate thresholds. The accuracy of change detection results can be affected by these elements, as well as the resolution of spatial and spectral images.

Research using change detection techniques should consider the elements that may affect the results. Moreover, researchers should provide a quantitative analysis of the results with their conclusions. Many of the studies reviewed failed to do this; as these studies compare the performance of change detection techniques, there was no ground reference for accuracy assessment. It is necessary to link the results obtained using change detection techniques with the real changes occurring on the ground. Some of the previous studies failed to consider the real changes.

Most of the change detection studies considered the impact of urban expansion on tropical and temperate environments; however, unfortunately, few studies addressed changes in desert environments. Certain elements, including urban expansion and natural factors, may lead to changes in a desert environment. These changes are sometimes challenging to detect using satellite imagery. Vegetation cover, for example, can be easily determined from satellite images in a forested area, but not in a desert area. Therefore, further studies are warranted to apply change detection techniques to desert environments.

## REFERENCES

1. A. Singh, "Review Article Digital Change Detection Techniques Using Remotely-Sensed Data," International Journal of Remote Sensing, Vol. 10, No. 6, 1989, pp. 989- 1003.

2. D. Lu, et al., "Change Detection Techniques," International Journal of Remote Sensing, Vol. 25, No. 12, 2004, pp. 2365-2401.

3. T. Ramachandra and U. Kumar, "Geographic Resources Decision Support System for Land Use, Land Cover Dynamics Analysis," Proceedings of the FOSS/GRASS Users Conference, Bangkok, September 2004.

4. J. R. E. Jensen, "Urban/Suburban Land Use Analysis," American Society of Photogrammetry, Falls Church, Virginia, Vol. 2, 1983, pp. 1571-1666.

5. J. Rogan and D. M. Chen, "Remote Sensing Technology for Mapping and Monitoring Land-Cover and Land-Use Change," Progress in Planning, Vol. 61, No. 4, 2004, pp. 301-325.

6. E. Brondizio, et al., "Land Use Change in the Amazon Estuary: Patterns of Caboclo Settlement and Landscape Management," Human Ecology, Vol. 22, No. 3, 1994, pp. 249-278.

7. T. Kuemmerle, et al., "Cross-Border Comparison of Land Cover and Landscape Pattern in Eastern Europe Using a Hybrid Classification Technique," Remote Sensing of Environment, Vol. 103, No. 4, 2006, pp. 449-464.

8. R. Pelorosso, et al., "Land Cover and Land Use Change in the Italian Central Apennines: A Comparison of Assessment Methods," Applied Geography, Vol. 29, No. 1, 2009, pp. 35-48.

9. R. B. Thapa and Y. Murayama, "Urban Mapping, Accuracy, & Image Classification: A Comparison of Multiple Approaches in Tsukuba City, Japan," Applied Geography, Vol. 29, No. 1, 2009, pp. 135-144.

10. W. Z. Michalak, "GIS in Land Use Change Analysis: Integration of Remotely Sensed Data into GIS," Applied Geography, Vol. 13, No. 1, 1993, pp. 28-44.

11. Q. Weng, "A Remote Sensing GIS Evaluation of Urban Expansion and Its Impact on Surface Temperature in the Zhujiang Delta, China," International Journal of Remote Sensing, Vol. 22, No. 10, 2001, pp. 1999-2014.

12. J. Rogan, et al., "Integrating GIS and Remotely Sensed Data for Mapping Forest Disturbance and Change," In: M. A. Wulder and S. E. Franklin, Eds., Understanding Forest Disturbance and Spatial Pattern: Remote Sensing and GIS Approaches, 2007, pp. 133-171.

13. D. A. Mouat, et al., "Remote Sensing Techniques in the Analysis of Change Detection," Geocarto International, Vol. 8, No. 2, 1993, pp. 39-50.

14. P. Deer, "Digital Change Detection Techniques in Remote Sensing," DTIC Document, 1995.

15. J. R. Jensen, "Introductory Digital Image Processing: A Remote Sensing Perspective," In: K. C. Clarke, Ed., 3rd Edition, Prentice Hall, The United States of America, 2005.

16. D. M. Muchoney and B. N. Haack, "Change Detection for Monitoring Forest Defoliation," Photogrammetric Engineering and Remote Sensing, Vol. 60, No. 10, 1994, pp. 1243-1251.

17. S. S. Podeh, et al., "Forest Change Detection in the North of Iran Using TM/ETM+ Imagery," Asian Journal of Applied Sciences, Vol. 2, No. 6, 2009, pp. 464-474.

18. M. Jahari, et al., "Change Detection Studies in Matang Mangrove Forest Area, Perak," Pertanika Journal of Science and Technology, Vol. 19, No. 2, 2011, pp. 307-327.

19. P. L. Rosin and T. Ellis, "Image Difference Threshold Strategies and Shadow Detection," Proceedings of the 6th British Machine Vision Conference, Citeseer, 1995.

20. P. A. Rogerson, "Change Detection Thresholds for Remotely Sensed Images," Journal of Geographical Systems, Vol. 4, No. 1, 2002, pp. 85-97.

21. R. Weismiller, et al., "Change Detection in Coastal Zone Environments," Photogrammetric Engineering and Remote Sensing, Vol. 43, No. 12, 1977, pp. 1533-1539.

22. T. L. Sohl, "Change Analysis in the United Arab Emirates: An Investigation of Techniques," Photogrammetric Engineering and Remote Sensing, Vol. 65, No. 4, 1999, pp. 475-484.

23. H. A. Afify, "Evaluation of Change Detection Techniques for Monitoring Land-Cover Changes: A Case Study in New Burg El-Arab Area," Alexandria Engineering Journal, Vol. 50, No. 2, 2011, pp. 187-195.

24. J. R. Jensen and D. Toll, "Detecting Residential Land-Use Development at the Urban Fringe," Photogrammetric Engineering & Remote Sensing, Vol. 48, No. 4, 1982, pp. 629-643.

25. D. Williams and M. Stauffer, "Monitoring Gypsy Moth Defoliation by Applying Change Detection Techniques to Landsat Imagery," Proceeding of the Sympsium on Vegetation Damage Assessment, American Society of Photogammetry, Falls Church, 1978, pp. 221-229.

26. M. K. Ridd and J. Liu, "A Comparison of Four Algorithms for Change Detection in an Urban Environment," Remote Sensing of Environment, Vol. 63, No. 2, 1998, pp. 95-100.

27. P. S. Chavez and D. J. MacKinnon, "Automatic Detection of Vegetation Changes in the Southwestern United States Using Remotely Sensed Images," Photogrammetric Engineering and Remote Sensing, Vol. 60, No. 5, 1994, pp. 571-583.

28. P. Pilon, et al., "An Enhanced Classification Approach to Change Detection in Semi-Arid Environments," Photogrammetric Engineering and Remote Sensing, Vol. 54, No. 12, 1988, pp. 1709-1716.

29. H. Alphan, "Comparing the Utility of Image Algebra Operations for Characterizing Landscape Changes: The Case of the Mediterranean Coast," Journal of Environmental Management, Vol. 92, No. 11, 2011, pp. 2961- 2971.

30. P. J. Howarth and G. M. Wickware, "Procedures for Change Detection Using Landsat Digital Data," International Journal of Remote Sensing, Vol. 2, No. 3, 1981, p. 277-291.

31. R. Nelson, "Detecting Forest Canopy Change Due to Insect Activity Using Landsat MSS," Photogrammetric Engineering and Remote Sensing, Vol. 49, 1983, pp. 1303- 1314.

32. A. Prakash and R. Gupta, "Land-Use Mapping and Change Detection in a Coal Mining Area—A Case Study in the Jharia Coalfield, India," International Journal of Remote Sensing, Vol. 19, No. 3, 1998, pp. 391-410.

33. S. Baronti, et al. "Principal Component Analysis for Change Detection on Polarimetric Multitemporal SAR Data," Geoscience and Remote Sensing Symposium, Surface and Atmospheric Remote Sensing: Technologies, Data Analysis and Interpretation, 1994.

34. G. M. Foody, et al., "Classification of Remotely Sensed Data by an Artificial Neural Network: Issues Related to Training Data Characteristics," Photogrammetric Engineering and Remote Sensing, Vol. 61, No. 4, 1995, pp. 391-401.

35. J. Jensen, et al., "Predictive Modelling of Coniferous Forest Age Using Statistical and Artificial Neural Network Approaches Applied to Remote Sensor Data," International Journal of Remote Sensing, Vol. 20, No. 14, 1999, pp. 2805-2822.

36. T. R. Allen and J. A. Kupfer, "Application of Spherical Statistics to Change Vector Analysis of Landsat Data: Southern Appalachian Spruce-Fir Forests," Remote Sensing of Environment, Vol. 74, No. 3, 2000, pp. 482-493.

37. C. Silapaswan, et al., "Land Cover Change on the Seward Peninsula: The Use of Remote Sensing to Evaluate the Potential Influences of Climate Warming on Historical Vegetation Dynamics," Canadian Journal of Remote Sensing, Vol. 27, No. 5, 2001, pp. 542- 554.

38. X. Liu and R. G. Lathrop, "Urban Change Detection Based on an Artificial Neural Network," International Journal of Remote Sensing, Vol. 23, No. 12, 2002, pp. 2513-2518.

39. L. Castellana, A. d'Addabbo and G. Pasquariello, "A Composed Supervised/Unsupervised Approach to Improve Change Detection from Remote Sensing," Pattern Recognition Letters, Vol. 28, No. 4, 2007, pp. 405-413.

40. W. Zhou, et al., "Object-Based Land Cover Classification of Shaded Areas in High Spatial Resolution Imagery of Urban Areas: A Comparison Study," Remote Sensing of Environment, Vol. 113, No. 8, 2009, pp. 1769-1777.

41. J. Miettinen, et al., "2010 Land Cover Map of Insular Southeast Asia in 250-m Spatial Resolution," Remote Sensing Letters, Vol. 3, No. 1, 2011, pp. 11-20.

42. Q. Zhang, et al., "Urban Built-Up Land Change Detection with Road Density and Spectral Information from MultiTemporal Landsat TM Data," International Journal of Remote Sensing, Vol. 23, No. 15, 2002, pp. 3057-3078.

43. Y. Liu, et al., "Analysis of Four Change Detection Algorithms in Bi-Temporal Space with a Case Study," International Journal of Remote Sensing, Vol. 25, No. 11, 2004, pp. 2121-2139.

44. D. Peijun, et al., "Fusion of Difference Images for Change Detection over Urban Areas," IEEE Journal of Selected Topics in Applied Earth Observations and Remote Sensing, Vol. 5, No. 4, 2012, pp. 1076-1086.

45. A. J. Peters, et al., "Drought Monitoring with NDVIBased Standardized Vegetation Index," Photogrammetric Engineering and Remote Sensing, Vol. 68, No. 1, 2002, pp. 71-75.

46. M. W. Schoppmann and W. A. Tyler, "Chernobyl Revisited: Monitoring Change with Change Vector Analysis," Geocarto International, Vol. 11, No. 1, 1996, pp. 13-27.

47. S. E. Franklin and B. Wilson, "Vegetation Mapping and Change Detection Using SPOT MLA and Landsat Imagery in Kluane National Park," Canadian Journal of Remote Sensing, Vol. 17, No. 1, 1991, pp. 2-22.

48. J. B. Thayn, "Assessing Vegetation Cover on the Date of Satellite-Derived Start of Spring," Remote Sensing Letters, Vol. 3, No. 8, 2012, pp. 721-728.

49. D. Zheng, et al., "Rates and Patterns of Landscape Change between 1972 and 1988 in the Changbai Mountain Area of China and North Korea," Landscape Ecology, Vol. 12, No. 4, 1997, pp. 241-254.

50. W. G. Kepner, et al., "A Landscape Approach for Detecting and Evaluating Change in a Semi-Arid Environment," Environmental Monitoring and Assessment, Vol. 64, No. 1, 2000, pp. 179-195.

51. J. Taylor, et al., "Monitoring Landscape Change in the National Parks of England and Wales Using Aerial Photo Interpretation and GIS," International Journal of Remote Sensing, Vol. 21, No. 13-14, 2000, pp. 2737-2752.

52. R. M. Lucas, et al., "Characterizing Tropical Secondary Forests Using Multi-Temporal Landsat Sensor Imagery," International Journal of Remote Sensing, Vol. 14, No. 16, 1993, pp. 3061-3067.

53. D. J. Hayes and S. A. Sader, "Comparison of ChangeDetection Techniques for Monitoring Tropical Forest Clearing and Vegetation Regrowth in a Time Series," Photogrammetric Engineering and Remote Sensing, Vol. 67, No. 9, 2001, pp. 1067-1075.

54. E. H. Wilson and S. A. Sader, "Detection of Forest Harvest Type Using Multiple Dates of Landsat TM Imagery," Remote Sensing of Environment, Vol. 80, No. 3, 2002, pp. 385-396.

55. E. J. Christensen, et al., "Aircraft MSS Data Registration and Vegetation Classification for Wetland Change Detection," International Journal of Remote Sensing, Vol. 9, No. 1, 1988, pp. 23-38.

56. J. Jensen, et al., "An Evaluation of the Coast Watch Change Detection Protocol in South Carolina," Photogrammetric Engineering and Remote Sensing, Vol. 59, No. 6, 1993, pp. 1039-1044.

57. C. Munyati, "Wetland Change Detection on the Kafue Flats, Zambia, by Classification of a Multitemporal Remote Sensing Image Dataset," International Journal of Remote Sensing, Vol. 21, No. 9, 2000, pp. 1787-1806.

58. J. L. Michalek, et al., "Multispectral Change Vector Analysis for Monitoring Coastal Marine Environments," Photogrammetric Engineering and Remote Sensing, Vol. 59, No. 3, 1993, pp. 635-641.

59. S. Berberoglu and A. Akin, "Assessing Different Remote Sensing Techniques to Detect Land Use/Cover Changes in the Eastern Mediterranean," International Journal of Applied Earth Observation and Geoinformation, Vol. 11, No. 1, 2009, pp. 46-53.

60. A. Singh and A. Harrison, "Standardized Principal Components," International Journal of Remote Sensing, Vol. 6, No. 6, 1985, pp. 883-896.

61. B. Tso and P. P. Mather, "Classification Methods for Remotely Sensed Data," 2nd Edition, CRC Press, Boca Raton, 2009.

62. D. C. Howell, "Chi-Square Test-Analysis of Contingency Tables," Women, Vol. 35, No. 3, 2009, pp. 28-83.

63. S. I. Gordon, "Utilizing LANDSAT Imagery to Monitor Land-Use Change: A Case Study in Ohio," Remote Sensing of Environment, Vol. 9, No. 3, 1980, pp. 189-196.

64. J. Jensen, et al., "Inland Wetland Change Detection Using Aircraft MSS Data," Photogrammetric Engineering and Remote Sensing, Vol. 53, No. 5, 1987, pp. 521-529.

65. M. U. H. Dimyati, et al., "An Analysis of Land Use/Cover Change in Indonesia," International Journal of Remote Sensing, Vol. 17, No. 5, 1996, pp. 931-944.

66. C. Song, et al., "Classification and Change Detection Using Landsat TM Data: When and How to Correct Atmospheric Effects?" Remote Sensing of Environment, Vol. 75, No. 2, 2001, pp. 230-244.

67. D. L. Civco, et al., "A Comparison of Land Use and Land Cover Change Detection Methods," Proceedings of the 2002 ASPRS Annual Convention, Washington DC, 22-26 April 2002.

68. J. Aguirre-Gutiérrez, et al., "Optimizing Land Cover Classification Accuracy for Change Detection, a Combined Pixel-Based and Object-Based Approach in a Mountainous Area in Mexico," Applied Geography, Vol. 34, 2012, pp. 29-37.

69. J. O. Sexton, et al., "Long-Term Land Cover Dynamics by Multi-Temporal Classification across the Landsat-5 Record," Remote Sensing of Environment, Vol. 128, No. 21, 2013, pp. 246-258.

70. J. F. Mas, "Monitoring Land-Cover Changes: A Comparison of Change Detection Techniques," International Journal of Remote Sensing, Vol. 20, No. 1, 1999, pp. 139- 152.

71. J. J. Schulz, et al., "Monitoring Land Cover Change of the Dryland Forest Landscape of Central Chile (1975- 2008)," Applied Geography, Vol. 30, No. 3, 2010, pp. 436-447.

72. X. Li and A. Yeh, "Principal Component Analysis of Stacked Multi-Temporal Images for the Monitoring of Rapid Urban Expansion in the Pearl River Delta," International Journal of Remote Sensing, Vol. 19, No. 8, 1998, pp. 1501-1518.

73. O. R. Abd El-Kawy, et al., "Land Use and Land Cover Change Detection in the Western Nile Delta of Egypt Using Remote Sensing Data," Applied Geography, Vol. 31, No. 2, 2011, pp. 483-494.

74. R. D. Macleod and R. G. Congalton, "A Quantitative Comparison of Change-Detection Algorithms for Monitoring Eelgrass from Remotely Sensed Data," Photogrammetric Engineering and Remote Sensing, Vol. 64, No. 3, 1998, pp. 207-216.

75. A. B. Brink and H. D. Eva, "Monitoring 25 Years of Land Cover Change Dynamics in Africa: A Sample Based Remote Sensing Approach," Applied Geography, Vol. 29, No. 4, 2009, pp. 501-512.

76. V. B. Rao and H. V. Rao, "C++ Neural Networks and Fuzzy Logic," MIS-Press, New York, 1993.

77. G. Foody and M. Arora, "An Evaluation of Some Factors Affecting the Accuracy of Classification by an Artificial Neural Network," International Journal of Remote Sensing, Vol. 18, No. 4, 1997, pp. 799-810

78. J. Benediktsson and J. Sveinsson, "Feature Extraction for Multisource Data Classification with Artificial Neural Networks," International Journal of Remote Sensing, Vol. 18, No. 4, 1997, pp. 727-740.

79. J. R. Jensen, F. Qiu and K. Patterson, "A Neural Network Image Interpretation System to Extract Rural and Urban Land Use and Land Cover Information from Remote Sensor Data," Geocarto International, Vol. 16, No. 1, 2001, pp. 21-30.

80. F. Qiu and J. Jensen, "Opening the Black Box of Neural Networks for Remote Sensing Image Classification," International Journal of Remote Sensing, Vol. 25, No. 9, 2004, pp. 1749-1768.

81. J. F. Mas, "Mapping Land Use/Cover in a Tropical Coastal Area Using Satellite Sensor Data, GIS and Artificial Neural Networks," Estuarine, Coastal and Shelf Science, Vol. 59, No. 2, 2004, pp. 219-230.

82. J. Y. G., et al., "A Review of Multi-Temporal Remote Sensing Data Change Detection Algorithms," The International Archives of the Photogrammetry, Remote Sensing and Spatial Information Sciences, Vol. 37, No. B7, 2008, pp. 757-762.

83. Y. Zhang, et al., "Hybrid Change Detection for Watershed Impervious Surface Using Multi-Time Remotely Sensed Data," IEEE International of Geoscience and Remote Sensing Symposium, Barcelona, 23-28 July 2007, pp. 1939-1942.

84. L. Wald, "Some Terms of Reference in Data Fusion," IEEE Transactions on Geoscience and Remote Sensing, Vol. 37, No. 3, 1999, pp. 1190-1193.

85. Y. Zeng, et al., "Image Fusion for Land Cover Change Detection," International Journal of Image and Data Fusion, Vol. 1, No. 2, 2010, pp. 193-215.

86. M. Xu, et al., "Decision Tree Regression for Soft Classification of Remote Sensing Data," Remote Sensing of Environment, Vol. 97, No. 3, 2005, pp. 322-336.

87. M. C. Hansen, et al., "Global Land Cover Classification at 1 km Spatial Resolution Using a Classification Tree Approach," International Journal of Remote Sensing, Vol. 21, No. 6-7, 2000, pp. 1331-1364.

88. X. Yang, et al., "Impacts of Land Use and Land Cover Changes on Evapotranspiration and Runoff at Shalamulun River Watershed, China," Hydrology Research, Vol. 43, No. 1-2, 2012, pp. 23-37.

89. R. Efe, et al., "Land Use and Land Cover Change Detection in Karinca River Catchment (NW Turkey) Using GIS and RS Techniques," Journal of Environmental Biology, Vol. 33, No. 2, 2012, pp. 439-447.

90. D. Tripathi and M. Kumar, "Remote Sensing Based Analysis of Land Use/Land Cover Dynamics in Takula Block, Almora District(Uttarakhand)," Journal of Human Ecology, Vol. 38, No. 3, 2012, pp. 207-212.

91. S. Gajbhiye and S. K. Sharma, "Land Use and Land Cover Change Detection of Indra River Watershed through Remote Sensing Using Multi-Temporal satellite Data," International Journal of Geomatics and Geosciences, Vol. 3, No. 1, 2012, pp. 89-96.

92. S. Reis, "Analyzing Land Use/Land Cover Changes Using Remote Sensing and GIS in Rize, North-East Turkey," Sensors, Vol. 8, No. 10, 2008, pp. 6188-6202.

93. P. P. Serra, et al., "Land-Cover and Land-Use Change in a Mediterranean Landscape: A Spatial Analysis of Driving Forces Integrating Biophysical and Human Factors," Applied Geography, Vol. 28, No. 3, 2008, pp. 189-209.

94. I. Abbas, et al., "Mapping Land Use-Land Cover and Change Detection in Kafur Local Government, Katsina, Nigeria (1995-2008) Using Remote Sensing and GIS," Research Journal of Environmental and Earth Sciences, Vol. 2, No. 1, 2010, pp. 6-12.

95. N. Nagarajan and S. Poongothai, "Effect of Land Use/ Land Cover Change Detection of Ungauged Watershed," World Applied Sciences Journal, Vol. 17, No. 6, 2012, pp. 718-723.

96. R. G. Congalton, "A Review of Assessing the Accuracy of Classifications of Remotely Sensed Data," Remote Sensing of Environment, Vol. 37, No. 1, 1991, pp. 35-46.

97. C. Conese and F. Maselli, "Use of Error Matrices to Improve Area Estimates with Maximum Likelihood Classification Procedures," Remote Sensing of Environment, Vol. 40, No. 2, 1992, pp. 113-124.

98. G. M. Foody, "Status of Land Cover Classification Accuracy Assessment," Remote Sensing of Environment, Vol. 80, No. 1, 2002, pp. 185-201.

99. R. Latifovic and I. Olthof, "Accuracy Assessment Using Sub-Pixel Fractional Error Matrices of Global Land Cover Products Derived from Satellite Data," Remote Sensing of Environment, Vol. 90, No. 2, 2004, pp. 153-165.

100. H. Liu and Q. Zhou, "Accuracy Analysis of Remote Sensing Change Detection by Rule-Based Rationality Evaluation with Post-Classification Comparison," International Journal of Remote Sensing, Vol. 25, No. 5, 2004, pp. 1037-1050.

101. C. Liu, et al., "Comparative Assessment of the Measures of Thematic Classification Accuracy," Remote Sensing of Environment, Vol. 107, No. 4, 2007, pp. 606-616.

102. P. P. L. Zimmerman, et al., "An Accuracy Assessment of Forest Disturbance Mapping in the Western Great Lakes," Remote Sensing of Environment, Vol. 128, No. 21, 2013, pp. 176-185.

103. A. Comber, et al., "Spatial Analysis of Remote Sensing Image Classification Accuracy," Remote Sensing of Environment, Vo. 127, 2012, pp. 237-246.

104. R. G. Congalton and K. Green, "Assessing the Accuracy of Remotely Sensed Data: Principles and Practices," 2th Edition, CRC Press, Boca Raton, 2008.

105. P. A. J. van Oort, "Interpreting the Change Detection Error Matrix," Remote Sensing of Environment, Vol. 108, No. 1, 2007, pp. 1-8.

106. P. P. Olofsson, et al., "Making Better Use of Accuracy Data in Land Change Studies: Estimating Accuracy and Area and Quantifying Uncertainty Using Stratified Estimation," Remote Sensing of Environment, Vol. 129, No. 15, 2013, pp. 122-131.

107. G. Banko, "A Review of Assessing the Accuracy of Classifications of Remotely Sensed Data and of Methods Including Remote Sensing Data in Forest Inventory," International Institute for Applied Systems Analysis, Laxenburg, 1998.

108. R. G. Congalton, et al., "Quality Assurance and Accuracy Assessment of Information Derived from Remotely Sensed Data," In: J. D. Bossler, Ed., Manual of Geospatial Science and Technology, CRC Press, London, 2001, pp. 349-361.

109. G. M. Foody, "Assessing the Accuracy of Land Cover Change with Imperfect Ground Reference Data," Remote Sensing of Environment, Vol. 114, No. 10, 2010, pp. 2271-2285.

# Chapter 7

# Development of a High-Resolution Wireless Sensor Network for Monitoring Volcanic Activity

**José Chilo[1], Andreas Schlüter[2] and Thomas Lindblad[3]**

[1] *University of Gävle S-80176 Gävle, Sweden*

[2] *Institute of Computer Science, Freie Universität Berlin, Germany*

[3] *Royal Institute of Technology, S-106 91 Stockholm, Sweden*

## 1. INTRODUCTION

This chapter describes the design of a high-resolution wireless sensor network to monitor infrasonic signals from volcanic activity. A prototype system is constructed and tested. The system is based on the ultra low power microcontroller MSP430 with the requirements of energy-awareness and high sensor node autonomy. The infrasonic signals are measured at 200 Hz using 12 bit resolution and the result is buffered on SD cards in case of a lack of bandwidth. The implementation of a cost-table driven network routing protocol allows a radio sleep schedule of almost 97% when no data has to be transmitted. Furthermore, the sensors need to be time-synchronized for later event localization. This work shows that it is feasible to have a synchronization accuracy of less than 1 ms using a GPS receiver that is powered on only a few seconds per hour.

In recent years the installation of infrasound sensors at seismic measuring stations has become common and now researchers can obtain large and heterogeneous infrasound signal data-sets generated in near real-time. However, most current infrasound stations are still using expensive infrasound microphones and traditional data acquisition systems which limit the deployment of new infrasound stations. To improve on this, we propose in this work a wireless data acquisition system based on FreeWave FGR09CSU 900 MHz radio modem and a Wireless sensor networks (WSN).

Infrasound is defined as the range of frequencies between 0.001-20 Hz. It is generated by a variety of events, both man-made and natural. Among the latter type, active volcanoes are efficient sources of infrasound. Volcanic eruptions are characterized by the acceleration of hot fluids from subsurface reservoirs into the atmosphere generating acoustic waves in the 1-20 Hz frequency range. Infrasonic airwaves produced by active volcanoes provide valuable insight into the eruption dynamics and related phenomena. Infrasound also provides a special opportunity for the comparison of eruptive activity among different volcanoes because atmospheric pressure records are mostly independent of site-specific propagation effects (Chilo 2008).

However, infrasound propagating long distances is a complex phenomenon. It is strongly influenced by the detailed temperature and wind profiles. The infrasonic signal detected by traditional infrasound systems contains the combination of the source's infrasound power spectrum and the distortions introduced by the atmosphere. In order to extract the source characteristics the data should be collected at close range: from a few meters to a few km distances. At short distances, the atmosphere is a homogeneous medium that preserves the infrasonic waveform as it is generated by the source. We need to seek new ways to enhance the capability of monitoring volcanic activity close to the source. Wireless sensor networks have the potential greatly benefit studies of volcanic activity.

A wireless sensor network is a collection of small devices having sensors, computational processing ability, wireless receiver and transmitter technology and a power supply (Culler, Estrin and Srivastava 2004). Typical WSNs communicate directly with centralized controller or a satellite, thus communication between the sensor and controllers is based on a single hop. Another kind of WSN could be a collection of autonomous nodes that communicate with each other by forming a multi-hop radio network and maintaining connectivity in a decentralized manner by forming an ad hoc network.

The last few years, the WSN has been used by a number of authors for volcanic eruptions monitoring. In references (Werner-Allen, Johnson, Ruiz, Lees and Welsh 2005, Werner-Allen, Welsh, Locrincz, Johnson, Marcillo, Ruiz and Lees 2006) a WSN was used together with infrasound microphones and seismometers for geophysical studies in the area of the volcano Tungurahua, Ecuador. The infrasound signals were sampled by 102 Hz, 10 bits resolution and transmitted over a 9 km wireless link to a remote base station. For the time synchronization a single GPS receiver was used in combination with the *Flooding Time Synchronization Protocol* (FTSP). The archived accuracy was 10 ms with an error of more than six milliseconds. The data transport was controlled by the remote base station.

In this approach we present a high-resolution WSN for long-term monitoring. The infrasonic signals are sampled and converted to digital data at a frequency of 200 Hz and a resolution of 12 bit. The proposed WSN requires a time stamp per infrasonic sample with an accuracy of one millisecond. Therefore, an algorithm was developed and evaluated which synchronizes the sensor nodes under the support of an equipped GPS device. The algorithm needs to be powered on the GPS device only a few seconds per hour. The collected data is handled under the concept of unlimited virtual data memory, whereby the data is swapped out to an SD card or, if the radio is switched on, transmitted towards the observatory. The used and evaluated routing protocol is the first implementation of the data-centric data dissemination protocol (D3). We extended this cost-based ad hoc routing protocol for the usage of radio time slots. This approach allows a sleep scheduled radio of almost 97% of the time.

The proposed WSN is planned to be deployed at University of San Agustin observatory station (ARE) in Southern Perú. The station is located 12.7 km south-east of Arequipa city, latitude S 16 27'56.67443", longitude W 71 29'35.23676" and elevation 2450 m. The ARE station provides a unique laboratory for studying regional infrasound and seismic wave propagation. The ARE station is located in the shadow of three giant volcanoes: Chachani (6075m), Misti (5821m) and Picchu Picchu (Chilo, Jabor, Liszka, Lindblad and Persson 2006). Furthermore, the most active volcano of Perú, volcano Ubinas, is situated 65 km from the ARE station which will be the focus for our infrasonic studies.

# 2. HARDWARE AND REQUIREMENTS

The platform used is the Modular Sensor Board MSB-430 shown in Figure 1 (left). In this figure the top part (MSB-430S) is the sensor module; the middle part (MSB-430) is the core module; and the bottom part (MSB-430T) is the base module. The base module MSB430T carries three AAA batteries, has a JTAG and serial/USB socket and is available with a GPS receiver FALCOM Smart Antenna FSA01. It should be noted that this platform can easily be exchanged to a more suitable platform. The features and capabilities of the MSB430 concerning our proposal are summarized in Table 1.

For infrasound recording the electret condenser element microphone MCE-200 from Panasonic was used. The details listed in Table 2 are given by the manufacturer. The mentioned frequency range is peculiar and seems not to cover the infrasonic range. Nevertheless, in contrast with the manufacturer specification, this microphone operates sufficiently in the infrasonic range according to the experiences of the authors (Chilo and Lindblad 2007).

**Figure 1.** Title Photo of MSB-430 (left) and the MCE-200 microphone with pertinent preamplifier and filter mounted on a PCB (right)

**Table 1**. MSB-430 features.

| | |
|---|---|
| **Microcontroller MSP430F1612** | 16bit RISC |
| | 100 kHz - 8MHz |
| | 55 kB Flash-ROM |
| | 5 kB RAM |
| | 256B Infomemory |
| **Transceiver Chipcon CC1020** | |
| | 868MHz |
| | 8.6 dBm, max 1km (tuneable to more than 5 km) |
| | 19.2 kbit/s using Manchester encoding |
| **Mass storage** | SD card (max 4 GB) |
| **ADC** | 12 Bit, unipolar [0-3 V] |
| **Sensors on board** | |
| | humidity and temperature sensor Sensirion SHT11 |
| **Supply voltage** | 2.7-3.6V |
| **Energy** | Active Mode: 330 µA at 1 MHz, 2.2 V |
| | Standby Mode: 1.1 µA |
| | Off Mode (RAM Retention): 0.2 µA |

The MCE-200 microphone is also available from Extronic AB (http://www.extronic.se/) mounted on a PCB, Fig 1, right part. A specially designed version includes a filter providing two bipolar output signals, one for infrasound and one for audible sound. The power supply is about 3-12 V. The PCB doesn't suit to be used as a sensor caused by high power consumption of more than 1 mA. We planned to use the microphone without this PCB, therefore, a tailor -made circuit needs to be designed including a low pass filter, a gain amplifier and an antialiasing filter.

**Table 2.** Nominal MCE-200 Microphone specifications.

| | |
|---|---|
| **Frequency range** | 20-16000 Hz |
| **Sensitivity** | 7.9mV/Pa/kHz ±2 dB |
| **Output impedance** | 1-2 kΩ |
| **Signal-to-noise** | ratio < 58 dB |
| **Couple capacitor** | 0.1-4.7µF |
| **Working temperature** | 0-40 °C |
| **Power supply** | 1.5-10V DC /0.5mA |

The following incomplete enumeration briefly sketches the requirements of the application domain.

1. An event classification requires the complete signal opening of each infrasonic record.
2. The record of an infrasonic event per microphone shall be automatically transported from volcano Ubinas to the University of San Agustin observatory station (ARE). The distance is about 65 km.
3. The operating time of the monitoring system is weeks or permanent.
4. The system shall automatically handle the failure of sensor nodes. The system will continue at least under a restricted operating mode following the malfunction of multiple nodes. It means that a complete system breakdown should be avoided.

The third requirement is in contradiction to the first one, because it forces a continuous AD conversion. Therefore, the sensor nodes will quickly run out of power. A trade-off is the continuous conversion by just a single node. The remaining nodes could use a more energy conserving comparator.

To analyze the wave propagation and in order to localize the infrasonic source six time synchronized signals recorded from different positions are required:

1. At least six microphones shall be deployed close to the volcano. They shall be spatially separated.
2. The chronological, accurate to a millisecond, and spatial position, accurate to 10 meters, of all records per microphone of an infrasonic event shall be available.

# 3. SYSTEM DESIGN

## 3.1. System Overview

The data transport requirements (about 65km distance) affect materially the hardware selection and distribution. The transmission range of the MSB-430 mounted radio goes about 5000 m, but only under special circumstances like a clear line of sight. Hence the node has to be combined in a multi hop network a more powerful transmitting devices must be used. The FreeWave FGR09CSU 900MHz radio modem (http://www.freewave.com) could be used for this task.

According to the manufacturer specifications the FGR09CSU modem reaches a range of 95 km in a clear line of sight. If no line of sight is possible, then either a modem can be used as a repeater or additional intermediates allow a more suited gateway position. An example deployment using the FreeWave modem is sketched in Figure 2. In the example seven infrasonic sensor nodes (green) are used for the data acquisition, one gateway node (magenta) is connected to a workstation located in the ARE observatory and one intermediate node (blue) pass the data from the sources to the sink, the gateway node, which finally delivers the data to a workstation.

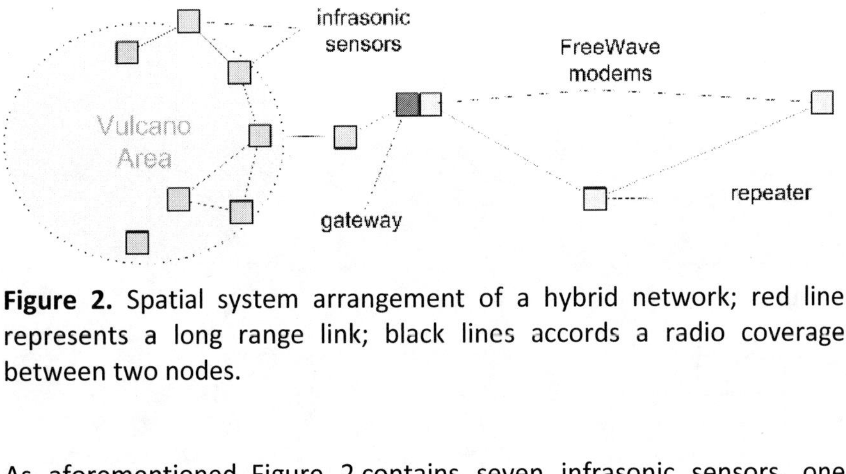


**Figure 2.** Spatial system arrangement of a hybrid network; red line represents a long range link; black lines accords a radio coverage between two nodes.

As aforementioned, Figure 2 contains seven infrasonic sensors, one more than needed. Not all sensors have to be simultaneously active. So, one can rest and maybe recover energy. Not every infrasonic sensor in the figure is inside radio coverage of an intermediate. In order to transport the data of the outer infrasonic sensors all of these must be intermediate nodes simultaneously. To assure sufficient CPU power for the data acquisition this double role needs to be controlled.

It is also vitally important that a failure of one or two nodes must not impact the whole system, thus the architecture shall actualize highly autonomous behaviour of the nodes not just for the day-to-day business but also for the handling of exceptions like malfunctions of some nodes. Furthermore, low power consumption and energy conservation strategies shall be taken into account, which is essential for a long term operation.

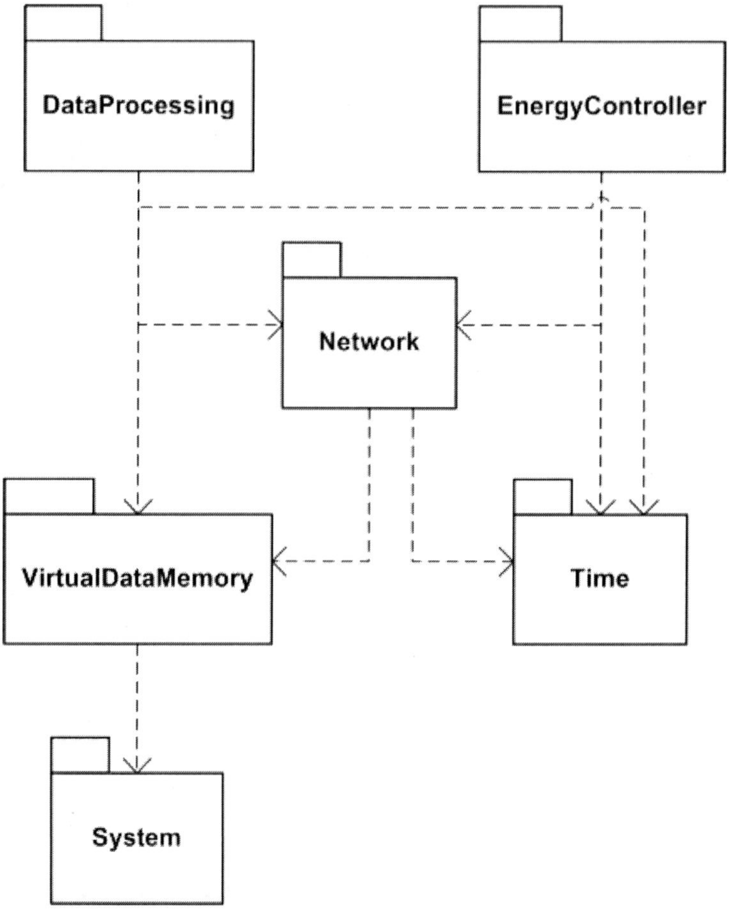


**Figure 3.** The diagram gives an overview of the system architecture. Arrows symbolize interface access.

### 3.2. Architecture

The applied method for the system decomposition covered both: the partitioning of the system according to the system functionality (distinction of concerns), and to map the components to the required hardware.

Figure 3 shows the component dependencies of a sensor node. Time synchronization using GPS and periodical signalization form the Time component. For instance, the analog-digital-converter (ADC) is controlled by a periodical signal. The converted data is evaluated by the DataProcessing component (DP) in order to detect events of interest. The component gathers all needed data for an infrasonic record including the time and position information provided by the Time component and hands it to the Network component. Both components, DP and Network, require major RAM parts, as they handle the same data. The VirtualDataMemory component (VDM) manages RAM and SD card blocks. The current voltage level is periodically watched by EnergyController. Depending to the remaining energy and in collaboration with the neighboring nodes the component set the node into a low-power mode by stopping the Network, the Time and indirectly the DP component. The hardware abstraction functionality is pooled within the System component.

# 4. TIME SYNCHRONIZATION

## 4.1. Hardware Mapping

The MSP430 version F1216 provides two timers, *Timer_A* and *Timer_B*. Timer_A has three (TACCR0–TACCR2), Timer_B has seven independent capture/compare registers (TBCCR0–TBCCR6). A timer register can be used in two different ways. Either the register's content is compared to the current counter value of the timer and creates an interrupt on equality (compare mode), or on an external signal, e.g. the rising edge of the GPS pulse, the current counter value is stored in the register (capture mode). Timer_A is configured to count the oscillations of the sub-main clock (SMCLK), which itself will be fed by the digitally controlled oscillator (DCO, Figure 4) and Timer_B counts the crystal watch oscillations (ACLK, 32768 Hz). In order to provide a stable main clock the DCO is watched and controlled according to the more reliable crystal oscillator (dcoChecker).

The second Timer_B register in compare mode is used to create the signal for the ADC. The third register is used to measure the GPS pps. In detail, the GPS signal captures the Timer_B counter value, stores it into the third register and wakes up the CPU, i.e. an interrupt service routine (ISR) of the Time component.

For scheduling tasks like a periodically GPS synchronization the functionality of software timers is needed. Software timers shall run either after a defined delay or at a defined point in time. For an accurate scheduling two values are required, the number of Timer_B cycles and the desired value of Timer_B, i.e. ACLK cycles. When the defined Timer_B cycles of a software timer expires, the desired watch oscillator cycles are written to the sixth compare register, which produces an interrupt if Timer_B reaches the value and the software timer is executed.

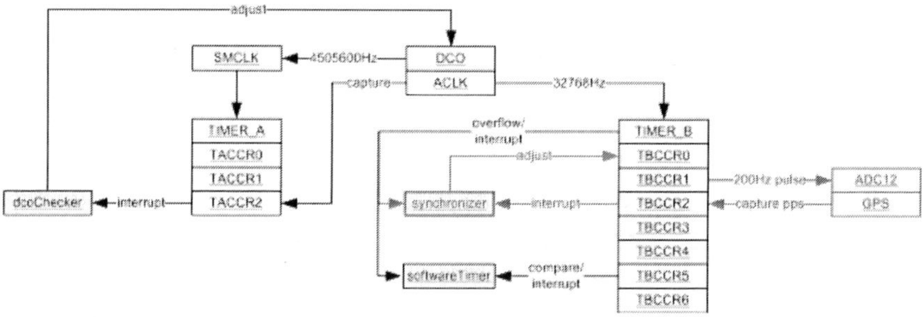


**Figure 4.** The diagram shows time synchronization via crystal watch. Green objects are interrupting service routines; red objects are available only on sensor nodes.

## 4.2. GPS-ACLK Synchronization

The idea of the ACLK synchronization is to let the timer count from zero to exactly the ACLK frequency decreased by one. Though, the exact ACLK frequency, about 32678 Hz, needs to be measured by using the GPS pulse-per-second (pps) as reference which provides an accuracy of 1 µs. The real ACLK frequency depends to the environment temperature. The interrupt TBCCR0 (for TBR=32767) and CAP2 (for pps capture) occur simultaneously. This fact complicates the deviation detection. The execution of an interrupt service routine may not follow the real chronological event order. Therefore, it is wise to move the pps capture reference from zero to (TBCCR0+1)/2. Thus, the interrupts are separated by 0.5 s, but only if the system is synchronous. In other words, if the Timer_B is synchronized, the pps is caught exactly by (TBCCR0+1)/2. Moreover, the ACLK ticks between the first and the second pps capture are counted, what should be close to 32768. This is the first synchronization stage.

To illustrate the procedure in detail we sketched a simplified example in Figure 5. In this figure each timer counting step is visible. One relevant detail is the fact that if the timer counts to TBCCR0 it altogether counts TBCCR0+1 ticks; as the timer value is in the range of [0..TBCCR0]. Before the first capture occurred, the timer was initialized by $f_{ACLK_1} = 9$ or TBCCR0 = 8. After two seconds $capture_1$=3 and $capture_2$ =1 are measured.

$$f_{ACLK_n} = (f_{ACLK_{n-1}} - capture_{n-1}) + capture_n \tag{1}$$

$$shift = capture_n + \left\lfloor \frac{f_{ACLK_n}}{2} \right\rfloor \tag{2}$$

Now, the more exact frequency can be computed by equation 1, with result

$f_{ACLK_2} = 7$. Before the register TBCCR0 is adjusted by $f_{ACLK_2}$, it is phase shifted, i.e. it is set to $shift = 4$. The shift is determined by the elapsed time since the last timer overflow plus the expected amount of ticks to the desired timer overflow, equation 2. This happens still in the ISR of the second capture.

On the next timer overflow (Figure 5, third TBIFG event) the timer border is set to the calculated value $f_{ACLK_1} - 1 = 6$. The third and fourth capture occurs exactly on TBR=3. The real time clock is synchronous. Furthermore, the local time is adjusted to the UTC time. The UTC time was written by the GPS device to the COM port between the first and second capture and must be mapped to the first capture; i.e. the first pps is the exact point in time given in the following NMEA record. The GPS device could be switched off after the second capture for a time period depending on the allowed time deviation.

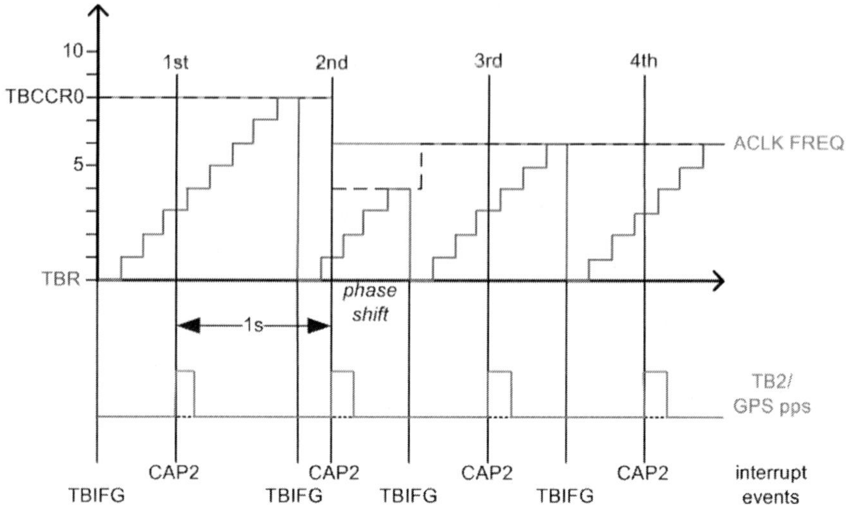


**Figure 5.** Example of phase shift and period correction via ACLK.

The allowed deviation of $\pm 0.5ms$ accords $\pm 5ms * f_{ACLK}\,ticks$, which is about 16 ticks. Even by a stable frequency in the worst case it could happen that each second the ACLK deviates almost one tick. It is caused by the fact, that the frequency is calculated as an integer and the fraction is dropped. In case of a fraction close to one the ACLK exceeds the allowed deviation after 16 s.

As a corrective the places after the decimal point for the rational ACLK frequency are approximated by measuring the average deviation for

*Δseconds*. Through the usage of the integer frequency $f_{ACLK}$ the deviation accords the fractional part of the rational ACLK frequency multiplied *Δseconds*. The fraction calculation is done within the second stage of synchronization. First after the third capture the GPS device is switched off and the captured values (captured ticks and local time in seconds) are stored. When for instance tens of minutes elapsed the GPS device is activated again and the fourth capture is awaited. Equation 3 calculates the average fraction, whereby $\Delta ticks_n = capture_n - capture_{n-1}$.

$$fraction_n = \frac{\Delta ticks_n}{\Delta seconds_n} \qquad (3)$$

The frequency correction can be simply done by adding uniquely Δ*ticksn* every Δ*secondsn* to TBCCR0 and restoring the original value of TBCCR0 one second later.

However, this approach would lead to unwanted steps within the time response. Much better is the method of equally distributing Δ*ticksn* with Δ*secondsn*. The resulting value for TBCCR0 is calculated on each timer overflow event like in equation 4. The variable *t* is the time in seconds after the last phase shift (*t*=0*s*).

As $fraction_n$ always is within the range (-1..1), the timer border *TBCCR0*(*t*) equals either $f_{ACLK_n} - 1$ or $f_{ACLK_n} - 1 \pm 1$. In other words: the correction of

$\Delta ticks_n$ within Δ*seconds* is implemented as an equal distribution of atomic corrections for the timer period.

$$TBCCR0(t) = f_{ACLK_n} - 1 + \lfloor (t+1) fraction_n \rfloor - \lfloor t fraction_n \rfloor \quad (4)$$

On the fifth pps the captured values needed for the next GPS adjustment are stored and the device is powered down for maximum 16* $\Delta seconds_n$ (on a constant frequency). One open point is the measuring of $fraction_n$ while concurrently the timer is adjusted by $fraction_{n-1}$. Equation 3 needs to be extended to equation 5 to take a concurrent adjustment into account.

$$fraction_n = \frac{\Delta ticks_n + \lfloor \Delta seconds_n fraction_{n-1} \rfloor}{\Delta seconds_n} \quad (5)$$

First tests showed a non acceptable result. A failure analysis discovered a dependency of the watch crystal on the environment temperature. Within a couple of hours the oscillator slowed down almost linear. The third synchronization stage forecasts the expected ACLK frequency taking the gradient of the frequency response into account.

The gradient is calculated in equation 6. The gradient goes into equation 4 and equation 5 to formequation 7 and equation 8.

$$gradient_n = \frac{(f_{ACLK_n} + fraction_n) - (f_{ACLK_{n-1}} + fraction_{n-1})}{\Delta seconds_n} \quad (6)$$

$$TBCCR0(t) = f_{ACLK_n} - 1 + \lfloor (t+1)\, fraction_n + (t+1)^2 gradient_n \rfloor - \lfloor t fraction_n + t^2 gradient_n \rfloor \tag{7}$$

$$fraction_n = \frac{\Delta ticks_n + \lfloor \Delta seconds_n fraction_{n-1} + \Delta seconds_n^2 gradient_{n-1} \rfloor}{\Delta seconds_n} \tag{8}$$

The synchronization stages are sketched in the activity diagram in Figure 6. Stage one happens after the second capture but only for the first time a node is activated. Between the third and fourth capture the GPS is powered down for a long time. For the best trade off between energy consumption and accuracy the time duration is determined by test series. On the fourth capture stage one and two are executed. The calculation of the gradient results zero, hence a prior fraction is missing in the current state. The next occurring capture is treated as the third capture again. The second time a *fourth* capture happens, the gradient can be determined.

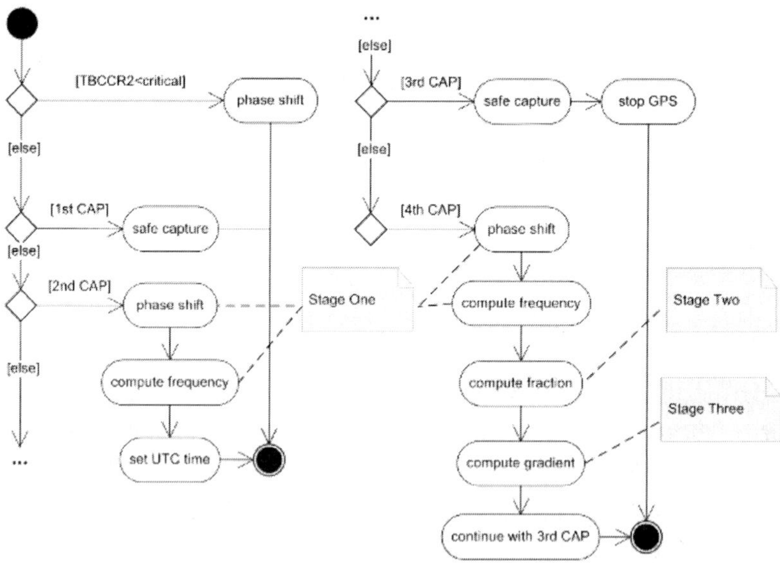


**Figure 6.** Final activity diagram of handling CAP2 events.

# 5. DATA ACQUISITION

## 5.1. Infrasound Data

The infrasound sensor is realized by the electret condenser element microphone (ECEM) MCE-200 from Panasonic. A PCB will connect the microphone to the ADC12. Therefore, the signal needs to be transformed to a unipolar signal up to 3V.

The complete data pipeline is arranged in a way that first the analog signal produced by the ECEM is filtered to pass only frequencies below the cutoff frequency and afterwards the signal is gained to the required voltage, Figure 7. This approach avoids an overmodulation by disturbing frequencies.

To date there are not ideal low-pass filter, i.e. passing the signal below 20 Hz and eliminating the frequencies above. Real filters are approximations, for instance the *Bessel*, *Chebyshev*, and *Butterworth*filter.

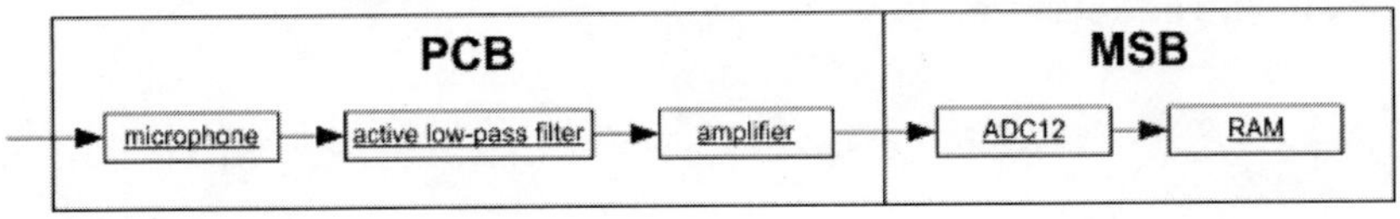


**Figure 7.** Data acquisition pipeline

A Butterworth filter was designed in our case. The quality of an active filter, for instance a steeper attenuation, is improved by cascading filter stages by combining multiple single filters. A *4th order*low-pass filter was realized which will require two operation amplifiers on the PCB.

## 5.2. Local Event Detection

To detect a supposed volcanic eruption a simple approach is watching the signal amplitude. If it passes a threshold one can assume an eruption occurred. The maximum of the amplitude is searched in time window $T_w$. For the detection of a 1 Hz signal $T_w$ must be at least 500 ms. If the threshold is to high, the amount of false negatives, i.e. missing an event, increases. If it is to low, more false positives will be detected, for instance short-term fluctuations or wind.

Short-term fluctuations can be smoothed out by calculating a moving average. The *EWMA-detector*implemented by (Werner-Allen, Johnson, Ruiz, Lees and Welsh 2005) supplied more reliable events than their implemented threshold based detector. The EWMA (exponential weighted moving average) function

$$average_t = average_{t-1} + \alpha(sample - average_{t-1})$$

computes the averages by fading older samples. The variable $\alpha \in [0..1]$ determines the fading factor of old samples. A high α fades off more. The elimination of long-term trends can be done by maintenance two averages, by name a short-term $average_s$ and a long-term average $average_l$. If $average_s$ exceeds $average_l$ by the amount of *div* an event is triggered.

The state machine in Figure 8 maps the distributed event detection. To trigger the data acquisition two independent events must occur within a time window. The elapsing of a timeout without an event always leads back to the initial state.

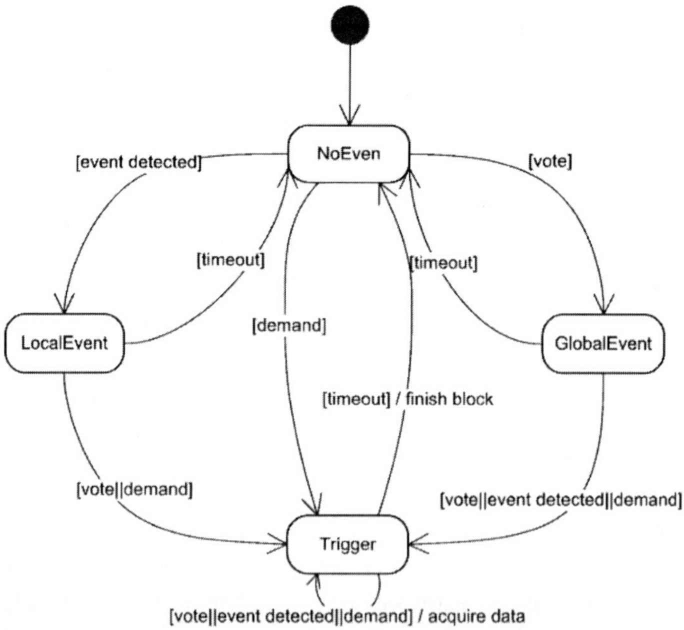


**Figure 8.** State machine for distributed event detection.

All state transitions are briefly described in Table 3. The demand message is put into brackets because of the following: if a sensor does not receive two votes it is probably to far away, that it could sample meaningful data, hence, the demand message could be discarded. However, the implementation efforts for this functionality are not high and may still finally benefit.

**Table 3.** List of state transitions used in Figure 8

| State Transition | Description |
|---|---|
| *local event* | local event detected |
| *vote* | dialog message received by a neighbour who detected an local event |
| *(demand)* | dialog message forces the start of data acquisition |
| *timeout* | time without any event elapsed |

## 6. DATA TRANSPORT

The implemented routing protocol D3 (Ditzel and Langendoen 2005) is a cost table and data centric routing protocol, i.e. no network addresses are used and the data is “floating” down a gradient. Therefore, only broadcasts are used. In respect to energy conservation, the medium is accessed in time slots, small ones for negotiation and large timeslots for the data transfer whereby only the participating nodes are active. The problems of hidden and exposed stations are no more solvable by the classical MACA protocol. Instead we used three different data timeslots: TXDATA for transmission, RXDATA for reception, and IDLE for a powered down radio. An example is given in Figure 9. All nodes with the same hop-count distance to the data-sink, i.e. cost, need to use the same slots.

Figure 10 shows the general idea of the D3 protocol. First the gateway, the sink, broadcasts an interest message, Figure 10(a). Each node updates its own cost, i.e. it stores the smallest received cost plus one and forwards one-time the INT message with the node's own cost. The data sink labels the interest message by a sequence number. If a node receives an INT message with a new sequence number, it overwrites its cost value even with a higher cost. In this way a changing node topology can be recognized by regular INT messages.

Before a node transmits a data message, it advertises the data by broadcasting an ADV message in the ADV slot before the next TXDATA slot, node E in Figure 10(b). The message contains the cost of the node (here two for node E) and the segment_key_t values of all unconfirmed packets in the memory. In the original D3 protocol the message also includes the time of data transmission, i.e. here the duration to the next data slot in ACLK ticks. However, the transmitted duration should be ideally the time between the first byte received by an interested node and the beginning of the data slot. To approach this value as close as possible the time of transmitting the first ADV byte needs to be measured by the initiating node and the duration should be computed according this value. This is only possible by transmitting the duration by a different message immediately after an ADV message; called*Time* message. To map a Time message to the responding ADV message, both messages need the node network address.

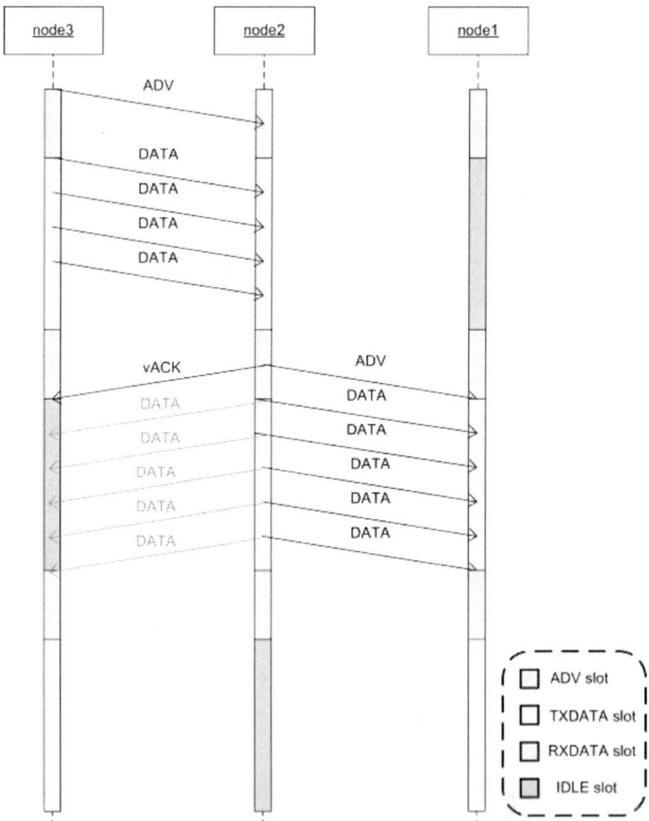


**Figure 9.** Sequence diagram of D3 timeslots.

Each neighbour with lower costs than the cost of a received ADV message schedules the data reception (node C and D). All other nodes just switch off their radio in the next RXDATA slot (node F and G). The data is transmitted at the beginning of the RXDATA slot, Figure 10(c).

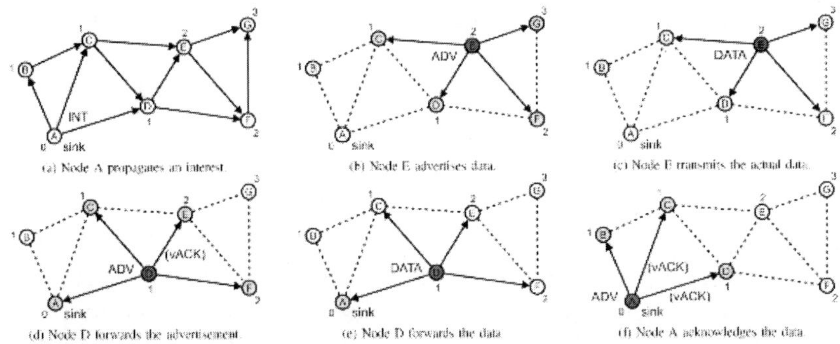

**Figure 10.** The D3 routing protocol. Initially the gateway floods an INT message. A node forwards data by advertising the new data and after transmitting the data itself.

The next data slot is an IDLE slot for the prior transmitter (node E) and a TXDATA slot for the prior receiving nodes (node C and D). Those nodes initiate a random backoff delay at the beginning of the ADV slot foregoing its TXDATA slot. The node for which the backoff delay expires first, node D inFigure 10(d), immediately broadcasts an ADV, which lets all other nodes with an advertisement intention cancel their backoff delay. Again, the message contains the nodes own cost (one for node D) and all segment identifications of the packets in the memory, which includes also the quite recently received packets. Furthermore the prior forwarding node (node E) takes that ADV message as a virtual acknowledgment (vACK).

Finally, the data message reaches the gateway, Figure 10(e). The gateway also advertises its data in its according ADV slot, Figure 10(f). Of course, in its following TXDATA slot it does not forward the data by the radio, but by the COM port to a gateway client.

The usage of explicit acknowledgments (eACK) is need fully to avoid the retransmission of plenty big data packets. The eACKs are transmitted immediately when a neighbour node advertises already confirmed packets.

Another scenario is the reseting of a node or the adding of a new node to the network. The ADV messages do not suffice for the integration of the node into the topology, because the following could happen: Let's assume a new node has the cost of five, but before the node doesn't receive an ADV message of a neighbour with the cost four, its cost is unknown. Now, this node receives an ADV message with the cost six and assumes a cost seven, i.e. the cost plus one. This lets also confirm all advertised packets in its history. In the case the node receives an ADV message with cost four; it updates its cost to five. If now the node who first advertised data needs to advertise the data twice, the new node assumes the data to be already confirmed and wrongly transmits an eACK. The data packets will be lost. A solution is the usage of interest request messages, which are transmitted the first time a node enters the network or frequently until the node is added to the network. All neighbours answer on an interest request with its last received interest message. However, the requesting node can use the ADV messages together with the time messages to synchronize to the network time slots, so it could transmit the request within an ADV slot.

# 7. RESULTS AND CONCLUSIONS

## 7.1. Data Processing Component

A prototype node is equipped with a specially designed PCB containing the infrasonic microphone, a 4th-order Butterworth active low-pass filter, and an amplifier. For the power supply a 3 V source is used. A function generator is connected to the couple capacitor and ground. It produces a sinus wave with an amplitude of $U_{i-max}$= 344 mV. The PCB output pin and ground is connected to a voltage oscillograph. The amplitude of the PCB output signal $U_{o-max}$ is measured. The output signal swings around 1.5 V.

The gain $G_{PCB}$ of the circuit depending to the input frequency $fi$ is computed by the equation 10,Figure 11, green curve.

$$G_{PCB}(f_i) = log(\frac{U_{o-max}(f_i) - 1.5V}{U_{i-max}})*10dB \quad (10)$$

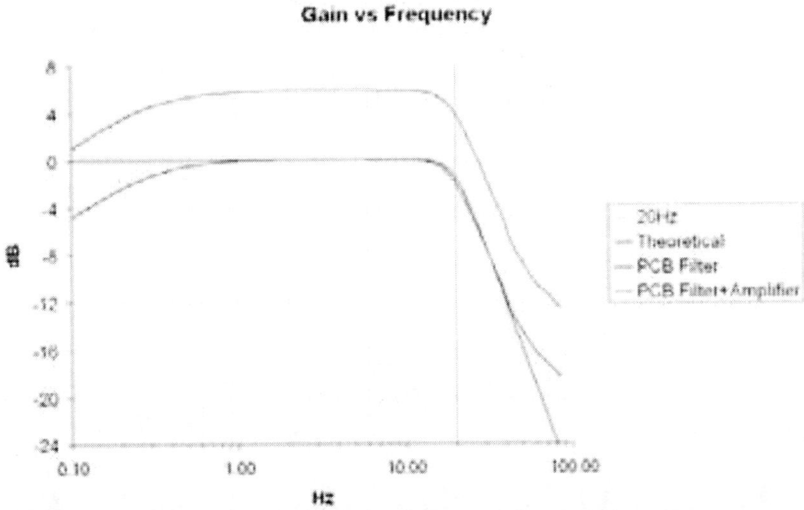


**FIGURE 11.** The response of the amplifier. Red, theoretical 4th-order Butterworth; blue, filter without amplifier and green, filter plus amplifier.

To measure the gain of the AC amplifier we used a frequency which is passed by the filter with unity gain. In other words we searched the output voltage maximum depending to the input frequency, what is by f=2.2 Hz and *Uo−max*

=2.844 Hz. The gain at this point is:

$$G_{amp} = \frac{2.844V - 1.5V}{U_{i-max}} = 3.95 \quad (11)$$

The calculated value is G=4.25. To get the gain

$G_{filter}$ assigned to the filter the amplifier gain must be undone, Figure 11, blue curve:

$$G_{filter}(f_i) = log(\frac{U_{o-max}(f_i) - 1.5V}{3.95U_{i-max}})*10db \quad (12)$$

The amplitude response of a 4th-order Butterworth filter is plotted for comparison, Figure 11 red curve. Two obvious deviations are visible. The first one below 1 Hz is explained by the high-pass effect of the couple capacitor and maybe even by the high-pass effect of the capacitor of the AC amplifier. The second one close to 100 Hz can be a measurement error caused by the fact that the output voltage in this range is nearly the offset DC of 1.5 V.

The cutoff frequency is defined as the frequency for which the filter returns $\sqrt{1/2}$ of the pass-band voltage. The cutoff frequency $f_c$ can be experimentally determined by finding the frequency which fulfils following equation:

$$log(\sqrt{1/2})*10dB = -1.51dB = G_{filter}(f_c) \tag{13}$$

The realized cutoff frequency is $f_{c-low}$=19 Hz. The cutoff frequency of the high-pass effect can be determined in the same way and is

$f_{c-high}$ =0.14 Hz which is acceptable.

## 7.2. Time Component

The three staged GPS synchronization allows a switched off GPS device for long periods ($t_{GPS\_OFF}$). The time period impacts not only the energy consumption but the RTC accuracy. To find the best trade-off between energy consumption and accuracy test series with different $t_{GPS\_OFF}$ values are accomplished. The board is simply made of a single sensor node equipped with a GPS device. It is configured to start the stage one synchronization immediately, stage two after ten minutes, and stage three after further ten minutes. Firstly now, the time period $t_{GPS\_OFF}$ is used. Hence, the GPS device needs between 45 s and about 165 s to provide a valid pps signal, the synchronization period is $t_{GPS\_OFF}$ plus the fluctuating GPS startup time.

Figure 12 shows the result for $t_{GPS\_OFF}$=45 min. The test runs more than 15 hours. One can extract different assumptions by interpreting the responses. The first fact is that within the first three hours the synchronization is unacceptable. The reason could be the temperature difference between the office and outside, thus the node needs some time to acclimatize to the outside temperature. Furthermore, the outside temperature declined in the evening hours.

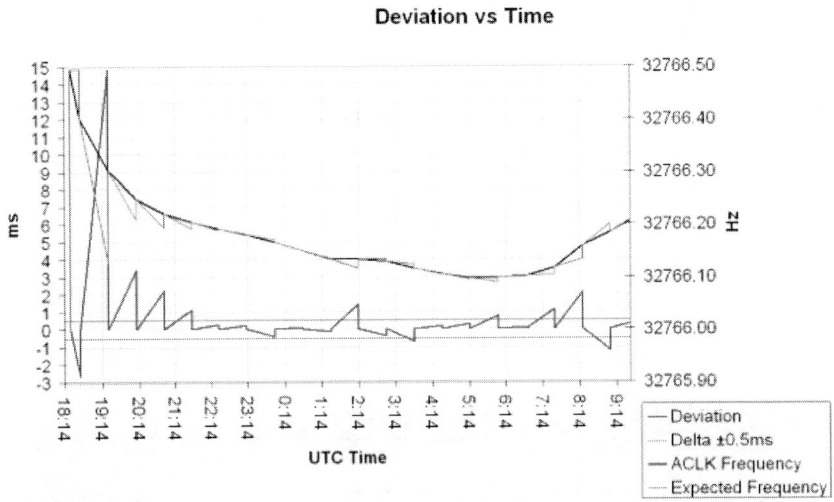


**Figure 12.** State 15 hour test result of the RTC deviation; three staged GPS synchronization happened about every 45 minutes.

Incontestable is the dependence of the deviation between the expected and the real frequency response to the RTC deviation. A bend in the frequency response leads to a deviation. However, the idea to take the second derivation into account would worsen the prediction of the frequency. One can appreciate this, if the expected frequency response is imaginary extended by the turns of the real curve. The mathematically unpredictable bends can maybe be physically predictable, if for instance the temperature is monitored. A future investigation of dependence of the frequency and temperature could pay off. Summarized, 33.7% of the time the RTC exceeds the allowed limit (ignoring the first three hours, 22.1%).

For the experimental assurance of the desired accuracy the time $t_{GPS\_OFF}$ needs to be defined according to the highest gradient, what is = 648 s, equation 14, ignoring the first three hour.

$$t_{GPS\_OFF} = 0.5ms / |gradient_{max}| = 648s \tag{14}$$

The average gradient, equation 15, of the deviation takes both into account: long periods of small deviations and the percentage part of peeks of high deviations. A fixing of the time $t_{GPS\_OFF}$ according to the average gradient, equation 16, allows an exceeding of the allowed deviation for a small time. For the average gradient without the first hour the result is

$t_{GPS\_OFF}$=2150*s*.

Very optimistic but energy conserving is to take only 75% of the smallest gradients into account. The resulting value according to the average gradient of 75% of the best values is $t_{GPS\_OFF}$=4168*s*.

$$\overline{gradient} = \frac{\sum_{min<i\leq max} \frac{|deviation_i|}{\Delta t_i}}{max - min} \tag{15}$$

$$t_{GPS\_OFF} = \frac{0.5ms}{\overline{\overline{gradient}}} \tag{16}$$

An adjustment every 30 minutes, Figure 13, doesn't show a considerable improvement. It needs two hours to acclimatize and to calculate the gradient to the realistic temperature response. After the first two hours the time of exceeding the limit deviation of 0.5 ms is about 19.9%. A shorter time period for the GPS adjustment doesn't solve sufficiently the deviation problem caused by the turns in the frequency response.

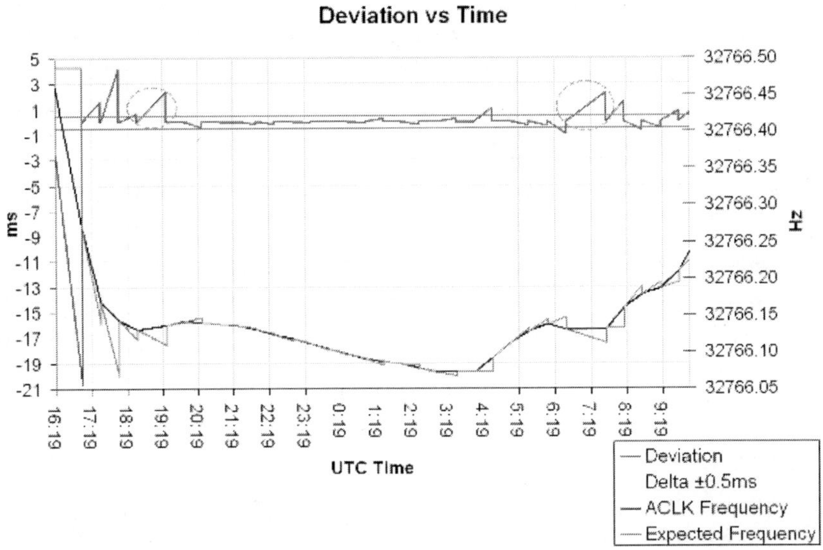


**Figure 13.** hour test result of the RTC deviation; three staged GPS synchronization happened about every 30 minutes; pink cycles - GPS timeout caused by rain.

For a complete evaluation of the measurement the knowledge of an outstanding fact is required. The pink cycles mark long time periods of a deficient GPS signal, which lasts respectively about an hour. Both GPS timeouts were caused by rain. Rain impedes the synchronization process twice. On the one hand, it induces GPS timeouts and on the other hand, it comes together with an increasing of temperature, which itself speeds up the ACLK. Nevertheless, excluding these values would result a violation of the accuracy requirements for about 12% of the time.

## 7.3. The Network Layer

The issues for the network layers are the capacity, the reliability, and the robustness. Two types of test scenarios were accomplished: the transmission of a continuous data stream and a sparse transmission of data records over a long time period.

For both tests the command *cost* was implemented on the gateway node in order to simulate a cost distribution. The first parameter of the command addresses a node. The second one specifies the desired cost of the node. By receiving a dialog message containing the command the destination node adjusts its cost accordingly. While the test application is running, the node ignores messages transmitted by nodes whose cost difference is greater than one.

Unfortunately, a second node equipped with a second GPS device was not timely available. Due to the fact that the network time slots are seeded by time synchronized nodes, the tests can only be done with a single data source. However, the intermediate nodes don't care for the packet originator and multiple data sources are not essentially required to test the functionality.

Four different configurations for the breadboard were realized. Different network depths and different numbers of nodes on the same level were tested. Of course the data sink was always on level zero. The data rates are measured by the GatewayClient.

**Table 4.** Test results of routing a continuous data stream.

| data rate [kbit/s] | Volume [kB] | cost level 1 | cost level 2 | cost level 3 |
|---|---|---|---|---|
| 2.35 | 530 | one data source | | |
| 2.26 | 502 | one intermediate | one data source | |
| 1.37 | 292 | two intermediates | one data source | |
| 2.09 | 524 | one intermediate | one intermediate | one data source |

Table 4 shows the data rates measured for the different network topologies. The packet loss rate (not listed in the table) for all topologies was 0%. The rate for the two node topology (line one in the table) complies exactly the theoretical value of equation 17 (one RAM block of 512 Bytes contains four network packets; one network packet carries a payload of 110 Bytes). The data rate is the measured value for two intermediates on the same cost level. Obviously, the small value is reasoned by collisions. This indicates a small value for the used backoff delay. Anyhow, the results still fulfil the requirements.

$$datarate_{MAX} = \frac{4blocks * 4packets * 110Byte}{3 * t_{ADV} + 3 * t_{DAT}} \approx 2.35kbit/s \tag{17}$$

By the following test case the reliability of the network synchronizations is measured. During the periods of no network traffic, no synchronization of the time slots happens. The network topology was linear with a depth of three, i.e. the data source was on cost level three. The size of the records was randomly created.

**Table 5.** Test results of routing periodically single records.

| test duration [min] | record separation [min] | volume [kB] | records |
|---|---|---|---|
| 124 | 15 | 51 | 8 |
| 185 | 30 | 33 | 7 |

A delay of 15 minutes between the record transmissions results a stable time slot behavior of the network, Table 5. However, for the delay of 30 minutes the networks got asynchronous after the 7th record. Another test with a delay of one hour (not listed) failed after the first record. It implies, that the time slot deviation after 30 minutes can exceed at least the duration of the half of the ADV slot duration, i.e. 94 ms or 3072 ACLK ticks. In other words, if the crystal oscillators of the intermediates deviates about 3.4 ticks per second, the ADV slot duration is exceeded after 30 minutes.

In order to increase the allowed deviation the ADV slots could simply be extended. Even, periodical transmissions of ADV messages would synchronize the nodes again. However, both approaches would increase the network energy consumption.

Another solution is the flooding of the measured frequency and its expected trend by the sensor nodes immediately after a sensor node synchronized to the GPS device. Therefore, it is assumed, that the oscillators of all nodes work under the same conditions.

Considering very long time periods without any network traffic, following strategy is promising: If the nodes got asynchronous it firstly matters in the case of transmission purposes. To synchronize again the acting node transmits its slot time multiple times with a delay, which assures that the message is received in at least one ADV slot by a child node. A delay of about the half of the ADV slot duration would suit. The acting node is *scanning* the time slots.

Though the unsatisfying result the implemented D3 routing protocol works in principal. A detailed analyzes and thereby the adjustment of the protocol parameters is the direction of future work, as well as large scale tests.

## 7.4. Conclusion and Outlook

The system presented in this chapter has the potential of a full-fledged application. The analysis of the requirements, as well as the design of the system architecture, was done using common and well proven software engineering techniques. Whereby, the balancing act between the reutilization and the easy interchange ability in opposite to the high application awareness succeeded for the most parts. Though the not entirely satisfying results and the open questions, the design is reasoned and self contained.

In order to connect the microphone to the MSB a sensor module was created. The circuit covers a fourth-order Butterworth low-pass filter and an amplifier. Furthermore, the signal is transformed from bipolar to unipolar. The comparison of the measured frequency response and the expected theoretical response shows acceptable results. A direction for future work is to design a digitally controlled amplifier. This would open the possibility to remotely adjust and fine tune the event detection on hardware level. In the same way, the problem of a missing infrasonic reference source for the best suiting amplifying factor could be avoided.

The time synchronization was one of the big challenges. We showed the feasibility of achieving the required accuracy by a rare use of the GPS device. The implemented approach comprises a high potential to get improved. In view of the measurement results, the prediction of the crystal oscillator frequency by taking the current temperature into account is promising. Even the usage of an external and more stable oscillator would be an enhancement. The maximum achievable accuracy is determined by the GPS receiver and lies in the range of microseconds.

The time synchronization greatly benefits the data acquisition. On the one hand, the synchronized hardware timer controls the sampling frequency through the pulse-width modulation, what assures the desired sampling frequency. On the other hand, the timer determines the beginning of the sampling. So, the mapping of the samples to the UTC time is reliably done.

The conflict of the high CPU demands of both, the data acquisition and the data transmission, is solved by applying a mutex for the CPU. Anyhow, considering a long-term usage it is recommended to implement a hardware trigger and start the data acquisition only if a threshold is passed. Therefore, the requirement of the signal opening needs to be discarded. A trade-off is the continuously sampling by just a single node.

The presented strategies for the local and the distributed event detection can be applied and expediently tested firstly, when several infrasonic sensor nodes are available. Again, the missing of an adequate reference signal complicates a contingently ongoing development. Future work should concentrate on the remote parameterizing of the thresholds. The event detection is an own field of research. An improvement here would payoff by reducing the power consumption of the network.

Another big challenge was the implementation of the routing protocol in respect to the time slots. The procedure was very time consuming and the test results didn't show a sufficient behaviour. Anyhow, we showed the functioning in principle and submitted possible adjustments. The most promising approach is the scanning of the time slots in the case of asynchrony. The benefit of the sleep scheduled radio for almost 97% of the time still reasons such an approach. A detailed analyze and proving of the routing protocol can fill an own study.

The implemented concept of the virtual data memory was highly optimized to the demands of the accessing components. The SD card read/write operations are only performed when it is absolutely required. For the data acquisition a free memory block is always assured. The Network component finds enough free memory in the reception time slot, while in the transmission time slot the memory is filled by the packets to be forwarded. In order to serve gateway requests of missing packets a history maps the packets to SD card addresses. The thereby required trade-off between the history size and the SD card address limitation can be solved by considering the swapping out the entire history to the SD card.

Finally, designing a circuit for the regeneration of chargeable batteries by solar panels could be the essential step towards a real long-term usage.

Wireless sensor networks present many exciting opportunities. The developed system is easily to customize in order to operate in different applications. For instance, the orthogonal usage of the infrasonic microphones can be used to measure the resonance frequency of high buildings. The implementation of a burglar alarm is another example, as well as tracing of moving objects by the time synchronized nodes. Nevertheless, exchanging the microphones to other sensors opens a broad spectrum of possibilities.

## REFERENCES

1. J. Chilo, 2008 Low-Frequency Signal Classification: Filtering and extracting features from infrasound data Verlag Dr. Muller, 978-3-63911-341-9
2. J. . Chilo, A. Jabor, L. Liszka, A. J. Eide, Lindblad, L. Persson, 2006 Infrasonic and Seismic Signals from Earthquake and Explosions in Arequipa, Peru, Proceedings of Western Pacific Geophysics Meeting, Beijing China, 2006.
3. J. Chilo, Th. Lindblad, 2007 A Low Cost Digital Data Acquisition System for Infrasonic Records, International Workshop on Intelligent Data Acquisition and Advanced Computing Systems, 3537 , IEEE Catalog 07EX1838C
4. D. Culler, D. Estrin, M. Srivastava, 2004 Overview of wireless sensor networks, IEEE Computer, Special Issue in Sensor Networks.
5. M. Ditzel, Langendoen, 2005 D3: Data-centric Data Dissemination in Wireless Sensor Networks, IEEE Computer Society, The European Conference on Wireless Technology.
6. G. Werner-Allen, J. Johnson, M. Ruiz, J. Lees, M. Welsh, 2005 Monitoring Volcanic Eruptions with a Wireless Sensor Network, Second European Workshop on Wireless Sensor Networks, EWSN'05.

7. G. Werner-Allen, M. Welsh, K. Locrincz, J. Johnson, O. Marcillo, M. Ruiz, J. Lees, 2006 Deploying a Wireless Sensor Network on an Active Volcano, IEEE Computer, Sensor-Network Applications.

# Chapter 8

# Mapping Forest Canopy Height over Continental China Using Multi-Source Remote Sensing Data

**Xiliang Ni [1], Yuke Zhou [2], Chunxiang Cao [1], Xuejun Wang [3], Yuli Shi [4], Taejin Park [5], Sungho Choi [5] and Ranga B. Myneni [5]**

[1] *State Key Laboratory of Remote Sensing Science, Institute of Remote Sensing and Digital Earth, Chinese Academy of Sciences, Beijing 100101, China*

[2] *State Key Laboratory of Resources and Environmental Information System, Institute of Geographical Sciences and Natural Resources Research, Chinese Academy of Sciences, Beijing 100101, China*

[3] *Survey Planning and Design Institute, State Forest Administration of China, Beijing 100714, China*

[4] *School of Remote Sensing, Nanjing University of Information Science and Technology, Nanjing 210044, China*

[5] *Department of Earth and Environment, Boston University, 675 Commonwealth Avenue, Boston, MA 02215, USA*

## ABSTRACT

Spatially-detailed forest height data are useful to monitor local, regional and global carbon cycle. LiDAR remote sensing can measure three-dimensional forest features but generating spatially-contiguous forest height maps at a large scale (e.g., continental and global) is problematic because existing LiDAR instruments are still data-limited and expensive. This paper proposes a new approach based on an artificial neural network (ANN) for modeling of forest canopy heights over the China continent. Our model ingests spaceborne LiDAR metrics and multiple geospatial predictors including climatic variables (temperature and precipitation), forest type, tree cover percent and land surface reflectance. The spaceborne LiDAR instrument used in the study is the Geoscience Laser Altimeter System (GLAS), which can provide within-footprint forest canopy heights. The ANN was trained with pairs between spatially discrete LiDAR metrics and full gridded geo-predictors. This generates valid conjugations to predict heights over the China continent. The ANN modeled heights were evaluated with three different reference data. First, field measured tree heights from three experiment sites were used to validate the ANN model predictions. The observed tree heights at the site-scale agreed well with the modeled forest heights ($R$ = 0.827, and RMSE = 4.15 m). Second, spatially discrete GLAS observations and a continuous map from the interpolation of GLAS-derived tree heights were separately used to evaluate the ANN model. We obtained $R$ of 0.725 and RMSE of 7.86 m and $R$ of 0.759 and RMSE of 8.85 m, respectively. Further, inter-comparisons were also performed with two existing forest height maps. Our model granted a moderate agreement with the existing satellite-based forest height maps ($R$ = 0.738, and RMSE = 7.65 m ($R^2$ = 0.52, and RMSE = 8.99 m). Our results showed that the ANN model developed in this paper is capable of estimating forest heights over the China continent with a satisfactory accuracy. Forth coming research on our model will focus on extending the model to the estimation of woody biomass.

**Keywords:** Tree height; Geoscience Laser Altimeter System (GLAS); Artificial neural Network (ANN); China Meteorological Data (CMD); Nadir bidirectional reflectance distribution function adjusted reflectance (NBAR)

## 1. INTRODUCTION

Forests play a key role in the global climate system and carbon cycle. Forests store carbon in their above- and below-ground biomass [1]. As an important predictor of forest biomass and carbon stock, the vertical structure of forests has been well monitored in previous studies [2,3,4]. Light detection and ranging (LiDAR) remote sensing, is useful in the large-scale investigations of forest structural attributes, such as forest canopy height [2,5,6,7,8,9,10].

In recent years, LiDAR instruments have demonstrated their capability to estimate forest heights [11]. For instance, the Geoscience Laser Altimeter System (GLAS) on board the Ice, Cloud and Land Elevation Satellite (ICESat) has provided within-footprint height measures at the global scale [2,5,12]. However, the lack of valid GLAS data in some regions is problematic. Because the GLAS data are spatially incomplete, multiple geospatial predictors were supplemented to obtain spatially-continuous forest height maps in recently-published approaches [2,3,5,13]. In addition, the GLAS data are highly sensitive to topographic features due to its large footprint size, which causes overestimations of forest height [14]. In this paper, in order to solve those two problems degrading the accuracy of GLAS-based forest height maps, we proposed a new approach based on the artificial neural network (ANN).

The principal method was to combine GLAS derived tree heights with ancillary variables in the ANN model and then to produce a wall-to-wall forest height map over the China continent. Selected geospatial predictors were a series of climate variables, elevation, vegetation cover and multispectral reflectance data from the Moderate Resolution Imaging Spectroradiometer (MODIS). We resampled all of the input predictors to obtain the final output map of forest heights at a 1-km spatial resolution.

## 2. MATERIALS AND METHODS

### 2.1. Data and Processing

#### 2.1.1. ICESat Data and Processing

The ICESat/GLAS is the first spaceborne LiDAR system, which was designed to obtain characteristics of the Earth's surface structures with unprecedented accuracy [15]. The ICESat/GLAS data can provide information related to land surface elevation with a spatial resolution of 70 m (ellipsoidal footprints) and 170 m spaced intervals [16]. The global ICESat/GLAS data are available for a period spanning from 2003 to 2008 [17]. Due to the short lifetime of GLAS, altimetry information was collected only in the following months: February–March, May–June, and October–November [11,18]. There are totally 15 GLAS data products including Level-1 and Level-2 that were disseminated by the NSIDC (National Snow and Ice Data Center). The GLA14 used in this study is an GLAS Level-2 Land Surface Altimetry product that provides information on surface elevations, laser footprint geolocation, waveform parameters and reflectance, as well as geodetic, instrument, and atmospheric corrections for range measurements. In previous studies, the GLA14 product has been used to estimate forest canopy heights within each footprint [1,19].

In this study, we obtained the GLAS data recorded from May to October (2003–2006), because these data represent the approximated growing season and thus, the best leaf-on condition of forests. In our analyses, GLAS footprints over non-forest area were excluded. We further screened invalid GLAS data using several preprocessing filters in order to reduce the impact of slope gradient, atmospheric forward scattering, signal saturation and cloud contamination [14,20,21]. The GLAS data were considered valid when footprints were located over the forest classes in the land cover (LC) map with >50% tree cover percent in the vegetation continuous field (VCF). Then, the quality of GLAS shots over the forested area was investigated using GLA14 waveform parameters regarding the cloud, signal saturation and atmospheric forward scattering [1,21,22]. Firstly, we removed the invalid GLAS footprints affected by signal saturation and atmospheric forward scattering based on the internal flags (“*FRir_qaFlag = 15*” and “*satNdx = 0*”) GLA14 product contained [20,21,22]. To further remove waveforms with signal dominated by low cloud, we just selected the waveforms based on the filter that was built by using the absolute difference (50 m) between ASTER DEM and the internal elevation value from GLA14 product [21]. The terrain slope gradient also impacts the quality of the GLAS data [23]. In order to minimize the bias resulting from the terrain slope, we filtered the GLAS shots where the terrain slope is higher than 10 degrees [24].

#### 2.1.2. Land Surface Reflectance

The land surface reflectance data used in this study were acquired from the MCD43B4 which is a product of the MODIS nadir bidirectional reflectance distribution function adjusted reflectance (NBAR) with a spatial resolution of 1 km [25]. The NBAR product represents the best characterization of surface reflectance over a 16-day period. In the study, seven spectral bands in the MCD43B4 averaged over the approximated growing season (June-September) between 2004 and 2006 were extracted at a 1-km spatial resolution.

### 2.1.3. Climate Data

Weather station-based climatic variables, precipitation and temperature, were derived from China Meteorological Data Sharing Service System (CMDSSS). There are total of 754 weather stations that can provide annual meteorological records. The precipitation and temperature data covered a time period from 2004 to 2006. For each weather station, 3-year averaged precipitation and temperature were obtained. Then, we interpolated these two climate variables using ordinary kriging to generate the gridded maps of annual average precipitation and temperature [26].

### 2.1.4. Ancillary Data

Ancillary data applied in this study involved the digital elevation model (DEM), LC, and VCF. The DEM data were obtained from the Advanced Spaceborne Thermal Emission and Reflection Radiometer (ASTER) Global DEM (GDEM V2) of which spatial resolution is 30 m [27]. The use of DEM data is first to remove invalid GLAS shots with a signal contaminated by low clouds and second to generate a gridded map of slope at the same spatial resolution of the GDEM V2 data (30 m). This continuous field of slope was used to correct the GLAS derived tree heights based on the terrain correction method [21].

The LC data were derived from a MODIS global LC product (MCD12Q1, 500-m grid). Additionally, the VCF data were also obtained from the MODIS (MOD44B, 250-m grid). We selected the International Geosphere-Biosphere Programme (IGBP) of the LC and tree cover percent of the VCF over China for the year 2005 [28].

### 2.1.5. Field-Measured Tree Heights

Field-measured tree heights were used to evaluate our modeled tree heights. We chose 92 plots from three experimental sites. These three field regions are located in Gansu (northwestern China), JiangXi (southern China) and Yunnan (southwestern China), respectively.

The location of the filed experimental regions and corresponding plots is shown in Figure 1. The detailed information on the field sites is delivered in Table 1. Plots in the field sites were selected according to the definition of forest area using LC and VCF in this study. The field measured tree heights were averaged for each plot.

**Table 1.** The detailed information on three field experiment regions.

| Sites | Number of Plots | Plot Size (m) | Acquisition Year | Forest Type | References |
|---|---|---|---|---|---|
| Dayekou, Gansu | 36 | 20 × 20, 25 × 25 | 2008 | *Picea Crassifolia* | [29] |
| Taihe, Jiangxi | 22 | 50 × 50 | 2012 | Masson pine, Slash Pine | |
| Puer, Yunnan | 34 | 15 × 15 | 2013 | *Pinus Kesiya*, Fir, Eucalyptus | |

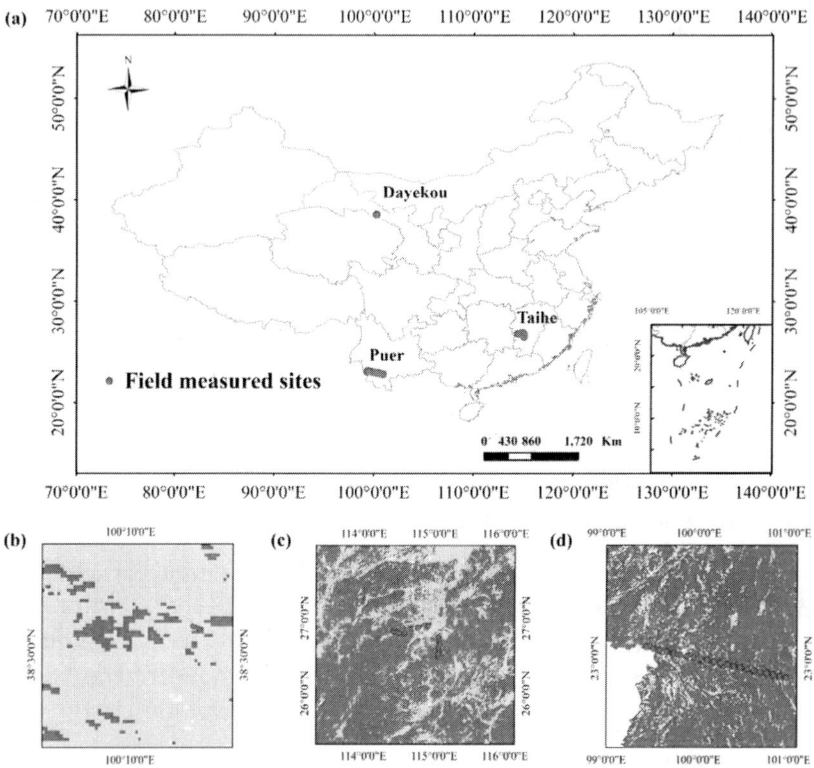


**Figure 1.** Field measurement sites. (**a**) Distribution of three field measurement regions; (**b**) Dayekou field sites in Gansu; (**c**) Taihe field sites in Jiangxi; (**d**) Puer field sites in Yunnan; (**b–d**) the background color map represents the MODIS land cover (LC) forest types.

## 2.2. Methods

### 2.2.1. GLAS Tree Height Estimation

The GLAS tree heights were used as training data to build the ANN model for tree height estimation and to validate model performances. GLAS height metrics were extracted based on the "decomposition of

GLAS waveforms into multiple Gaussian distribution curves" as described in previous studies [18,19,30]. The direct method was used to estimate canopy height based on the vertical difference between the signal start point and the ground peak. The signal start point was provided in the GLA14 product (i_SigBegOff). Also, the GLA14 product contains up to six Gaussian peaks for each GLAS waveform (i_gpCntRngOff). The distance between "i_SigBegOff" and "i_gpCntRngOff" represents the waveform metric RH100 (100% energy return height) of GLAS-derived tree height estimation [9,17]. Because the RH100 can be potentially affected by the topographic gradient [20], we applied the topographic correction approach [14,21] to obtain more accurate GLAS tree heights. The calculation formula of GLAS-derived tree heights based on topographic correction is shown as Equation (1),

$$H_{GLAS} = ( D_{SigBegOff} - D_{gpCntRngOff}) - \frac{d * \tan\theta}{2} \quad (1)$$

where $H_{GLAS}$ represents the GLAS derived tree heights, $D_{sigBegoff}$ and $D_{gpcntRngoff}$ separately represent the values of the beginning signal and the ground peak of the GLAS full-waveform, $d$ is the GLAS footprint size of 70 m [3,16], and $\vartheta$ is the topographic slope.

### 2.2.2. Tree Height Modeling

We employed the ANN algorithm to combine the GLAS-derived tree heights with the geospatial predictors. The neural network algorithms are products of artificial intelligence as black-box models, and the first prototype neural network was proposed in 1943 [31]. To date, more than 30 different neural network models have been developed [32], which have been widely used in various fields [33]. Here, we used the feed-forward neural network (FFNN) algorithm for the forest height predictions. The selected model consisted of 11 neurons in the input layer, 11 neurons in the hidden layer and 1 neuron in the output layer. The 11 parameters in the input layer include the LC class, VCF tree cover percent, temperature, precipitation, and seven MODIS NBAR bands. All of the 11 parameters in the input layer are related to forest growth, distribution or the characteristics' expression. The neuron in the output layer refers to the modeled forest canopy heights. Figure 2 depicts the schematic diagram of the ANN model proposed in this study.

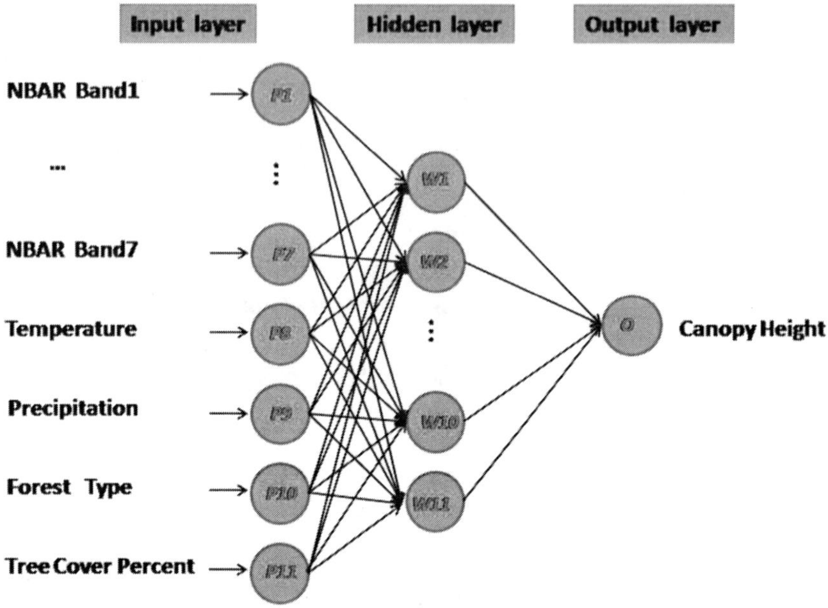


**Figure 2.** The schematic diagram of the ANN model proposed in this study. There are 11 neurons in the input layer, 11 neurons in the hidden layer and 1 neuron in the output layer.

To train the ANN model, we applied the back-propagation (BP) process algorithm to train the neural networks [34]. For each pixel, we trained the FFNN to estimate forest canopy heights. Based on the temperature and precipitation values, 15 pairs of training data were chosen to train the FFNN while 5 validation pairs remained to prevent the over-fitting of the FFNN. The selection discipline of training and validation data pairs is to obtain 20 GLAS shots that are located in the most similar climatic condition factor (temperature and precipitation) based on Equation (2),

$$D_i = (\frac{T_{pix} - T_i}{T_{max} - T_{min}})^2 + (\frac{P_{pix} - P_i}{P_{max} - P_{min}})^2 \qquad (2)$$

where $D_i$ is the distance of the climate variable difference between the reference pixel of the canopy height estimation map and the *i*-th GLAS shot; $T_{pix}$ is the temperature value of the reference pixel; $T_i$ is the temperature value of *i*-th GLAS shot; $T_{pix}$ is the precipitation value of the reference pixel; $P_i$ is the precipitation value of the *i*-th GLAS shot; $T_{max}$ and $P_{max}$ are respectively the maximum values of temperature and precipitation values over continental China, and $T_{min}$ and $P_{min}$ are respectively the minimum values of temperature and precipitation values over the continental China.

After selecting the training and validation data pairs for a pixel, the network gets trained by the training data until the FFNN model loses its best performance given validation data. Then, the trained FFNN model was used for estimating the forest canopy height over the target pixel.

### 2.2.3. Error Analysis

To assess the performance of the canopy height model in the study, we prepared several reference datasets. These datasets mainly include field-measured tree heights, GLAS-derived tree heights and existing forest height products. The root mean square error (RMSE) was calculated using the following Equation (3):

$$RMSE\ (m) = \sqrt{\frac{\sum_{i=1}^{N}(H_{Predicted} - H_{Ref})^2}{N}} \qquad (3)$$

where, $H_{predicted}$ is the predicted heights derived from the model; and $H_{Ref}$ is the validation heights derived from the reference datasets which were used to validate the predicted tree heights.

### 2.2.4. Calibration and Comparison with Existing Canopy Height Products

In this study, we present three forms of validation. One is the field calibration of the modeled tree heights, which was based on the field measurements mentioned in Section 2.1.5. For each field plot, we averaged all tree heights to obtain the representative height of this plot. The pixel on the modeled height map that is closest to the field plot centroid was chosen for the validation. The second is that the modeled heights were directly evaluated with the footprint-level GLAS heights (footprint *vs.* modeled pixel). The third is wall-to-wall map validation from the interpolated GLAS tree height map. Here, we averaged the GLAS tree heights from 20 GLAS shots selected according to Equation (2) to produce a full gridded map. Then, pixel-to-pixel comparisons (FFNN *vs.* averaged GLAS tree heights) were performed.

Lastly, we compared our modeled tree height map with two existing forest height products. These two reference products are named $H_{Simrad}$ and $H_{Ni}$, respectively, following their creators Simard and Ni [2,3]. Our modeled data and two reference products commonly provide forest canopy heights over the continental China, but their modeling strategies are significantly different. We also calculated and discussed the differences between our modeled height map and the two reference products in a view of the different canopy height modeling approaches.

# 3. RESULTS AND DISCUSSION

## 3.1. Canopy Height Map in China

A contiguous map of canopy heights with a spatial resolution of 1 km over the continental China was generated from the FFNN using the gridded geospatial predictors (Figure 3a). This map (mean = 36.44 m, Figure 3a) showed a good consistency with the spatial pattern of GLAS tree heights (mean = 30.00 m, Figure 3b).

From the estimation result of canopy heights over China, we can see that relatively tall trees were growing in the central and southern regions (Sichuan, Hubei, Yunnan and Chongqing). The trees distributed in the Northern China show obviously lower heights than those in Southern regions. Specially, the forests distributed in Heilongjiang and East of Inner Mongolia province, which are close to the border of Russia, have the lowest tree heights. The varying distribution of tree heights over China likely captures climate gradients. However, further assessment is still needed for exploring the underlying patterns of the tree height distribution.

## 3.2. Ground Validation and Error Analysis

92 field plots were used in comparison with the FFNN-modeled tree heights. The validation showed a good agreement between the modeled and field-measured tree heights (Figure 4a). In order to show the outperformance of our model with respect to previous studies, we also evaluated $H_{Simrad}$, $H_{Ni}$ and the GLAS tree height map

produced from 20 GLAS shots selected according to Equation (2). From these comparison results (Figure 4b–d), we can see that the modeled heights have better accuracy than other approaches.

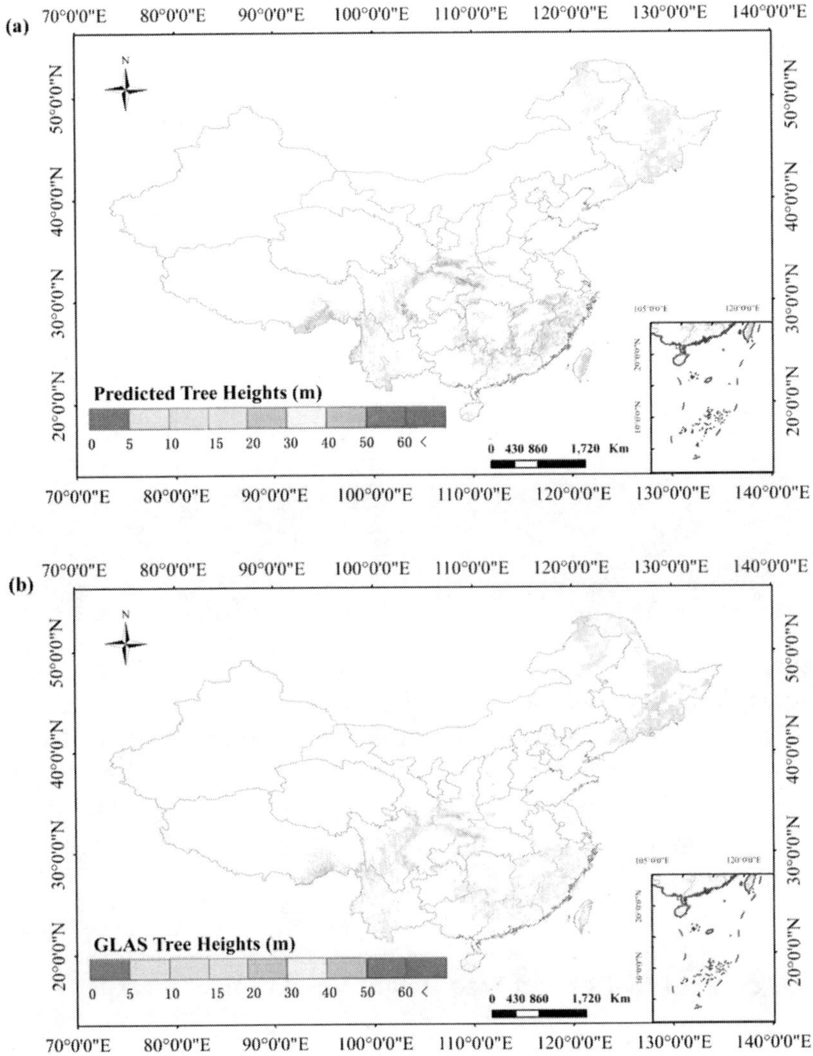


**Figure 3.** Maps of modeled and averaged GLAS-derived tree heights over continental China at 1-km spatial resolution. (**a**) The modeled height map using the trained ANN model proposed in this paper; (**b**) GLAS tree height map. The GLAS tree height map is made from the interpolation of the averaged GLAS-derived tree heights.

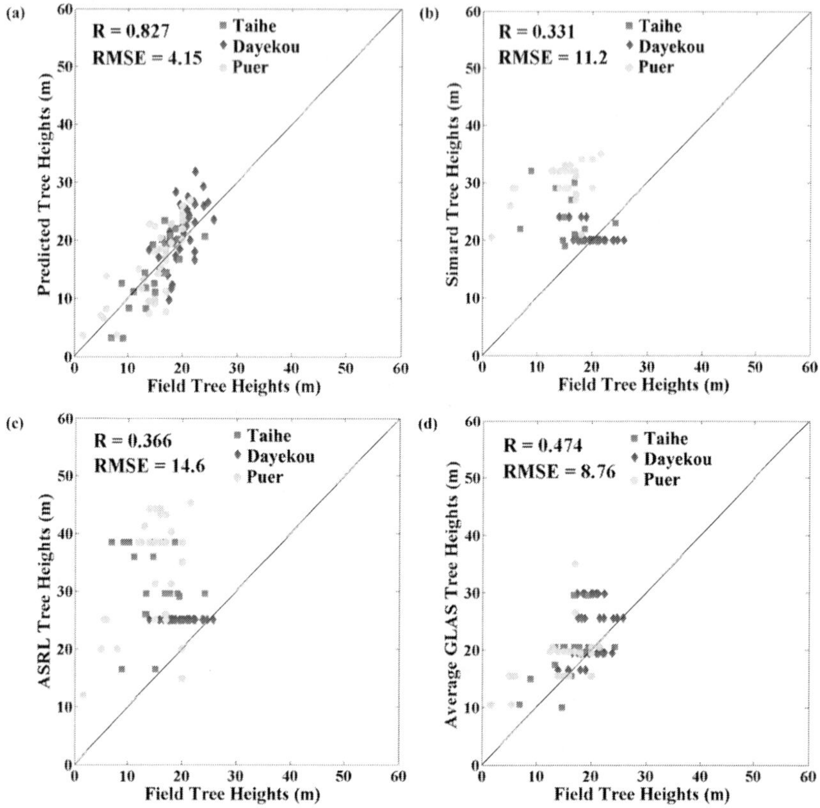


**Figure 4.** Ground validation result of the modeled tree heights, $H_{Simrad}$, $H_{Ni}$and GLAS tree height map. (**a**) The comparison between modeled tree heights and field measured tree heights; (**b**) The comparison between $H_{Simrad}$and field measured tree heights; (**c**) The comparison between $H_{Ni}$ and field measured tree heights; (**d**) The comparison between averaged GLAS tree heights and field measured tree heights.

### 3.3. Actual GLAS-Derived Tree Height Validation and Error Analysis

The modeled heights from the FFNN at the continental China scale were also evaluated using the GLAS-derived heights. This validation was based on two types of GLAS metrics, including the footprint-level GLAS heights and a full gridded GLAS height map.

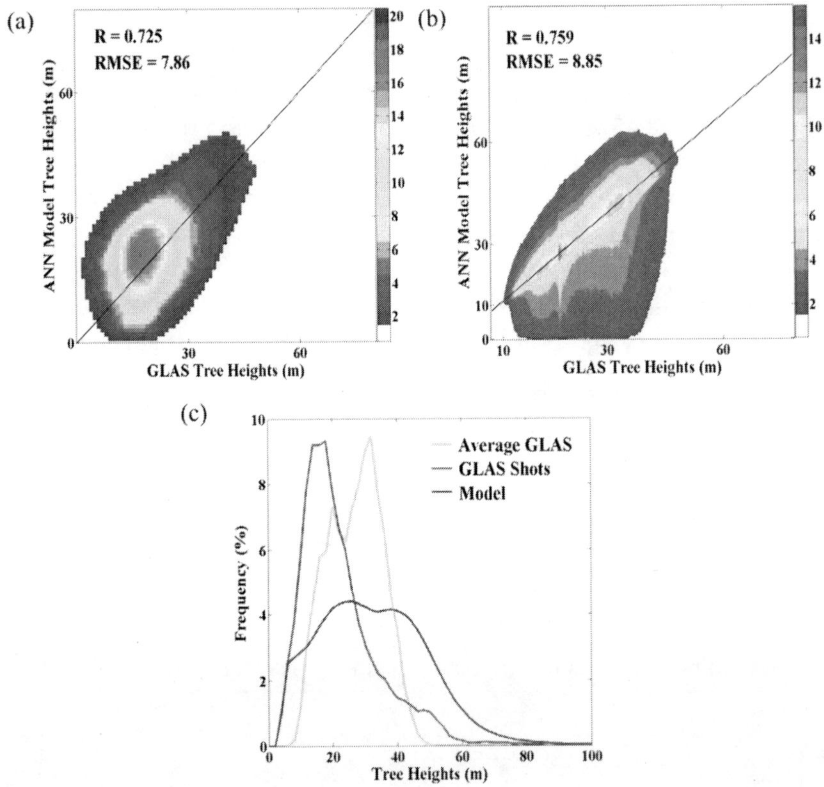


**Figure 5.** The comparison results between modeled tree heights and actual GLAS derived tree heights. (**a**) The comparison between modeled tree heights and discrete GLAS tree heights; (**b**) the comparison between modeled tree heights and averaged GLAS tree heights; (**c**) statistical characteristics of modeled tree heights, discrete GLAS tree heights and averaged GLAS tree heights.

Both types of validation showed a consistency agreement between the FFNN modeled and GLAS height metrics ($R$ = 0.725, RMSE = 7.86 m for the footprint-to-pixel level; $R$ = 0.759, RMSE = 8.85 for the pixel-to-pixel level; Figure 5). However, Figure 5a depicts some difference between the model tree heights and footprint-to-pixel-level tree heights derived from GLAS shots. The main reason is that tree heights model in this study were trained from 20 GLAS shots which were selected according to Equation (2), including 15 training data and five validation data. As for one GLAS shot location ($l_{pix}$), these 20 GLAS shots for training and validating the tree height model should have the most similar temperature and precipitation with the GLAS shot location ($l_{pix}$). Additionally, the GLAS shot of location $l_{pix}$ was included in the 20 GLAS shots. In general, modeled tree height should be similar to the GLAS shot height of location $l_{pix}$. When the temperature and precipitation of the GLAS shot location ($l_{pix}$) have a significant difference from the other 19 GLAS shots, the corresponding modeled tree height might be different from the GLAS shot height of location $l_{pix}$. Conversely, the modeled tree height is still similar to the mean value height of these 20 GLAS shots, which can explain the better consistency agreement inFigure 5b than in Figure 5a.

In order to further evaluate the modeled tree heights, we demonstrated the spatial distribution of deviations between the modeled and GLAS-derived heights. As shown in the map depicting the difference at the pixel-to-pixel level (Figure 6a), most forested pixels showed a moderate predictive power but overestimations were dominant over some regions.

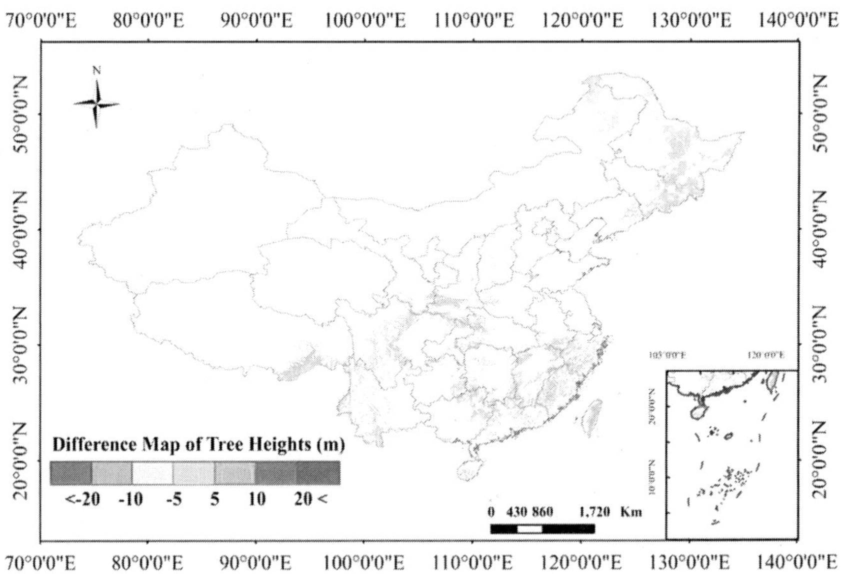


**Figure 6.** The difference map of modeled tree heights with averaged GLAS tree heights.

The overestimation of tree heights tended to distribute over some special area where temperature is relatively low and precipitation is relatively high (e.g., east of Heilongjiang and Jilin, southeast of Tibet). The overestimation may be caused by the tree height model, which is more sensitive to the precipitation than temperature during the model training process. In addition, relatively high precipitation may have bigger impacts on forest trees' growing and characteristics' expression (leaf reflectance of forest tree) than a relatively low temperature, which might affect the accuracy of modeled tree heights. The specific reasons will be explored in a further study.

## 3.4. Comparison with Existing Tree Height Map

### 3.4.1. Comparison with Simard Tree Heights Map

Based on the definition of forest land area in this study, we obtained the pixel-to-pixel comparison result between our modeled heights and the $H_{simard}$. The difference map (Figure 7a) showed a moderate agreement. The differences were less generally 10 m.

From the comparison result ($R$ = 0.738, RMSE = 7.65 m; Figure 7b), we can see a very good consistency between modeled tree heights and the HSimard when tree heights are less than 40 m. It is noteworthy that there are some obvious discrepancies existing where modeled tree heights are higher than 40 m. The disagreement between modeled tree heights and the HSimard for trees greater than 40 m might be caused by the following two reasons. On the one hand, the HSimard Simard tree heights are is the regional average height which made the forest height prediction result likely being less than 40 m. On the other hand, the overestimation of the modeled tree heights above 40min some places also appears to be an obvious disagreement with the HSimard.

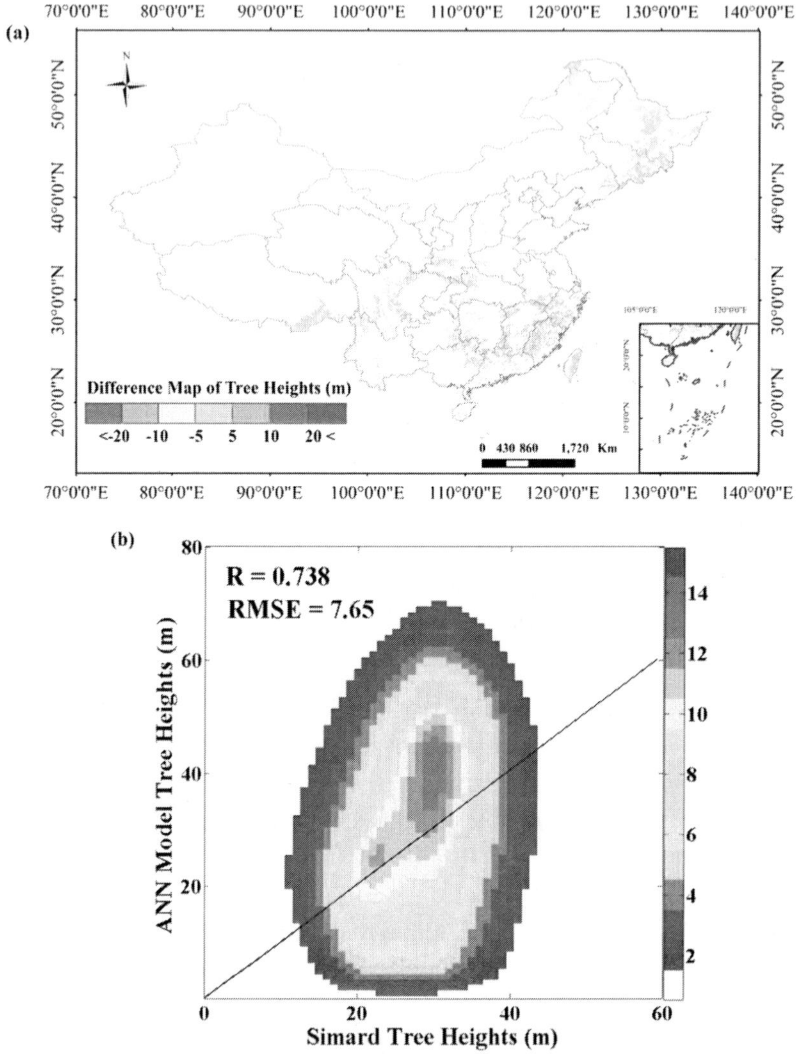


**Figure 7.** The comparison result between modeled tree heights and Simard tree heights. (**a**) The difference map of modeled tree heights with Simard tree heights; (**b**) the comparison between modeled tree heights and Simard tree heights.

### 3.4.2. Inter-Comparison with Ni Tree Heights

Based on the climate zones defined by Ni [3], the comparison between the modeled heights and the $H_{Ni}$ was conducted (Figure 8). As depicted in Figure 8a, the modeled heights and the $H_{Ni}$ generally agreed. For the most of forest land area, the difference between the modeled heights and the $H_{Ni}$ was less than 10 m. However, some regions in Southern China had relatively more significant difference, which might be caused by higher $H_{Ni}$ values in this region.

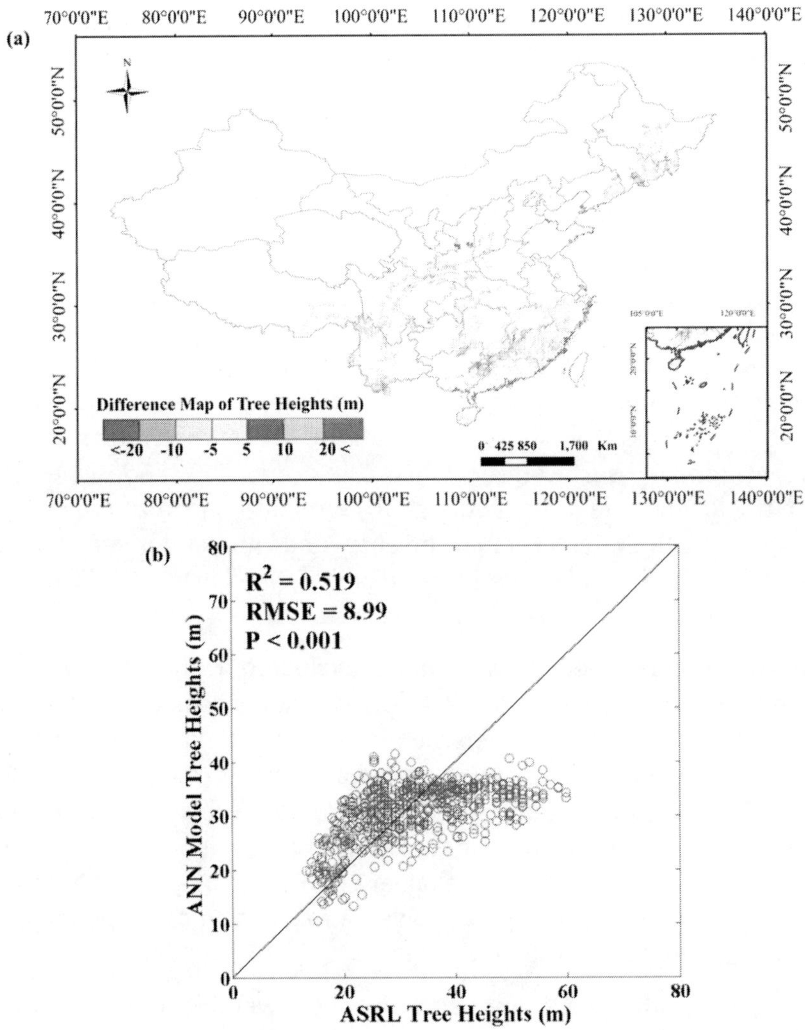


**Figure 8.** The comparison result between the modeled tree heights and the$H_{Ni}$ s. (**a**) The difference map of the modeled tree heights against $H_{Ni}$; (**b**) the comparison between the modeled tree heights and $H_{Ni}$.

The same comparison result was shown in Figure 8b ($R^2$ = 0.519, and RMSE = 8.99 m). In the case that the heights are less than 40 m, there is a good agreement between the modeled heights and the $H_{Ni}$. However, when the heights are higher than 40 m, the difference became larger. The main reason for these obvious errors is that in this process of comparison, we calculated the mean value of the modeled heights in climate zones, while the $H_{Ni}$ is the maximum tree heights predicted from the optimized allometric scaling and resource limitation (ASRL) model as the height of the climate zone.

## 4. CONCLUDING REMARKS

The objective of this work was to predict the large-scale spatial pattern of forest heights over the continental China. The GLAS derived heights were used as an input for the model and as a result, a spatially-continuous forest height map was produced with a spatial resolution of 1000 m. Additional geospatial predictors were climatic variables (temperature and precipitation), forest type, tree cover percent and land surface reflectance. The approach was designed to train the artificial neural network (ANN) model to minimize the errors between GLAS-derived actual heights and model-derived forest heights. For each pixel of the forested land defined in the study, there was one ANN model to be trained and tested.

The ANN-modeled heights were firstly evaluated at the site scale. In three field experimental sites, regression analysis was performed to assess the correlation between the modeled and field-measured tree heights. The comparison showed a good agreement ($n$ = 92, $R$ = 0.827, and RMSE = 4.15 m). Meanwhile the modeled heights were assessed at the GLAS footprint-to-pixel level. We used more than thirty thousand GLAS shots for this evaluation. The comparison between the modeled and GLAS heights showed a reasonable correspondence ($R$ =0.725, and RMSE = 7.86 m).

In addition, the ANN modeled heights were compared with continuous forest height maps over the China continent. The comparison was firstly conducted with the interpolated GLAS height map. Their difference map showed that there were relatively small errors in the most Chinese forested lands (<10 m), and the ANN model tended to predict little higher tree heights than the gridded GLAS heights. Furthermore, the modeled tree heights were validated using the published products of tree heights (RMSE = 7.65 m, $R$ = 0.738 for $H_{Simard}$; $R^2$ = 0.519, RMSE = 8.99 for $H_{Ni}$). According to the difference maps between the modeled and existing products, we obtained a small overestimation compared with the $H_{Simard}$ and an underestimation compared with the $H_{Ni}$. These discrepancies could be mainly attributed to some inherent limitations of various tree height estimation approaches.

This paper reported the ANN model for the extensive forest canopy height predictions. The training method of the ANN model for each pixel successfully compensated for the limitations of previous tree height estimation approaches by using a single model in large-scale regions, which can greatly improve the accuracy of forest tree height estimation to some extent. As shown in the results, the trained ANN model in this article demonstrated a good performance in the estimation of tree heights over the continental China.

Nevertheless, the ANN model is a black-box model, which cannot describe the growth mechanism of trees. For this reason, the trained ANN model in this paper still yielded ambiguous results in some regions. These ambiguous results were possibly caused by the uncertainties of input data in the areas with complex geographic and environmental conditions, or by the inaccurate training data due to the topographic effects. Further studies should focus on reducing the impacts of environment and topography on the uncertainties of tree height estimation, and extending the approach to biomass prediction.

## ACKNOWLEDGMENTS

The authors would like to thank the three anonymous reviewers whose comments significantly improved this manuscript. This study was partially funded by the Free Inquiry/Young Talent program of the State Key Laboratory of Remote Sensing Science (Grant No. 14ZY-02). This work was also supported by a grant from the National Natural Science Foundation of China (General Program No. 41171307 and Innovation Group Program No. 41421001), the Key Programs of the Chinese Academy of Sciences (Grant No. KZZD-EW-07 and KZZD-EW-TZ-17) and the Non-profit Industry Financial Program of the Ministry of Water Resources (Grant No. 1261430110032).

## AUTHOR CONTRIBUTIONS

The analysis was performed by Xiliang Ni, Yuke Zhou, Chunxiang Cao and Yuli Shi. All authors contributed with ideas, writing and discussions.

## REFERENCES


1. Hese, S.; Lucht, W.; Schmullius, C.; Barnsley, M.; Dubayah, R.; Knorr, D.; Neumann, K.; Riedel, T.; Schröter, K. Global biomass mapping for an improved understanding of the $CO_2$ balance—The Earth observation mission Carbon-3D. *Remote Sens. Environ.* **2005**, *94*, 94–104.

2. Simard, M.; Pinto, N.; Fisher, J.B.; Baccini, A. Mapping forest canopy height globally with spaceborne LiDAR. *J. Geophys. Res.: Biogeosci.* **2011**, *116*.

3. Ni, X.; Park, T.; Choi, S.; Shi, Y.; Cao, C.; Wang, X.; Lefsky, M.A.; Simard, M.; Myneni, R.B. Allometric scaling and resource limitations model of tree heights: Part 3. Model optimization and testing over continental China. *Remote Sens.* **2014**, *6*, 3533–3553.

4. Bellassen, V.; Delbart, N.; Le Maire, G.; Luyssaert, S.; Ciais, P.; Viovy, N. Potential knowledge gain in large-scale simulations of forest carbon fluxes from remotely sensed biomass and height. *For. Ecol. Manag.* **2011**,*261*, 515–530.

5. Baghdadi, N.; Le Maire, G.; Fayad, I.; Bailly, J.S.; Nouvellon, Y.; Lemos, C.; Hakamada, R. Testing different methods of forest height and aboveground biomass estimations from ICESat/GLAS data on Eucalyptus plantations in Brazil. *IEEE J. Sel. Top. Appl. Earth Obs. Remote Sens.* **2014**, *7*, 290–299.

6. Pang, Y.; Lefsky, M.; Sun, G.Q.; Ranson, J. Impact of footprint diameter and off-nadir pointing on the precision of canopy height estimates from spaceborne LiDAR. *Remote Sens. Environ.* **2011**, *115*, 2798–2809.

7. Fayad, I.; Baghdadi, N.; Bailly, J.S.; Barbier, N.; Gond, V.; El Hajj, M.; Fabre, F.; Bourgine, B. Canopy height estimation in French Guiana with LiDAR ICESat/GLAS data using principal component analysis and random forest regressions. *Remote Sens.* **2014**, *6*, 11883–11914.

8. Lefsky, M.A. A global forest canopy height map from the Moderate Resolution Imaging Spectroradiometer and the Geoscience Laser Altimeter System. *Geophys. Res. Lett.* **2010**, *37*.

9. Saatchi, S.S.; Harris, N.L.; Brown, S.; Lefsky, M.; Mitchard, E.T.A.; Salas, W.; Zutta, B.R.; Buermann, W.; Lewis, S.L.; Hagen, S.; *et al*. Benchmark map of forest carbon stocks in tropical regions across three continents. *Proc. Natl. Acad. Sci. USA* **2011**, *108*, 9899–9904.

10. Wulder, M.A.; White, J.C.; Nelson, R.F.; Næsset, E.; Ørka, H.O.; Coops, N.C.; Hilker, T.; Bater, C.W.; Gobakken, T. Lidar sampling for large-area forest characterization: A review. *Remote Sens. Environ.* **2012**,*121*, 196–209.

11. Van Leeuwen, M.; Nieuwenhuis, M. Retrieval of forest structural parameters using LiDAR remote sensing.*Eur. J. For. Res.* **2010**, *129*, 749–770.

12. Schutz, B.; Zwally, H.; Shuman, C.; Hancock, D.; Di Marzio, J. Overview of the ICESat mission. *Geophys. Res. Lett.* **2005**, *32*.

13. Shi, Y.L.; Choi, S.; Ni, X.L.; Ganguly, S.; Zhang, G.; Duong, H.V.; Lefsky, M.A.; Simard, M.; Saatchi, S.S.; Lee, S.; *et al*. Allometric scaling and resource limitations model of tree heights: Part 1. Model optimization and testing over continental USA. *Remote Sens.* **2013**, *5*, 284–306.

14. Lee, S.; Ni-Meister, W.; Yang, W.; Chen, Q. Physically based vertical vegetation structure retrieval from ICESat data: Validation using LVIS in White Mountain National Forest, New Hampshire, USA. *Remote Sens. Environ.* **2011**, *115*, 2776–2785.

15. Abshire, J.B.; Sun, X.; Riris, H.; Sirota, J.M.; McGarry, J.F.; Palm, S.; Yi, D.; Liiva, P. Geoscience laser altimeter system (GLAS) on the ICESat mission: On-orbit measurement performance. *Geophys. Res. Lett.***2005**, *32*.

16. Gong, P.; Li, Z.; Huang, H.; Sun, G.; Wang, L. ICEsat GLAS data for urban environment monitoring. *IEEE Trans. Geosci. Remote Sens.* **2011**, *49*, 1158–1172.

17. Harding, D.J.; Carabajal, C.C. ICESat waveform measurements of within-footprint topographic relief and vegetation vertical structure. *Geophys. Res. Lett.* **2005**, *32*.

18. Sun, G.; Ranson, K.; Kimes, D.; Blair, J.; Kovacs, K. Forest vertical structure from GLAS: An evaluation using LVIS and SRTM data. *Remote Sens. Environ.* **2008**, *112*, 107–117.

19. Neuenschwander, A.L.; Urban, T.J.; Gutierrez, R.; Schutz, B.E. Characterization of ICESat/GLAS waveforms over terrestrial ecosystems: Implications for vegetation mapping. *J. Geophys. Res.: Biogeosci.* **2008**, *113*.

20. Zhang, G.; Ganguly, S.; Nemani, R.R.; White, M.A.; Milesi, C.; Hashimoto, H.; Wang, W.; Saatchi, S.; Yu, Y.; Myneni, R.B. Estimation of forest aboveground biomass in California using canopy height and leaf area index estimated from satellite data. *Remote Sens. Environ.* **2014**, *151*, 44–56.

21. Choi, S.; Ni, X.L.; Shi, Y.L.; Ganguly, S.; Zhang, G.; Duong, H.V.; Lefsky, M.A.; Simard, M.; Saatchi, S.S.; Lee, S.; *et al*. Allometric scaling and resource limitations model of tree heights: Part 2. Site based testing of the model. *Remote Sens.* **2013**, *5*, 202–223.

22. Chen, G.; Hay, G.J. An airborne LiDAR sampling strategy to model forest canopy height from Quickbird imagery and GEOBIA. *Remote Sens. Environ.* **2011**, *115*, 1532–1542.

23. Duncanson, L.I.; Niemann, K.O.; Wulder, M.A. Estimating forest canopy height and terrain relief from GLAS waveform metrics. *Remote Sens. Environ.* **2010**, *114*, 138–154.

24. Lefsky, M.A.; Harding, D.J.; Keller, M.; Cohen, W.B.; Carabajal, C.C.; Espirito-Santo, F.D.; Hunter, M.O.; de Oliveira, R. Estimates of forest canopy height and aboveground biomass using ICESat. *Geophys. Res. Lett.* **2005**, *32*.

25. Román, M.O.; Schaaf, C.B.; Woodcock, C.E.; Strahler, A.H.; Yang, X.; Braswell, R.H.; Curtis, P.; Davis, K.J.; Dragoni, D.; Goulden, M.L.; *et al*. The MODIS (Collection V005) BRDF/albedo product: Assessment of spatial representativeness over forested landscapes. *Remote Sens. Environ.* **2009**, *113*, 2476–2498.

26. Krige, D.G. A Statistical Approach to Some Mine Valuations and Allied Problems at the Witwatersrand. Master's Thesis, University of Witwatersrand, Johannesburg, South Africa, 1951.

27. NASA. *Land Processes Distributed Active Archive Center (LP DAAC), ASTER L1B*; USGS/Earth Resources Observation and Science (EROS) Center: Sioux Falls, SD, USA, 2001.

28. Ni, X.L.; Shi, Y.L.; Choi, S.H.; Cao, C.X.; Myneni, R.B. Estimation of tree heights using remote sensing data and an allometric scaling and resource limitations (ASRL) model. In Proceedings of the 2012 IEEE International Geoscience and Remote Sensing Symposium (IGARSS), Munich, Germany, 22–27 July 2012; pp. 7248–7251.

29. He, Q.C.E.; An, R.; Li, Y. Above-ground biomass and biomass components estimation using LiDAR data in a coniferous forest. *Forests* **2013**, *4*, 984–1002.

30. Boudreau, J.; Nelson, R.F.; Margolis, H.A.; Beaudoin, A.; Guindon, L.; Kimes, D.S. Regional aboveground forest biomass using airborne and spaceborne LiDAR in Québec. *Remote Sens. Environ.* **2008**, *112*, 3876–3890.

31. McCulloch, W.S.; Pitts, W.H. A logical calculus of the ideas immanent in neural nets. *Bull. Math. Biophys.* **1943**, *5*, 115–133.

32. Samardak, A.; Nogaret, A.; Janson, N.B.; Balanov, A.G.; Farrer, I.; Ritchie, D.A. Noise-controlled signal transmission in a multithread semiconductor Neuron. *Phys. Rev. Lett.* **2009**, *102*, 226802.

33. Maier, H.R.; Dandy, G.C. Neural networks for the prediction and forecasting of water resources variables: A review of modelling issues and applications. *Environ. Model. Softw.* **2000**, *15*, 101–124.

34. Wang, L.X.; Mendel, J.M. Back-propagation fuzzy systems as nonlinear dynamic system identifiers. In Proceedings of the IEEE 1992 International Conference on Fuzzy Systems, San Diego, CA, USA, 8–12 March 1992; pp. 1409–1418.

# Chapter 9

# Remote Sensing Study of Glacial Change in the Northern Patagonian Icefield

**Lucy Dixon, Shrinidhi Ambinakudige**

*Department of Geosciences, Mississippi State University, Starkville, MS, USA*

## ABSTRACT

The Patagonian Icefield has the largest temperate ice mass in the southern hemisphere. Using remote sensing techniques, this study analyzed multi-decadal glacial retreat and expansion of glacier lakes in Northern Patagonia. Glacial boundaries and glacier lake boundaries for 1979, 1985, 2000, and 2013 were delineated from Chilean topographic maps and Landsat satellite images. Aster stereo images were used to measure mass balance from 2007 to 2012. The highest retreat was observed in San Quintin glacier. The area of glacier lakes increased from 13.49 $km^2$ in 1979 to 65.06 $km^2$ in 2013. Four new glacier lakes formed between 1979 and 2013. Between 2007 and 2012, significant glacial thinning was observed in major glaciers, including HPN1, Pared Norte, Strindberg, Acodado, Nef, San Quintin, Colonia, HPN4, and Benito glaciers. Generally, ablation zones lost more mass than accumulation zones.

**Keywords:** Patagonia, Glaciers, South America, ASTER

# 1. INTRODUCTION

Glaciers are key indicators for assessing climate change [1] -[3] . Beginning in the nineteenth century, glaciers in many parts of the world retreated significantly, which was a clear indicator of climate warming [3] - [7]. Therefore, an accurate estimation of mass balance, glacial retreat, and glacier lake expansion in mountain glaciers will contribute to the understanding of the global impacts of climate change. Such data will also help develop an early warning system for Glacier Lake Outburst Floods (GLOF), which can devastate vulnerable downstream localities.

Further, documentation of glacier measurements in major glacial regions is not comprehensive. For example, the European Alpine glaciers have been well documented using both in situ and remote sensing methods. The Himalayan glaciers have also attracted more attention in recent decades. Conversely, South American glaciers are not studied as intensively. Like other glacier regions, the reduction in glacier mass is also very evident in Patagonia [8]. However, documentation of the status of glaciers in the Chilean Andes has been very limited.

The glacier mass balance is the total loss or gain in a glacier's mass over the course of a hydrological year [9]. There are both direct and indirect methods to measure mass balance. The direct method involves an in-situ collection of glacier data to measure retreat and mass balance taken at lower ablation regions [10] [11]. In recent years, remote sensing satellite images have helped researchers study even the world's most inaccessible glaciers [12] - [15]. The indirect method of mass balance measurement requires the use of remote sensing techniques, which involves the use of satellite images, aerial photos, and photogrammetric methods [15] [16].

In this study, we first analyzed the spatial and temporal positional shifts in the terminus of the glaciers in the North Patagonian Icefield (NPI). We also analyzed the expansion of glacier lakes from 1979 to 2013. Then, using Aster satellite stereo images from 2007 to 2012, we measured the mass balance of some of the major glaciers in the NPI.

# 2. STUDY AREA

The Chilean cryosphere can be divided into four parts based on climate and glacier characteristics: 1) the Arid North (18˚S to 32˚S); 2) the

Central Andes (32°S to 38°S), which has high precipitation in the winter and dry summers; 3) Southern Chile (38 to 41°S), which has an active volcano region; and 4) the Northern and Southern Patagonian ice fields (41°S), which has abundant precipitation [14]. The NPI and the Southern Patagonia Icefield (SPI) together have the largest temperate ice mass in the southern hemisphere [17]. The temperatures of these temperate glaciers are always at melting point throughout the glacial area [18]. These glaciers respond to climate changes after only short periods [18]. Therefore, such glaciers must be studied in order to understand the dynamics of climate change.

Patagonian glaciers have long been neglected for mass-balance studies, although researchers have undertaken other glacial studies in the area [17] . Due to the remoteness, inaccessibility, and bad weather of Patagonia, field-based mass-balance studies are lacking [18]. Because glaciers in this region have not been as closely studied, researchers are uncertain about the degree that these glacier-melts contribute to sea level rise [19]. Glaciers are the most prominent water source in this area; therefore, the lack of records of water balances and annual mean discharges in the region is of great concern [20]. In recent years, however, several studies have been initiated to study glacier changes in this region [21] [22] and have clearly shown glacial retreat and thinning in the Patagonian ice field.

The NPI is in the Central Andes region [23]. The NPI receives abundant precipitation: about 2000 to 11,000 mm of water equivalent per year [19]. The eastside of NPI, however, receives precipitation of only approximately 1500 mm per year [24] . The precipitation on the east side is predominantly attributable to mid-latitude storms that move westward between 40°S and 50°S, as well as the mean latitudinal locations of the storm tracks, which closely follow the axis of the westerly winds in the middle and upper troposphere [25] .

The west wind becomes very strong and nearly permanent when the mean air temperature at sea level decreases gradually from 13.7°C at latitude 37°23'S to 6.5°C at latitude 53°10'S [23] . The NPI is also characterized by the Föhn winds, which are dry, warm, and down-sloping winds. These winds lead to dry conditions and high temperatures on the eastern side of the icefield. They also increase the precipitation gradient in the west-east direction [26] . But precipitation in the NPI shows no significant seasonality [27] . The elevation of the limit between rainfall and snowfall during precipitation events is the main climatic factor that causes glacier fluctuations in the NPI [23] .

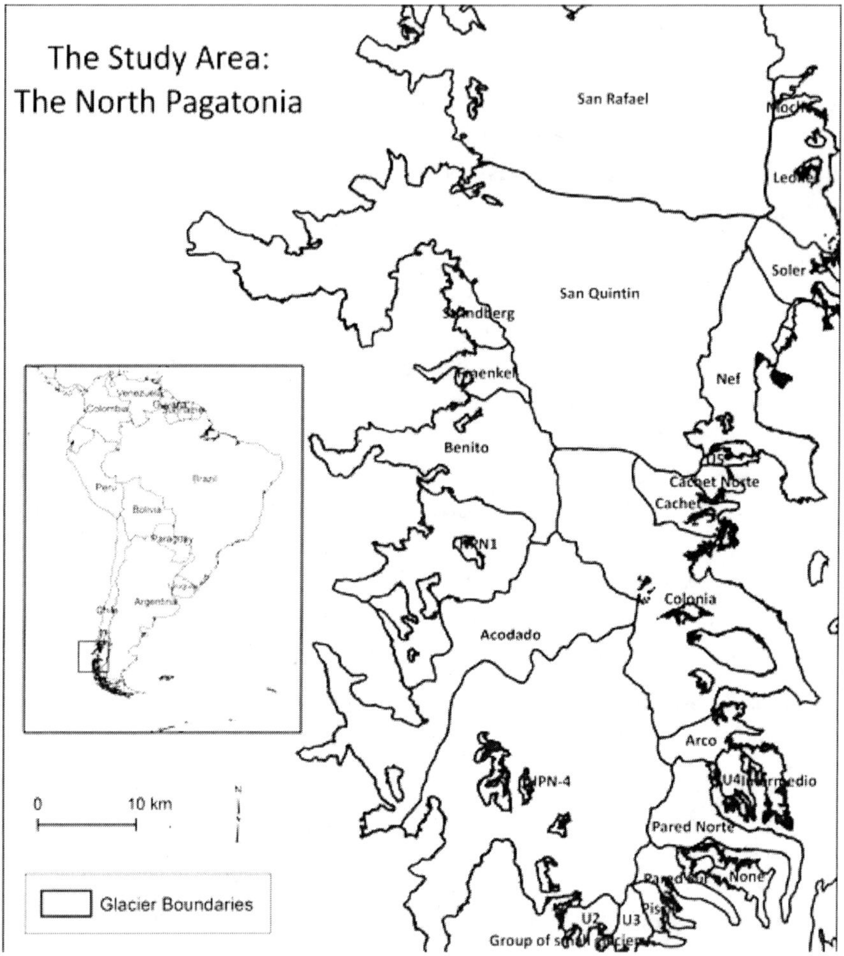


**Figure 1**. The study area.

The NPI is located between 46°30'S and 47°30'S, and it stretches approximately 100 km along 73°30'W, covering a total area of 4200 km$^2$ (Figure 1). It has a width of approximately 71 km in the west-east direction, from the tongues of San Quintin glacier to Soler glacier [21]. Most of the glaciers lie at altitudes of 1000 m in the west and 1500 min the east. A northwest-to-southeast line of nunataks divides the northeast icefield from the southwest icefield. The NPI has twenty-eight outlet glaciers [17]. San Quintin and San Rafael glaciers are the largest glaciers in the NPI, each with an area of about 760 km$^2$, followed by Steffan and Colonial glaciers [28]. San Rafael is the only tidewater-calving glacier in the NPI. Most of the other glaciers have freshwater-calving fronts [29] . Both calving-glacier types respond directly to climate change [30].

Equilibrium Lines of Altitude (ELA) of about 900 to 1350 m separate the NPI into a 2578 km$^2$accumulation zone and a 1550 km$^2$ ablation zone [28] . The summit of Mount San Valentin, at the northeast end of the ice field, has the highest elevation, with 4032 ± 1 m above sea level. The tidal water of Laguna San Rafael, located on the western side of the ice field, has a minimum altitude at sea level [21]. Due to climatic conditions, topography, and the nunataks that divide the icefield, the western outlet glaciers are larger and more active than the eastern glaciers [21] [31]. Over the last 60 years, [32] indicated a general retreat of 21 glaciers in the NPI. After the 1990s, there was an increase in the rate of retreats, with the western side experiencing greater retreat than the eastern side [33].

Glacier Lake Outburst Floods (GLOF) are a disastrous discharge of large volumes of water, caused by the breaking of moraine-dammed lakes [34]. Glacier lakes form from melting glacier ice and snow. Higher rates of melting have increased the threat of glacier lake outburst in the world's high mountain ranges [35] . Various natural phenomena, such as avalanches, earthquakes, or even the weight of the water itself, break the moraine dams. Some of the most devastating GLOF occurred in the Peruvian Andes. Researchers [36] have described the natural hazards associated with the GLOF of Palcacocha Lake in Cordillera Blanca. In 2000, a rock fall partially displaced a moraine-dammed lake in Lago Calafate glacier and caused a GLOF in the Northern Patagonia Ice field [37] .

## 3. DATA AND METHODS

We acquired topographic maps of the study area with a 1:50,000 scale created by the Istituto Geografico Militar of Chile (IGM) in 1979. The contour interval in these maps is 50 m. The vertical random error of these maps is 17 m [38]. The National Imagery and Mapping Agency [39] estimated the horizontal error in the maps as 15 m. The maps were projected to WGS 1984 UTM Zone 18S.

To delineate the glaciated area from the non-glaciated area, we digitized the glacier boundaries on the ASTER 2012 and 2007 images in Arc GIS 10.1. We digitized the region that overlapped between the 2012 and 2007 images and used the resulting image for the mass balance analysis. Fourteen glaciers fell within the overlapping region: San Quintin, Fraenkel, Strindberg, Colonia, Acodado, HPN 1, HPN-4, Cachet, Cachet Norte, Nef, Arco, Benito, Pared Norte, and U5 (Figure 1).

We used three Landsat images and 1:50,000 topographic maps to measure the retreat of the NPI glaciers and to analyze the expansion of glacier lakes. The Landsat images were from 1985 (Landsat 5 TM), 2000 (Landsat 7 ETM+), and 2013 (Landsat 8). We downloaded the Landsat images from the USGS EROS website (http://earthexplorer.usgs.gov). Both image types were re-projected to WGS 84 UTM Zone 18S.

To clearly delineate the glacier lakes, we created the Normalized Differenced Water Index (NDWI). We also visually inspected the images of the lakes using various band combinations. The NDWI uses (equation 1) - the (Near Infrared) NIR and Blue band to differentiate water bodies from other features [35].

$$NDWI = \frac{B_{NIR} - B_{blue}}{B_{NIR} + B_{blue}} \quad (1)$$

Using the NDWI and the Landsat images, we digitized the glacier lakes in the NPI and in the topographic map. We observed a significant increase in the area of the glacier lakes between 1979 and 2013.

For the mass balance estimation, we used two ASTER images, one from April 4, 2007 and one from March 18, 2012. We used ASTER satellite data because it has the stereoscopic capability to generate Digital Elevation Models (DEMs) [40]. ASTER bands 3N and 3B can be used in a DEM creation. Bands 3N and 3B are the Nadir and the backward-looking telescopes on the NIR wavelength [40]. These bands have a wavelength range of 0.78 to 0.86 μm. For this study, we geo referenced each of the ASTER images to WGS84 datum, UTM Zone 18S.

## DEM Creation

We used the Ortho Engine extension of PCI Geomatica 2012 software to create DEMs. To create DEMs of 15 m resolution, we collected 50 well-distributed Ground Control Points (GCPs) and 100 well-distributed Tie Points (TPs) from the 2007 and 2012 images. During the GCPs and TPs collections, we gave priority to non-glaciated regions, because, in the absence of a natural disaster, such regions hardly change. Glaciated areas, shaded areas, and cloud-covered areas were excluded from GCPs, as such areas are highly variable and could introduce inaccuracies to the DEM. The root mean square error (RMSE) after the GCPs collection was 2.14 m for the 2012 image, with an X RMSE of 1.81 m and a Y RMSE of 1.13 m. The RMSE for the 2007 image was 1.37 m, with an X RMSE of 1.03 m and a Y RMSE of 0.90 m. We computed the change in elevation values by finding the difference in pixel-by-pixel DEM values between the 2012 image and the 2007 image. We converted the elevation difference image to points. We excluded shaded areas, steep slopes, and clouded areas from the analysis. The density of ice used to calculate mass balance was 900 $kg \cdot m^{-3}$ [9]. The mass balance estimate was presented as meter water equivalent per year. The elevation of the master DEM (2007) was compared with the elevation of the 1:50,000 topographic maps. The correlation of over 70 well-distributed random points from topographic map elevations and the DEM elevations was very high ($b = 0.91$). Considering the lower resolution of the topographic map, DEM values seem to be highly accurate.

We expected the non-glaciated region to have a zero-elevation difference value, because there should be no change in elevation unless there is a natural disaster or a vertical error in the DEM. We used elevation differences between 2007 and 2012 DEMs in non-glaciated areas to calculate the uncertainty in the mass balance estimation in the glaciated areas.

We computed uncertainties in the elevation difference in glacier and non-glacier areas using individual Standard Error (SE) [41]. We calculated SE in the non-glaciated area using the equation 2, where n is the number of included pixels.

$$\mathrm{SE} = \frac{\mathrm{STDV\ non-glac}}{\sqrt{n}} \qquad (2)$$

In accordance with the law of error propagation, we used the SE and the mean difference of elevation (MED) of the non-glaciated area to calculate uncertainty (equation 3) in glaciated areas [41].

$$e = \sqrt{(\mathrm{SE})^2 + (\mathrm{MED})^2} \qquad (3)$$

## 4. RESULTS

We measured the retreat of nine glaciers in the NPI that occurred between 1979 and 2013 (Figure 2). These glaciers are San Rafael, San Quintin, Colonia, Acodado, HPN-1, Nef, Strindberg, Fraenkel, and Pared Norte. We measured retreat by the change in position of the tongue of the glacier during the study period [21] , as seen in Figure 2.

Table 1 shows the extent of retreat of all nine glaciers. San Quintin retreated the most, 4.5 km. This glacier tongue was intact in 1979 but by 2013 had retreated to form a new lake. Another glacier tongue in San Quintin retreated by 3.03 km. San Rafael retreated about 3.3 km between 1979 and 1987 [21]. San Quintin, conversely, retreated by only approximately 0.5 km within the same time period. In general, all glaciers in the study area retreated, which is attributable to atmospheric warming in the region [21]. The rate at which these glaciers retreat is a direct response to climate, because these are temperate glaciers and they respond rapidly to change in temperature [23]. Due to the lack of weather data in the region, however, we could not analyze long-term temperature changes.

The number of glacier lakes in the study area increased significantly between 1979 and 2013. In 1979 there were 9 glacier lakes; in 1985 there were 10; in 2000 there were 11; and by 2013, there were 13 glacier lakes. The glacier lake area increased from 13.49 $km^2$ in 1979 to 65.06 $km^2$ in 2013 (Figure 3).

San Quintin Glacier Lakehad the maximum expansion of area, from 5.42 $km^2$ to 42.37 $km^2$. This expansion is the result of the creation of a new glacier lake at the tongue of San Quintin. A study [42] observed a similar trend between 1945 and 2011. They also found that this drastic increase in San Quintin Lake compared to other glaciers. Table 2 shows the number of lakes and their total areas.

In the non-glaciated area, we selected 522,513 random points to estimate the uncertainty in the mass balance estimation. The average elevation difference in the non-glaciated area was 2.82 m, with a standard deviation of 8.99 m and a standard error of mean of 0.012 m. Based on these parameters, we estimated uncertainty in mass balance of the glaciated area to be ±0.51 m.w.e.$a^{-1}$. In glaciers, we estimated mass balance for accumulation and ablation zones separately. Table 3 presents mass balance values from 2007 to 2012.

To estimate mass balance, we used only the areas of glaciers that fall within the boundaries of the satellite images. In general, ablation zones had higher loss of mass than accumulation zones (Table 3). The HPN1 glacier ablation zone showed the highest loss of mass, with −4.76 ± 0.51 m.w.e.$a^{-1}$, followed by Pared Norte (−4.58 ± 0.51 m.w.e.$a^{-1}$), the Strindberg ablation zone (−3.28 ± 0.51 m.w.e.$a^{-1}$), the Acodado ablation zone (−3.00 ± 0.51 m.w.e.$a^{-1}$), the Nef ablation zone (−2.33 ± 0.51 m.w.e.$a^{-1}$), the Fraenkel ablation zone (−1.86 ± 0.51 m.w.e.$a^{-1}$), the San Quintin ablation zone (−1.80 ± 0.51 m.w.e.$a^{-1}$), the Nef accumulation zone (−1.77 ± 0.51 m.w.e.$a^{-1}$), the Colonia ablation zone (−1.52 ± 0.51 m.w.e.$a^{-1}$), the HPN-4 ablation zone (−1.27 ± 0.51 m.w.e.$a^{-1}$), the HPN-4 accumulation zone (−1.25 ± 0.51 m.w.e.$a^{-1}$), Acodado (−1.21 ± 0.51 m.w.e.$a^{-1}$), and the Benito ablation zone (−1.14 ± 0.51 m.w.e.$a^{-1}$). Accumulation zones of HPN1 accumulation (−0.89 ± 0.51 m.w.e.$a^{-1}$), Strindberg (−0.74 ± 0.51 m.w.e.$a^{-1}$), San Quintin (−0.72 ± 0.51 m.w.e.$a^{-1}$), and Fraenkel (−0.23 ± 0.51 m.w.e.$a^{-1}$) also showed negative mass balance, but these values are close to uncertainty values. Accumulation zones in Cachet Norte, Arco, and U5 glaciers had positive mass changes; however, other accumulation zones also showed positive mass changes, but the estimates were close to the uncertainty in the model estimation.

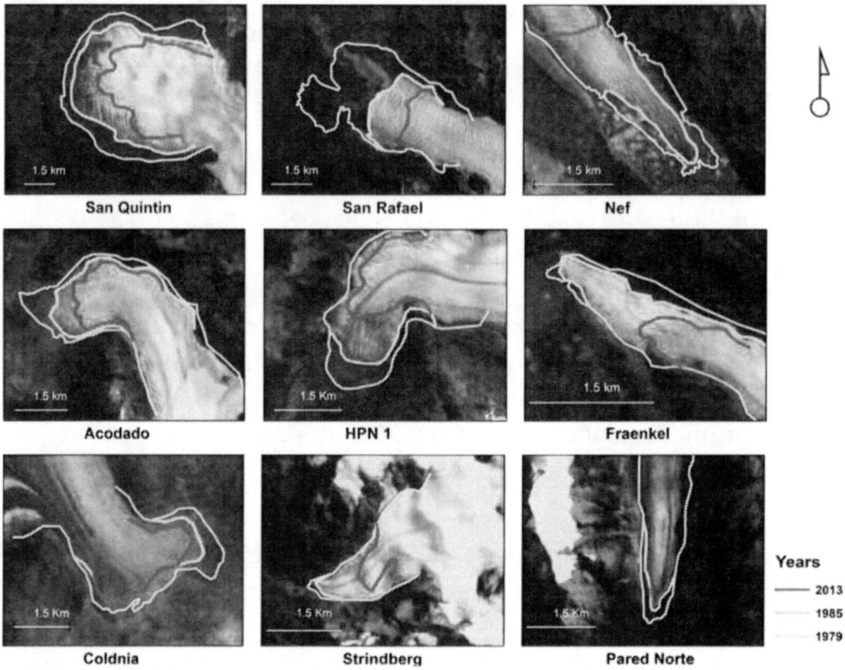


**Figure 2**. Retreat of nine major glaciers in the NPI.

**Table 1**. Retreat of glaciers between 1979 and 2013.

| Glacier | Retreat (km) |
|---|---|
| San Quintin | 4.50 |
| San Rafael | 3.92 |
| Nef | 3.82 |
| Acodado | 3.40 |
| HPN-1 | 2.43 |
| Fraenkel | 2.19 |
| Colonia | 2.13 |
| Strindberg | 1.48 |
| Pared Norte | 0.83 |

**Table 2**. Glacier lakes area for 1979 and 2013.

| Glacier Lake | 1979 Area (km$^2$) | 2013 Area (km$^2$) |
|---|---|---|
| San Quintin | 5.42 | 42.37 |
| Fraenkel | 1.48 | 2.41 |
| Benito | 0.39 | 1.00 |
| Acodado | 1.80 | 9.38 |
| Cachet | 2.50 | 4.30 |
| Nef | 1.90 | 5.60 |

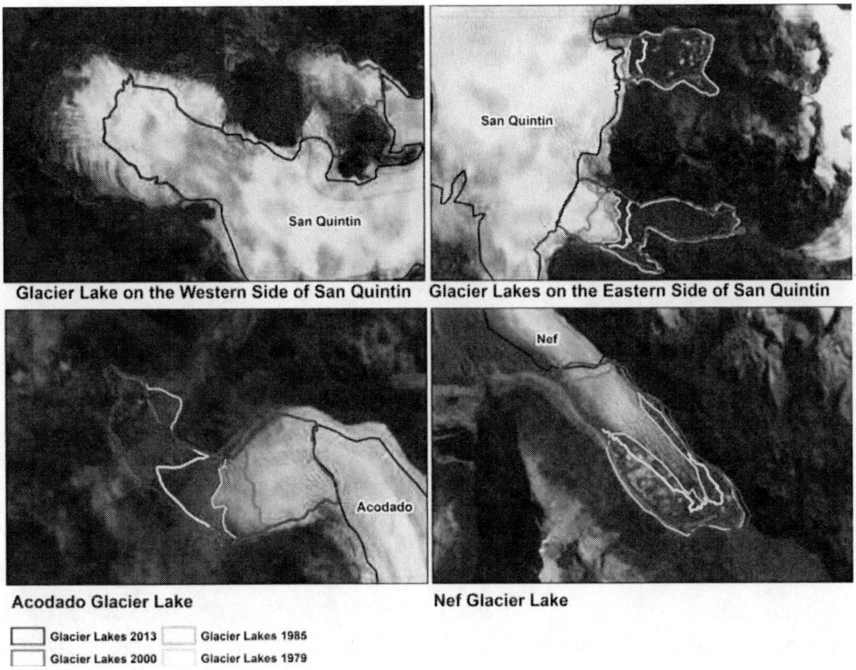


**Figure 3**. Expansion of glacier lakes.

**Table 3**. Mass balance of glaciers between 2007 and 2012.

| *Glaciers* | *Area* (km$^2$) | *Average Elevation Difference* (m.a$^{-1}$) | *STD* (m.a$^{-1}$) | *Specific Mass Balance* (m.w.e.a$^{-1}$) |
|---|---|---|---|---|
| Acodado Accumulation | 190.96 | −1.21 | 3.14 | −1.09 ± 0.51 |
| Acodado Ablation | 44.82 | −3.00 | 4.15 | −2.70 ± 0.51 |
| Arco | 10.31 | 0.84 | 3.19 | 0.75 ± 0.51 |
| Benito-Accumulation | 108.91 | 0.43 | 2.51 | 0.38 ± 0.51 |
| Benito Ablation | 71.12 | −1.14 | 2.26 | −1.03 ± 0.51 |
| Cachet | 46.86 | 0.54 | 3.90 | 0.49 ± 0.51 |
| Cachet Norte accumulation | 10.75 | 1.38 | 3.51 | 1.24 ± 0.51 |
| Cachet Norte Ablation | 1.04 | 0.14 | 2.50 | 0.13 ± 0.51 |
| Colonia Accumulation | 246.50 | 0.08 | 3.61 | 0.07 ± 0.51 |
| Colonia Ablation | 24.47 | −1.52 | 3.48 | −1.37 ± 0.51 |
| Fraenkel Accumulation | 24.99 | −0.23 | 2.35 | −0.21 ± 0.51 |
| Fraenkel Ablation | 9.74 | −1.86 | 2.79 | −1.67 ± 0.51 |
| HPN1 Accumulation | 102.31 | −0.89 | 2.79 | −0.80 ± 0.51 |
| HPN1 Ablation | 67.62 | −4.76 | 3.48 | −4.28 ± 0.51 |
| HPN-4 Accumulation | 264.55 | −1.25 | 2.94 | −1.13 ± 0.51 |
| HPN-4 Ablation | 5.98 | −1.27 | 3.41 | −1.15 ± 0.51 |
| Nef Accumulation | 86.52 | −1.77 | 4.57 | −1.59 ± 0.51 |
| Nef Ablation | 7.54 | −2.33 | 3.04 | −2.09 ± 0.51 |
| Pared Norte Accumulation | 5.19 | −4.58 | 7.47 | −4.13 ± 0.51 |
| San Quintin Accumulation | 410.10 | −0.72 | 2.99 | −0.65 ± 0.51 |
| San Quintin Ablation | 107.81 | −1.80 | 3.00 | −1.62 ± 0.51 |
| Strindberg Accumulation | 23.83 | −0.74 | 1.90 | −0.66 ± 0.51 |
| Strindberg Ablation | 3.24 | −3.28 | 2.38 | −2.95 ± 0.51 |
| U5 | 6.63 | 0.68 | 3.42 | 0.61 ± 0.51 |

Note: Area is not the actual area of the glaciers, but it represents only the area that was covered in the image.

# 5. DISCUSSION AND CONCLUSIONS

The results of this study clearly indicated general retreat, expansion of glacier lakes, and loss of mass in NPI. However, other studies have found the advancement of glacial tongues in some glaciers in the previous decade. For example, San Quintin glacier advanced in 1993 and then retreated between 1996 and 2000 [43]. Our study showed that the retreat is continuing in San Quintin. Overall, there was a 4.5 km retreat in the San Quintin glacier tongue between 1979 and 2013. A prior study observed the retreat of San Rafael glacier stopped in 1991 [44]. However, that study found a 3.92 km retreat in San Rafael glacier during the study period.

A study [29] observed a mass balance loss in all the glaciers in the NPI between 1975 and 2001. These average ice-loss rates have more than doubled to an equivalent sea level rise in the North and South Patagonian Icefields between 1995 and 2000 [19]. Similarly, this study observed negative mass balance in most glaciers. This study also found that accumulation zones are losing mass at a higher rate than ablation zones.

Glaciers on the western side of the nunataks experienced more retreat than glaciers on the eastern side of the nunataks. Another study [28] referred to orographic factors as the cause of this pattern. He further explained that the eastern side of the NPI receives less snowfall compared to the western side, because the NPI lies in the zone of the westerlies. The eastern side thus receives more snow from the western side through drift snow that is blown by wind. Due to this drift snow, the eastern side accumulates more snow and experiences less retreat in the NPI. A study [45] estimated the thinning rate of HPN-1 glacier as −0.7 m.w.e.$a^{-1}$ between 1945 and 1975; another study [29] estimated the thinning to be −4.0 m.w.e.$a^{-1}$ between 1975 and 2001; and we recorded a negative mass loss of HPN-1 to be −4.76 ± 0.51 m.w.e.$a^{-1}$ in the ablation zone and −0.80 ± 0.51 m.w.e.$a^{-1}$ in the accumulation zone.

Although these results were not validated with in-situ data for either mass balance or retreat, similar results were observed by [18] [19] [28] [45] for mass balance studies and [21] [29] [41] for glacier retreat and lakes. The results of this study and previous studies observed overwhelming retreat of glaciers and loss of mass in the NPI. These results must be taken seriously, as these conditions are linked to climate fluctuations in the region.

## CITE THIS PAPER

LucyDixon,ShrinidhiAmbinakudige, (2015) Remote Sensing Study of Glacial Change in the Northern Patagonian Icefield. *Advances in Remote Sensing*,**04**,270-279.

## REFERENCES

1. Ambinakudige, S. (2014) Glaciers. In: Warf, B., Ed., Oxford Bibliographies in Geography, Oxford University Press, New York.
2. Bolch, T., Buchroithner, M., Pieczonka, T. and Kunert, T. (2008) Planimetric and Volumetric Glacier Changes in Khumbu Himalaya since 1962 Using Corona Landsat TM and ASTER Data. Journal of Glaciology, 54, 592-600.
3. IPCC. (2013) Summary for Policymakers. In: Stocker, T.F., Qin, D., Plattner, G.-K., Tignor, M., Allen, S.K., Boschung, J., Nauels, A., Xia, Y., Bex, V. and Midgley, P.M., Eds., Climate Change 2013: The Physical Science Basis. Contribution of Working Group I to the Fifth Assessment Report of the Intergovernmental Panel on Climate Change, Cambridge University Press, Cambridge, UK and New York, USA.
4. Ambinakudige, S. (2010) A Study of the Gangotri Glacier Retreat in the Himalayas Using Landsat Satellite Images. International Journal of Geoinformatics, 6, 7-12.
5. Ambinakudige, S. and Joshi, K. (2015) Multi-Decadal Changes in Glacial Parameters of the Fedchenko Glacier in Tajikistan. International Journal of Advanced Remote Sensing and GIS, 4, 911-919.
6. Barry, R.G. (2006) The Status of Research on Glaciers and Global Glacier Recession: A Review. Progress in Physical Geography, 30, 285-306.
7. Liu, K.C., Reese, A. and Thompson, L.G. (2005) Ice-Core Pollen Record of Climatic Changes in the Central Andes during the Last 400 yr. Quaternary Research, 64, 272-278.
8. Casassa, G., Haeberli, W., Jones, G., et al. (2007) Current Status of Andean Glaciers. Global and Planetary Change, 59, 1-9.
9. Paterson, W. (1994) The Physics of Glaciers. 3rd Edition, Pergamon, Oxford.
10. Nishida, K., Satow, K. Aniya, M., Casassa, G. and Kadota, T. (1995) Thickness Change and Flow of Tyndall Glacier Patagonia. Bulletin of Glacier Research, 13, 29-34.

11. Rott, H., Rack, W., Skvarca, P. and De Angelis, H. (2002) Northern Larsen Ice Shelf, Antarctica: Further Retreat after Collapse. Annals of Glaciology, 34, 277-282.

12. Frey, H., Machguth, H., Huss, M., Huggel, C., Bajracharya, S., Bolch, T., Kulkarni, A., Linsbauer, A., Salzmann, N. and Stoffel, M. (2014) Estimating the Volume of Glaciers in the Himalayan-Karakoram Region Using Different Methods. The Cryosphere, 8, 2313-2333.

13. Bhambri, R., Bolch, T., Chaujar, R. and Kulshreshtha, S. (2011) Glacier Changes in the Garhwal Himalaya, India, from 1968 to 2006 Based on Remote Sensing. Journal of Glaciology, 57, 543-556.

14. Pellicciotti, F., Ragettli, S., Carenzo, M. and McPhee, J. (2014) Changes of Glaciers in the Andes of Chile and Priorities for Future Work. Science of the Total Environment, 493, 1197-1210.

15. Racoviteanu, A., Arnaud, Y., Williams, M. and Ordonez, J. (2008) Decadal Changes in Glacier Parameters in the Cordillera Blanca, Peru Derived from Remote Sensing. Journal of Glaciology, 54, 499-510.

16. Oerlemans, J. (2005) Extracting a Climate Signal from Glacier Records. Science, 308, 675-677.

17. Warren, C.R. and Sugden, D.E. (1993) The Patagonian Icefields: A Glaciological Review. Arctic and Alpine Research, 25, 316-331.

18. Aniya, M., Sato, H., Naruse, R., Skvarca, P. and Casassa, G. (1996) The Use of Satellite and Airborne Imagery to Inventory Outlet Glaciers of the Southern Patagonia Icefield South America. Photogrammetric Engineering and Remote Sensing, 62, 1361-1369.

19. Rignot, E., Rivera, A. and Casassa, G. (2003) Contribution of the Patagonia Icefields of South America to Sea Level Rise. Science, 302, 434-437.

20. Escobar, F., Vidal, F. and Garin, C. (1992) Water Balance in the Patagonia Icefield. In: Natuse, R., Ed., Glaciological Researches in Patagonia, Japanese Society of Snow and Ice, Nagoya, 109-119.

21. Lopez, P., Chevallier, P., Favier, V., Pouyaud, B., Ordenes, F. and Oerlemans, J. (2010) A Regional View of Fluctuations in Glacier Length in Southern South America. Global and Planetary Change, 71, 85-108.

22. Willis, M.J., Melkonian, A.K., Pritchard, M.E. and Ramage, J.M. (2012) Ice Loss Rates at the Northern Patagonian Icefield Derived Using a Decade of Satellite. Remote Sensing of Environment, 117, 184-198.

23. Lliboutry, L. (1998) Glaciers of Wet Andes. In: Ferrigno, W.M.J., Ed., Satellite Image Atlas of Glaciers of the World, US Government Printing Office, Washington DC, I109-I206.

24. Enomoto, H. and Nakajima, C. (1985) Recent Climate-Fluctuations in Patagonia. In: Nakajima, C., Ed., Glaciological Studies in Patagonia Northern Icefield 1983-1984, Data Center for Glacier Research, Japanese Society of Snow and Ice, Nagoya, 7-14.

25. Garreaud, R., Vuille, M., Campagnucci, R. and Marengo, J. (2009) Present-Day South American Climate. Paleogeography, Paleoclimatology, Paleoecology, 281, 180-195.

26. Fujiyoshi, Y., Kondo, H., Inoue, J. and Yamada, T. (1987) Characteristics of Precipitation and Vertical Structure of Air Temperature in the Northern Patagonia. Bulletin of Glacier Research, 4, 15-23.

27. Carrasco, J., Casassa, G. and Rivera, A. (2002) Meteorological and Climatological Aspects of the Southern Patagonia Icefield. In: Casassa, G., Sepulveda, F. and Sinclair, R., Eds., The Patagonian Icefields: A Unique Natural Laboratory for Environmental and Climate Change Studies, Kluwer Academic/Plenum Publishers, New York, 29-65.

28. Aniya, M. (1988) Glacier Inventory for the Northern Patagonia Icefield Chile and Variations 1944/45 to 1985/86. Arctic and Alpine Research, 20, 179-187.

29. Rivera, A., Benham, T., Casassa, G., Bamber, J. and Dowdeswell, J. (2007) Ice Elevation and Areal Changes of Glaciers from the Northern Patagonia Ice Field Chile. Global and Planetary Change, 59, 126-137.

30. Warren, C. and Aniya, M. (1999) The Calving Glaciers of Southern South America. Global and Planetary Change, 22, 59-77.

31. Aniya, M. and Enomoto, H. (1986) Glacier Variations and Their Causes in the Northern Patagonia Icefield Chile since 1944. Arctic and Alpine Research, 18, 307-316.

32. Aniya, M. (2007) Glacier Variations of Hielo Patagonico Norte, Chile for 1944/45-2004/05. Bulletin of Glaciological Research, 24, 59-70.

33. Aniya, M. (1992) Glacier Variation in the Northern Patagonia Icefield Chile between 1985/86 and 1990/91. Bulletin of Glaciological Research, 10, 83-90.

34. Yamanda, T. and Sharma, C. (1993) Glacier Lakes Outburst Floods in the Nepal Himalaya. In: Young, G.J., Ed., Snow and Glacier Hydrology (Proceedings of an International Symposium, Kathmandu, 16-21 November 1992), IAHS No. 218, 319-330.

35. Huggel, C., Kaab, A., Haeberli, W., Teysseire, P. and Paul, F. (2002) Remote Sensing Based Assessment of Hazards from Glacier Lake Outbursts: A Case Study in the Swiss Alps. Canadian Geotechnical Journal, 39, 316-330.

36. Vilimek, V., Zapata, M.L., Klimes, J., Patzelt, Z. and Santillan, N. (2005) Influence of Glacial Retreat on Natural Hazards of the Palcacocha Lake Area, Peru. Landslides, 2, 107-115.

37. Harrison, S., Glasser, N., Winchester, V., Haresign, E., Warren, C. and Jansson, K. (2006) A Glacial Lake Outburst Flood Associated with Recent Mountain Glacier retreat, Patagonian Andes. The Holocene, 16, 611-620.

38. Falkner, E. (1995) Aerial Mapping: Methods and Applications. CRC Press Inc., Boca Raton, 322.

39. NIMA (1997) Department of Defense World Geodetic System 1984: Its Definition and Relationships with Local Geodetic Systems. Third Edition, NIMA TR8350.2, National Imagery and Mapping Agency, Bethesda, MD.

40. Hirano, A., Roy, W. and Lang, H. (2003) Mapping from ASTER Stereo Image Data: DEM Validation and Accuracy Assessment. ISPRS Journal of Photogrammetry and Remote Sensing, 57, 356-370.

41. Bolch, T., Pieczonka, T. and Benn, D.I. (2011) Multi-Decadal Mass Loss of Glaciers in the Everest Area (Nepal Himalaya) Derived from Stereo Imagery. The Cryosphere, 5, 349-358.

42. Loriaux, T. and Casassa, G. (2012) Evolution of Glacial Lakes from the Northern Patagonia Icefield and Terrestrial Water Storage in a Sea-Level Rise Context. Global and Planetary Change, 102, 33-40.

43. Harrison, S., Warren, R.C.R., Winchester, V. and Aniya, M. (2001) Onset of Rapid Calving and Retreat of Glacier San Quintin, Hielo Patagónico Norte, Southern Chile. Polar Geography, 25, 54-61.

44. Winchester, V. and Harrison, S. (1996) Recent Oscillations of the San Quintín and San Rafael Glaciers Patagonian Chile. Geografiska Annaler, 78, 35-49.

45. Aniya, M. (1999) Recent Glacier Variations of the Hielos Patagonicos South America and Their Contribution to Sea-Level Change. Arctic, Antarctic and Alpine Research, 31, 165-173.

# Chapter 10

# Water Stress Estimation from NDVI-Ts Plot and the Wet Environment Evapotranspiration

**Daniela Girolimetto*, Virginia Venturini**

*Centre for Water and Environmental Studies, Universidad Nacional del Litoral, Santa Fe, Argentina*

## ABSTRACT

In this work we present a new simple index to estimate water stress (WS) for different types of surfaces, from remotely sensed data. We derive a WS index, named $WSI_{Ew}$, modifying the Water Deficit Index (WDI) proposed by Moran et al. by using the wet environment evapotranspiration ($E_w$) instead of the potential evapotranspiration ($E_{pot}$) concept. Jiang and Islam model was used to simulate actual evapotranspiration (ET) and Priestley and Taylor equation to estimate $E_w$. The $WSI_{Ew}$ results were compared to ground observations of ET, precipitation (PP), soil temperature ($T_{soil}$) and soil moisture (SM) in the Southern Great Plains-EEUU. Preliminary results suggest the method is sensitive to the water status of different surfaces. However, the $WSI_{Ew}$ would range from 0 to 0.7, having a value of 0.4 for a dry surface with 5% of SM. The methodology is operationally simple and easy to implement since it requires only information from remote sensors.

**Keywords:** water stress; water status; evapotranspiration; remote sensing

# 1. INTRODUCTION

The future of food depends largely on the water availability and strategic planning of the water resources. Therefore, to aid farmers to optimize the water uses is critical to maximize food production for the World. Nowadays vegetation WS indexes are extensively used in assisting farmer to maximize the crop yield when optimizing the irrigation system [1]. The need to monitor large areas motivated the development of WS indexes based on remotely sensed data. Indeed, the advent of the thermal infrared sensors allowed surface temperature ($T_s$) to be monitored and related to vegetation water deficits [2-5]. The correlation between surface temperature and water stress is based on the assumption that as a crop transpires, the leaves cool the air below them and the air temperature ($T_a$) drops [6]. Examples of successful water stress indexes are the Critical Temperature Variability (CTV) [5,7], the Crop Water Stress Index (CWSI) [3] and the Water Deficit Index (WDI) [8], among others.

Moran et al. [8] provided the theoretical derivation of the WDI based on the relative evaporation, defined as the ratio between actual evapotranspiration (ET) and potential evapotranspiration ($E_{pot}$). One of the main assumptions of the WDI is that soil and vegetation exchange energy and they cannot be analyzed separately [8]. This concept is particularly interesting for remote sensing applications where mix pixels, i.e. soil + vegetation, provide a single signal in every electromagnetic wavelength. The calculation is based on the vegetation index (VI) and $T_s$ space, introduced by Price [9]. The most common VI used is the Normalized Difference Vegetation Index (NDVI) [10-12]. The IV-$T_s$ space has also been used to estimate ET [13-15].

In this work we propose a new form of the WDI based on Jiang and Islam [16], and Priestley and Taylor [17] methods. The theoretical background provided by Moran et al. [8] is consistent to Jiang and Islam [16] methodology to estimate ET, therefore they are linkable in a single index called $WSI_{Ew}$.

# 2. METHODOLOGY

The background methodologies are the WDI published by Moran et al. [8] and Jiang and Islam [16] method. In this session both methods are briefly presented before we present the rationale behind the new index.

## 2.1. The WDI Index

Moran et al. [8] discussed the validity of the CWSI theory for partially vegetated areas. The derivation of their new index is based on the interpretation of the trapezoidal shaped IV-($T_s$-$T_a$) plot, where $T_a$ is the air temperature.

**Figure 1** exemplifies the borders of the trapezoid. The segment 1 - 3 represents the cold edge, with well watered conditions and vegetation cover ranging from bare soil to fully vegetated areas (NDVI ranging from 0 - 1). The segment 2 - 4, the warm edge, represents dry surfaces with vegetation varying from bare soil to full canopies. The segment 1 - 2 represents fully vegetated areas. Point 1 corresponds to well watered vegetation. Point 2 characterizes dry vegetation. The segment 3 - 4 represents bare soil [18]. Given the value of $T_s$-$T_a$ at any point in the trapezoid (i.e. point B in **Figure 1**), the evaporation can be obtained by from the distances AB and BC.

The WDI considers two important assumptions associated to the relationship between IV and the difference $T_s$-$T_a$. First, the authors assume that the difference $T_s$-$T_a$ is linearly related to percentage of vegetated area and the canopy and soil temperatures. Another important statement made by the authors is that given a certain net energy ($R_n$), the temperature of the foliage and soil are linearly related to the transpiration and evaporation respectively. Therefore, the variations in $T_s$-$T_a$ would be associated to ET. Thus, for a partially vegetated area,

$$\mathrm{WDI} = 1 - \left(\mathrm{ET}/E_{\mathrm{pot}}\right) \tag{1}$$

where ($ET/E_{pot}$) is the relative evaporation. $E_{pot}$ is associated to a surface with unlimited water supply. The authors computed it with Penman-Monteith´s equation [19] assuming the vegetation resistance ($r_{cp}$) closes to zero, without being canceled out.

Moran et al. [8] asserted that the VI has to be sensitive to the canopy variations and insensitive to spectral changes in soil background. Hence, the soil-adjusted vegetation index (SAVI) was the selected VI, and the linear edges of the SAVI-$T_s$ space were adjusted to determine $ET/E_{pot}$. Details of the WDI are provided in Moran et al. [8].

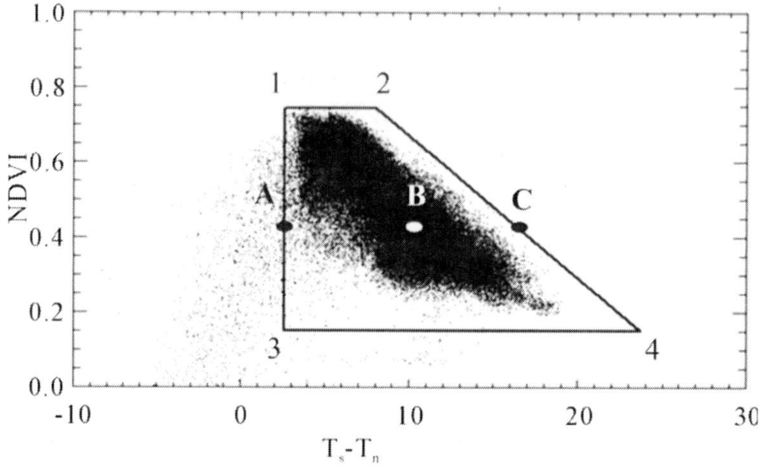


**Figure 1**. Trapezoidal diagram NDVI-($T_s$-$T_a$) for date 04/05/ 2011.

## 2.2. Jiang and Islam Method

The Jiang-Islam's interpretation of the NDVI-$T_s$ relationship provides the basis to estimate ET by modifying Priestley and Taylor's equation [17]. Jiang and Islam introduced a coefficient to account for unsaturated areas, which replaced the original Priestley and Taylor's coefficient (α). The resulting modified equation is,

$$ET_{J\text{-}I} = \phi\left[\frac{\Delta}{\Delta+\gamma}\right](R_n - G) \tag{2}$$

where f is Jiang-Islam's parameter, Δ is the slope of the saturation vapor pressure curve, γ is the psychrometric constant, $R_n$ is the net radiation at the surface level = and G is soil heat flux.

The parameter f varies from zero, for a dry bare soil surface, to α for a saturated or well vegetated surface, i.e. it becomes equal to Priestley and Taylor's equation. This parameter f is calculated by a simple two-step linear interpolation between the sides of the NDVI-$T_s$ triangle, as shown in **Figure 2**.

Jiang and Islam interpreted the upper edge, with high temperatures and low values of f, as the minimum value of ET for each class of NDVI, while the cold edge, associated with low $T_s$ and maximum values of f, represents maximum ET rate. Therefore, the value of $\phi$ vary within the limits of the triangle. Thus, NDVI-$T_s$ plot is applied to derive $\phi$ by using the normalized temperature,

$$\phi_i = \alpha \frac{T_{max} - T_i}{T_{max} - T_{min}} \tag{3}$$

where $T_{max}$ and $T_{min}$ are the maximum and minimum $T_s$ for a given vegetation class and $T_i$ is the radiometric temperature for a given pixel.

In practice, the value of $T_{max}$ is the temperature obtained extrapolating the upper edge to intersect the $T_s$ axis (**Figure 2**) for a NDVI = 0, while $T_{min}$ is obtained as the average $T_s$ of those pixels identified as water, i.e. with NDVI < 0. A full description of f calculation can be found in Jiang and Islam [16].

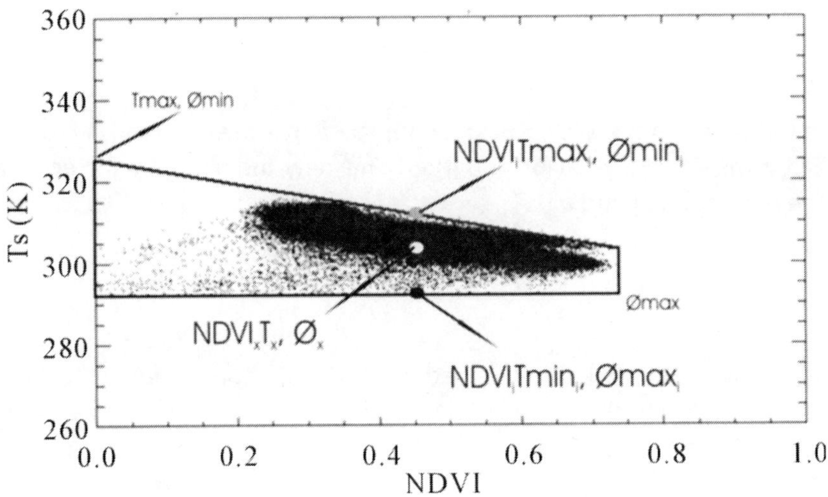


**Figure 2**. NDVI-$T_s$ triangle space with upper and lower bonds and Jiang-Islam's parameters for date 04/05/2011.

### 2.3. The New Water Status Index (WSIEw)

Moran et al. [8] related the WS with ET/$E_{pot}$ (see equation (1)), where $E_{pot}$ was defined from the PenmanMonteith's equation [19] suggesting that the $R_n$, $T_a$, vapor pressure deficit (VPD) and the soil moisture (SM) are the main controlling factors for the stomata opening [2,4,8]. One of the main assumption of this model is that $r_{cp}$ tends to zero. This

$E_{pot}$ definition is closer to a wet environment evapotranspiration ($E_w$) concept than to a truly potential concept, where the energy is maximum, the surrounding air is dry while the surface is saturated [20]. Hence, for a given atmospheric condition and unlimited water supply, the soil+ plant complex will evapotranspire at its maximum rate according to the available energy. In this case, the maximum rate of ET is given by $E_w$ [17,21-23].

The vegetation stress is mainly caused by deficit of moisture at the root zone. In the absence of watering, the moisture content in the root zone will be reduced as a result of crop intake. In turn, water stress causes the closure of the stoma of the plants and hence a reduction in the transpiration rate. Then, the ratio $ET/E_w$ is a good indicator of evapotranspiration deficit [24]. Thus, $E_w$ can replace the $E_{pot}$ in equation (1), and the WDI (see equation (1)) can be written as,

$$\mathrm{WDI} = 1 - \frac{\mathrm{ET}}{E_w} \tag{4}$$

In this new form of WDI, ET can be replaced by the equation (2) [16] and $E_w$ by the Priestley-Taylor equation. The new index, $WSI_{Ew}$ in terms of the parameters f and a is:

$$WSI_{Ew} = 1 - \frac{\phi}{\alpha} \tag{5}$$

replacing f by equation 3,Traducir del: españolEscribe texto o la dirección de un sitio web, o bien,traduce un documento.CancelAlpha the $WSI_{Ew}$ becomes,

$$WSI_{Ew} = 1 - \frac{T_{\max} - T_i}{T_{\max} - T_{\min}} = \frac{T_i - T_{\min}}{T_{\max} - T_{\min}} \tag{6}$$

where $T_i$ is the radiometric temperature for a given pixel and $T_{max}$ and $T_{min}$ are the Jiang and Islam's parameters.

Operational robust methods could be achieved with purely remotely sensed data, as we propose here. In this case, $T_a$ and the VPD are not explicitly required to compute the new index. The $WSI_{Ew}$does not require the understanding of the crop type biophysical functions under specific climates. The $WSI_{Ew}$ could be calculated from data recorded from current satellite missions, such as NOAA series, EOS-Terra and EOS-Aqua. In this work we applied to $WSI_{EW}$ to the Southern Great Plains (SPG), using MODIS images and comparing the results with observations.

# 3. STUDY AREA AND DATA

## 3.1. Study Area

The SGP region of US is a flat terrain, heterogeneous land cover with seasonal variation in temperature and humidity. It extends over the State of Oklahoma and southern part of Kansas, running from longitude 95.5°W to 99.5°W and from latitude 34.5°N to 38.5°N.

This region has relatively extensive and well distributed coverage of ground stations, maintained by the Atmospheric Radiation Measurement (ARM) program. The stations are widely distributed over the whole domain (**Figure 3**). E8 and E22 are located in a grazed rangeland region, E4 in an ungrazed rangeland area, E13 is positioned in a region with pasture and wheat, E7, E9, E15, E20 and E27 are located in pastures. E18 and E19 are in ungrazed pasture area, E12 is located in a native prairie, E10 is in alfalfa, E16 is in wheat region and E2 is in grass region.

## 3.2. Data and Images

The Energy Balance Bowen Ratio (EBBR) system compute 30-min estimates of the sensible and latent heat vertical fluxes at the local scale. Flux estimates are calculated from observations of net radiation, soil surface heat flux, and the vertical gradients of temperature and relative humidity. The instruments and measurement applications are well established and have been used for validation purposes in many studies [22,23,25,26]. Further information about the ARM EBBR data and methodology is available at http://www.arm.gov.

MODIS is one of the instruments on board EOS-Terra and EOS-Aqua satellites http://modis.gsfc.nasa.gov/ [27,28].

Daytime MODIS-Aqua images for nine days in years 2009, 2010 and 2011 in spring and summer with at least 82% of the study area free of clouds were selected. **Table 1** summarizes the image information including date, day of the year, satellite overpass time and image quality. The product MYD02 and MYD11 were used in this work. MYD02 provides corrected radiance, reflectance and geolocations for 36 bands and MYD11 provides $T_s$ images on a daily base [23,29,30].

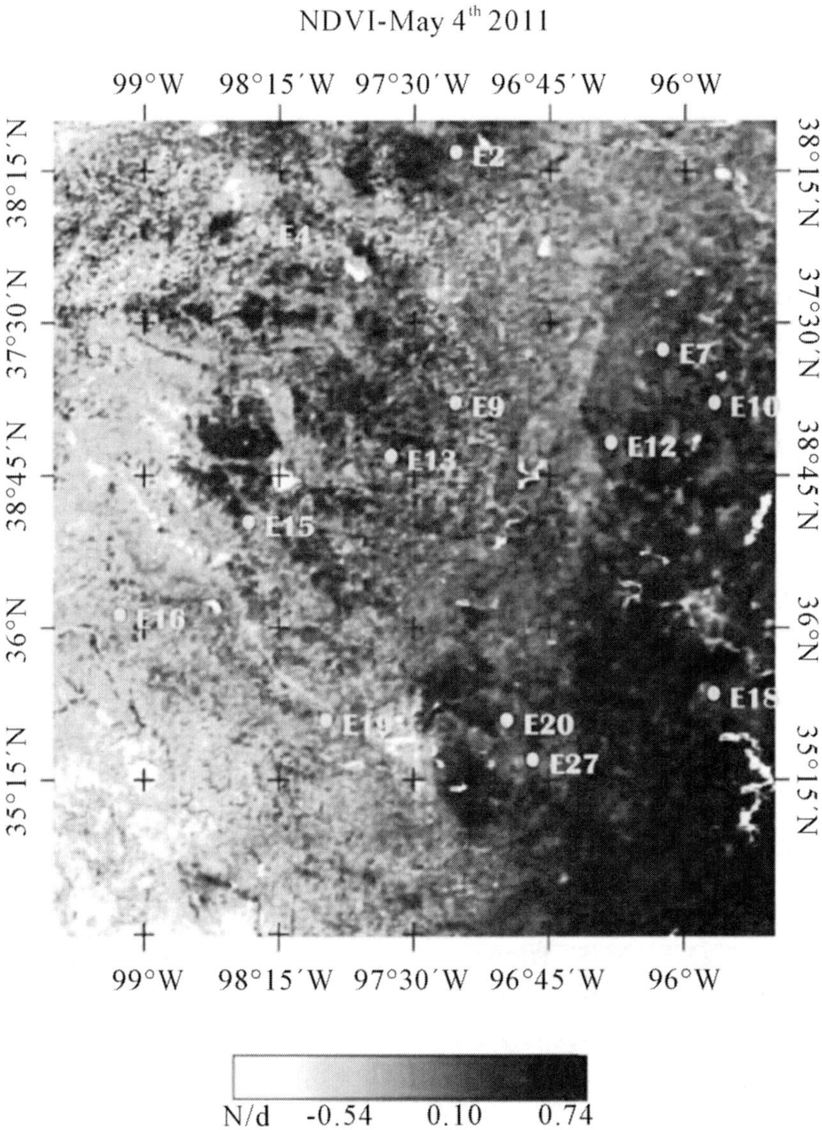


**Figure 3**. Southern Great Plains and ground station locations.

**Table 1**. Date day of the year overpass time and image quality of the nine study days.

| Date | Day of the year (DOY) | Overpass time (UTC) | Image quality (% clouds) |
|---|---|---|---|
| April 6th 2011 | 96 | 19:30 | 15 |
| May 4th 2011 | 124 | 19:55 | 1 |
| May 26th 2011 | 146 | 19:20 | 2 |
| May 29th 2011 | 149 | 19:50 | 1 |
| June 5th 2011 | 156 | 19:55 | 14 |
| April 10th 2010 | 100 | 19:40 | 5 |
| June 4th 2010 | 155 | 19:45 | 18 |
| June 5th 2009 | 156 | 19:25 | 14 |
| August 22nd 2009 | 234 | 19:35 | 11 |

# 4. RESULTS

## 4.1. Preprocessing

The MODIS images were georeferenced from the Latitude and Longitude associate to each pixel. The study area was pulled out of each image and geographically projected in a grid of 445 columns by 445 rows, with

pixels of approximately 1 km resolution. The reflectance of the red band (R) and near-infrared band (NIR) were obtained at the top of the atmosphere. The R and NIR images were used to obtain the NDVI,

$$NDVI = \frac{NIR - R}{R + NIR} \tag{7}$$

The NDVI-$T_s$ spaces were plotted to obtain Jiang-Islam's parameters ($T_{max}$ and $T_{min}$) and then obtain the $WSI_{Ew}$ by equation (6). Then, the $WSI_{Ew}$ was validated with field data and contrasted with variables that are commonly associated to water stress.

Given that the WS is not directly measurable, we computed the WS with equation (4) using ET ground observations ($ET_{obs}$) and Pristley and Taylor's equation as follows,

$$WS_{obs} = 1 - \frac{ET_{obs}}{\alpha \frac{\Delta}{\Delta + \gamma}\left(R_n^{Obs} - G^{Obs}\right)} \tag{8}$$

where $R_n^{Obs}$ is the net radiation observed at the bowen ratio stations, $G^{Obs}$ is the ground data of soil head flux, D and g were computed from observed air temperature and atmospheric pressure.

$WSI_{Ew}$ was compared with the $WS_{obs}$ and WDI obtained from NDVI-($T_s$-$T_a$) space. Finally, in order to analyze the applicability of $WSI_{Ew}$ we compared it with different variables associated to the stress.

## 4.2. WS Results

The regional statistic of $WSI_{Ew}$, i.e. maximum, mean and standard deviation, for each of the days are shown in **Table 2**.

The regional minimum is always equal to 0.0, since Jiang and Islam's method [16] requires free water pixels to estimate $T_{min}$. The minimum temperature represents ET = $E_w$, i.e. no-stress condition; thus, no-stress pixels are always present in this methodology. The regional maximum ranges from 0.50 and 0.72. The mean values of $WSI_{Ew}$ vary from 0.14 to 0.42. For all the study days, the standard deviation is lower than 0.12 suggesting little regional dispersion around the mean of $WSI_{Ew}$. Wang et al. [12] analyzed the phenological cycle of winter wheat under irrigation in the North China Plain. The authors associated CWSI values lower than 0.34 to no stress conditions. Kar and Kumar [31] analyzed the CWSI in peanut crops under irrigation. They found CWSI values between 0.61 and 0.63 just before the application of irrigation. Hence, the regional $WSI_{Ew}$ obtained here are comparables with those published by other authors. How-ever further analysis is needed to characterize the $WSI_{Ew}$ valid range.

**Table 2**. regional statistics $WSI_{Ew}$ (values maximum, average and standard deviation).

| Día | Average | Max | Standard Deviation |
|---|---|---|---|
| April 6th 2011 | 0.30 | 0.54 | 0.080 |
| May 4th 2011 | 0.42 | 0.72 | 0.098 |
| May 26th 2011 | 0.23 | 0.68 | 0.118 |
| May 29th 2011 | 0.27 | 0.60 | 0.099 |
| June 5th 2011 | 0.34 | 0.66 | 0.114 |
| April 10th 2010 | 0.32 | 0.70 | 0.068 |
| June 4th 2010 | 0.26 | 0.52 | 0.101 |
| June 5th 2009 | 0.23 | 0.50 | 0.090 |
| August 22nd 2009 | 0.14 | 0.51 | 0.067 |

Results of $WSI_{Ew}$ and $WS_{obs}$ were compared. The bias was obtained as $\sum(WS_{obs} - WSI_{Ew})/n$ and RMSE as $(\sum(WS_{obs} - WSI_{Ew})^2/n)^{0.5}$, where n is the number of observations. These statistics show bias of 0.05 and RMSE of 0.120, which represents approximately 24% of the mean WS (assuming a mean value of 0.5).

It is not evident how to validate the stress indexes from the literature revised here. For instance, Colaizzi et al. [32] compared the WDI and the soil water deficit index (SWDI). They found values of RMSE lower than 0,143 (29% of the mean) and bias lower than 0.112, consistent with the results presented here. In both cases, the RMSE are of the same order indicating that the method's errors are about 30% of the mean.

The comparison between WDI and $WSI_{Ew}$ remarks the differences between both models, with a bias of 0.28 and RMSE of 0.27 (54%). These results could be due to the separation between the bulk of the pixels and the dry border in the trapezoid (see **Figure 1**). These differences would be related to the differences in the $E_w$ and $E_{pot}$ concepts already explained in Section 2.3.

The $WSI_{Ew}$ results were also compared with records of different variables that index water stress.**Figure 4** presents the $WSI_{Ew}$ versus observed SM (at 5 cm below the surface).

There is a well defined inverse relationship between the SM and the $WSI_{Ew}$ (**Figure 4**) with a coefficient of determination ($R^2$) equal to 0.52. The dissimilar spatial resolution of both data sets may be the cause of the dispersion observed in **Figure 4**. Certainly the $WSI_{Ew}$ is calculated as a result of the signal from 1 km × 1 km pixel while the ancillary data are representative of few meters around the point station [30]. Nevertheless, values of SM of 5%, characteristic of low moisture in most soils, correspond to $WSI_{Ew}$ of about 0.45 while a SM of 30%, i.e. a well watered surface, could be associated to a $WSI_{Ew}$ of 0.2.

**Figure 5** displays $WSI_{Ew}$ versus $T_{soil}$ (at 5 cm below the surface). These relationship showed a $R^2$= 0.72. It should be noted that f is a dimensionless temperature, what may explain the relationship with the soil temperature presented in **Figure 5**. Patel et al. [33] examined the potential of using canopy-air temperature difference ($T_c$-$T_a$) for assessing the crop water status. The authors correlated $T_c$-$T_a$ with SM (at 15 cm) with a $R^2$ = 0.59. Fensholt and Sandholt [34] found a $R^2$ = 0.48 when comparing the Shortwave Infrared Water Stress Index (SIWSI) with soil moisture observed in situ.

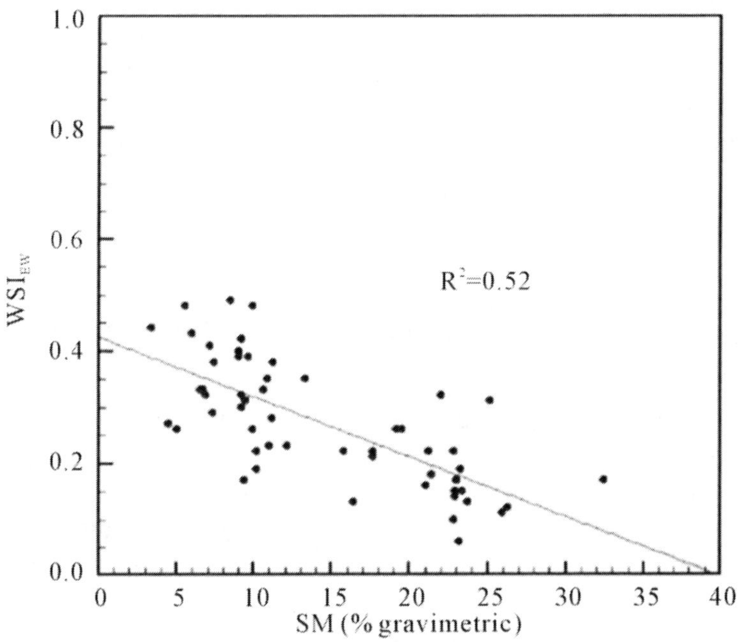


**Figure 4**. $WSI_{Ew}$ vs observed SM (at 5 cm).

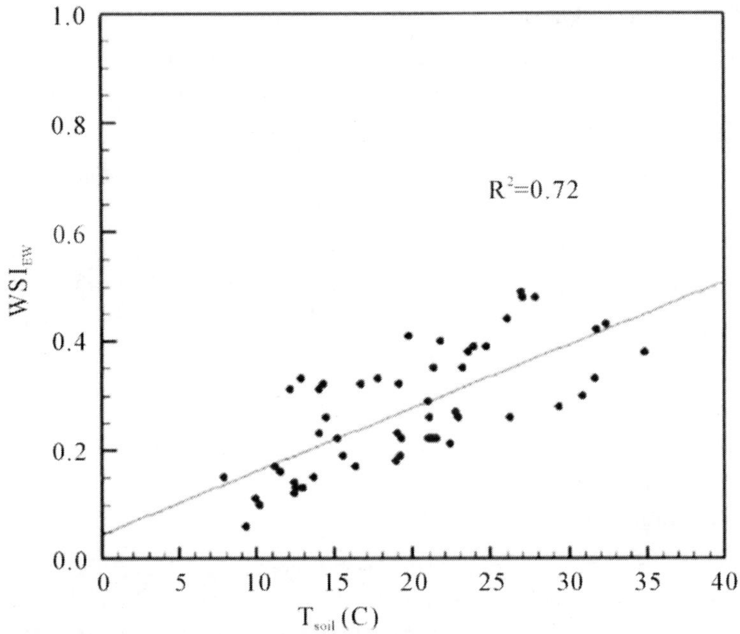


**Figure 5**. $WSI_{Ew}$ vs observed Tsoil (at 5 cm).

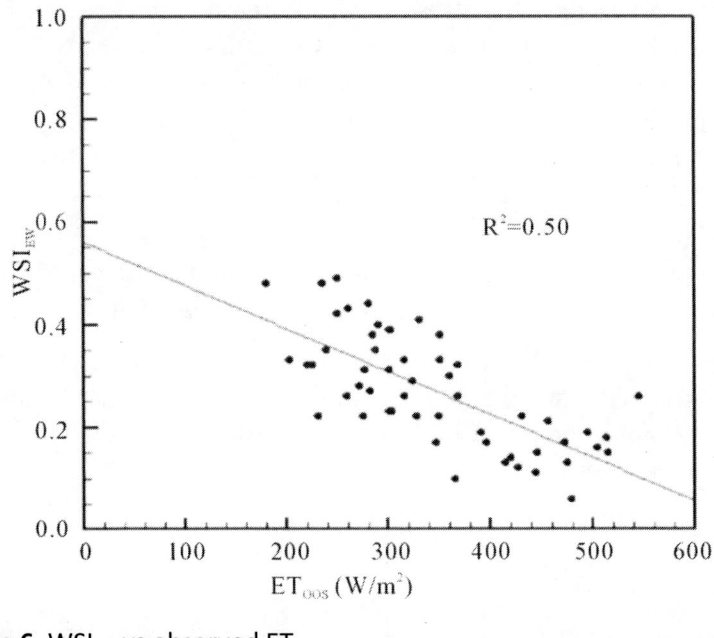


**Figure 6**. $WSI_{Ew}$ vs observed ET.

The inverse relationship between observed $ET_{obs}$ and calculated $WSI_{Ew}$ presented in **Figure 6** is noteworthy, given that the ET records are independent to $ET_{J-I}/E_{w}$. In general, the $WSI_{Ew}$ seems to capture the surface water stress condition well.

The $WSI_{Ew}$ was compared with the rainfall accumulated five days prior to the analyzed dates (PP). The comparison is shown in **Figure 7** where the decrease of WS when rainfall increases is observed ($R^2 = 0.62$). As the surface gets humid more water is available for ET and the ratio $ET/E_{pot}$ tends to one. Thus, the stress decreases with the precipitation increase.

Wang et al. [12] studied the wheat in the plain of China. They associated the wheat low stress to high rainfall events. The authors concluded that a good wheat harvest was obtained for CWSI lower than 0.34, representing a not-stress condition.

**Figure 8** shows the comparison of $WSI_{Ew}$ with soil moisture and PP for station E12. The SM observed at 5 cm is supplied by PP and represent the roots water availability [34], thus any reduction of PP, causes a decrease in the SM and it is expected to cause an increase of the $WSI_{Ew}$.

Finally, **Figure 9** shows the $WSI_{Ew}$ map for the SGP for the May $4^{th}$ 2011, where we observed the stress would decreases from West to East. In general, the mountain areas present $WSI_{Ew}$ of 0.7 and the prairies at the East show values of $WSI_{Ew}$ about 0.3. The black areas are cloudy masked pixel [16,30].

## 4.3. Sensitivity Analysis

$T_{max}$ is the main parameter of the index presented in this paper, thus the sensitivity of the $WSI_{Ew}$to variation of $T_{max}$ is analyzed here.

Different methods to draw the warm and cold edge of the triangle [10,11] would render different values of $T_{max}$ in Jiang-Islam method. Thus, we apply the First Order Analysis to estimate the effect of $T_{max}$ on $WSI_{Ew}$ variance.

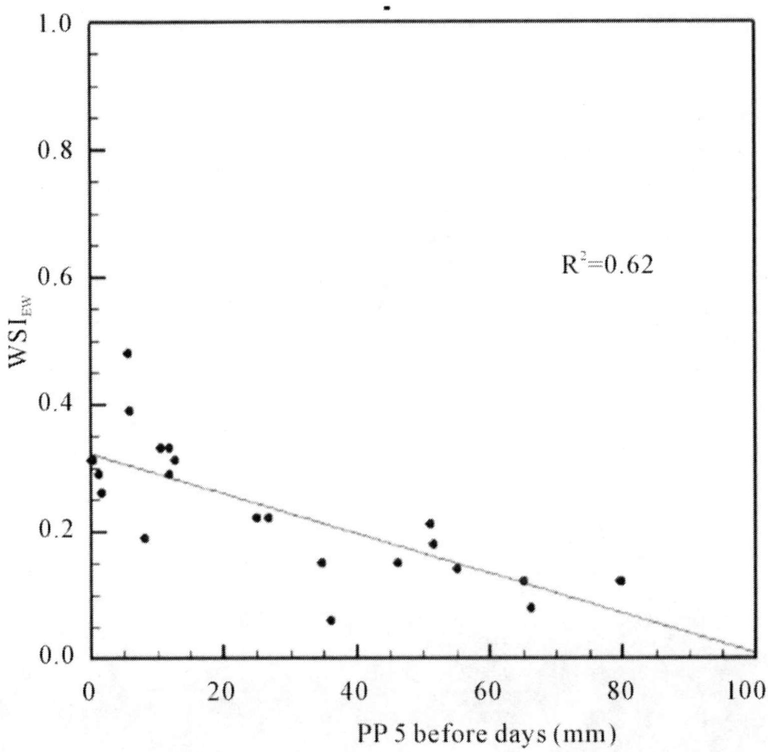


**Figure 7**. $WSI_{Ew}$ vs PP (mm).

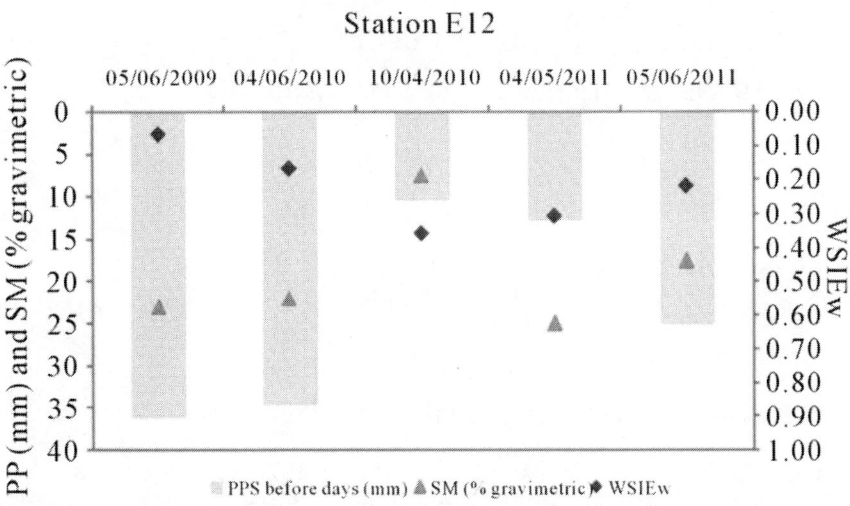


**Figure 8**. $WSI_{Ew}$ vs SM and PP for station E12.

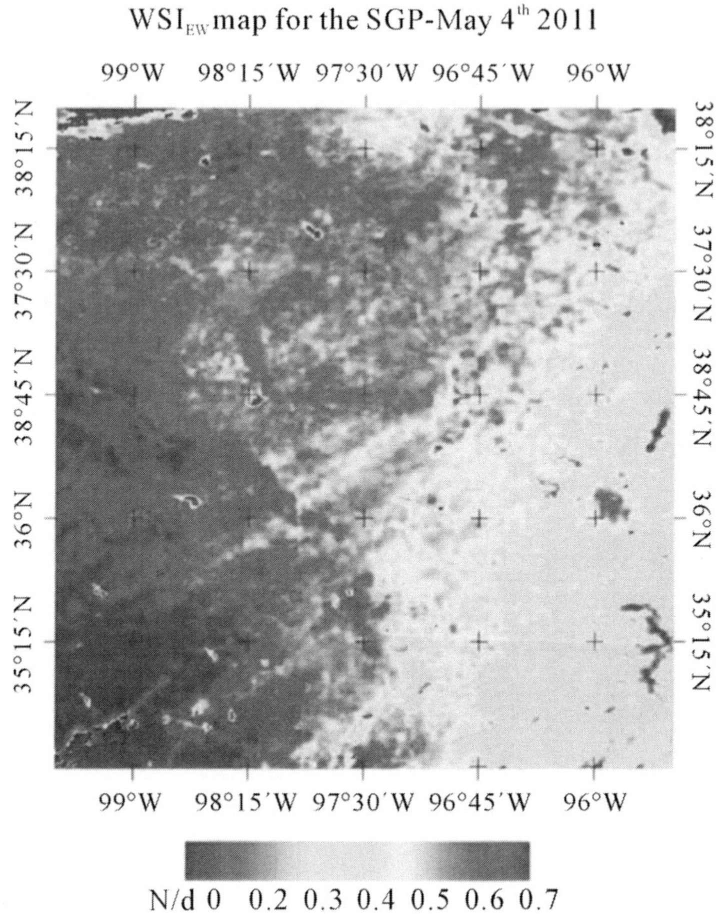


**Figure 9**. $WSI_{Ew}$ map for the SGP for the May 4th 2011.

$$S^2_{WSI_{E_w}} = \left( \frac{T_{min} - T_i}{\left(T_{max} - T_{min}\right)^2} \right)^2 S^2_{T\,max} \tag{9}$$

The results of applying equation (9) with differents $T_{min}$, $T_{max}$, $T_i$ statistic and assuming that $S^2_{T_{max}}$ varies from 1% to 15%, suggested that $S^2_{WSI_{E_w}}$ varies from 2.5% to 30% approximately. These results indicate that variations up 10% of $T_{max}$ would cause errors of about 15% in $WSI_{Ew}$.

## 5. DISCUSSION

Several methods have been developed to determine the WS from observations of the $T_s$ [2-5]. The index proposed by Moran et al. [8], based on the relative evaporation concept, is widely applied all over the world [18,32, 35]. In this work we presented a modification of WDI proposed by Moran et al. [8]. The new index replaces the $E_{pot}$ used in the original formulation for the $E_w$ concept. Given that $E_w$ £ $E_{pot}$ [20], the ratio $ET/E_{pot}$ is smaller than $ET/E_w$, yielding two WS indexes that point out different causes of vegetation stress. The $WSI_{Ew}$ would only consider the stress due to water shortness in the root zone while the WDI would also reflect the stoma closure cause by the atmospheric and radiation effects.

The WDI required field information (observations of $R_n$, G, wind speed, $T_a$) to establish the end-points that define the trapezoid (**Figure 1**(a)). If such information was not available, it would be necessary to guess them from each analyzed image; i.e. indentifying groups of pixels that represent the natural conditions of the endpoints. Guessing the extreme points creates uncertainty in the interpretations of the diagrams [35]. The $WSI_{Ew}$ also requires determination of the triangle borders, however in this case, the source of error would be mainly associated with definition of $T_{max}$.

The $WSI_{Ew}$ is another "thermal index" with the form of a normalized temperature, i.e.

$$WSI_{Ew} = (T_i - T_{min})/(T_{max} - T_{min})$$

[30]. $T_{max}$ is the temperature for a hypothetical dry bare soil pixel and $T_{min}$ is the temperature for a saturated pixel. It must be noted that the $T_{max}$ is not the maximum thermal infrared of the image or maximum canopy temperature; on contrast, it is a "hypothetical" maximum temperature. It is usually found in the literature thermal stress indices calculated as a normalized $T_s$, the difference between the methods is given the extreme temperatures estimation. For instance, Galleguillos et al. [18] used the WDI to derive the daily ET. The authors assumed that $T_a$ is constant in the study area and they estimated the maximum and minimum $T_s$ from the energy balance. Wang et al. [35] made used of the WDI to estimate soil moisture. These authors explored the $T_s$-EVI

(Enhanced Vegetation Index) space to determine the minimum and maximum values of $T_s$.

The relationship between the $WSI_{Ew}$ and SM ($R^2 = 0.52$) provided interesting information. In**Figure 3**(a) we observe that a soil with 5% of moisture matches a $WSI_{Ew}$ of about 0.45. On the other extreme, a SM of 30% agrees with a $WSI_{Ew}$ of 0.1. In other words, the $WSI_{Ew}$ would not get close to 1 although the SM indicates a dry surface.

We compare the $WSI_{Ew}$ with the rainfall accumulated five days prior to the analyzed dates, observing that the WS decreases as precipitation increases, however there is no sufficient analysis about irrigation fields and precipitation spatial distribution in this work to deepen the conclusions about the $WSI_{Ew}$ range of variation.

The advantage of this new formulation is that there is no need of crop-field observations, yet overestimated $T_{max}$ values might yield an underestimated water stress condition. Thus, a $T_{max}$significantly larger than the actual dry soil temperature, would not index the stress with a $WSI_{Ew}$value close to 1, as it would be expected by the end-users. In general, the $WSI_{Ew}$ shows significant correlations with water stress indicators.

The $WSI_{Ew}$ is applicable to different satellite missions and requires minor image processing. The new index can be very useful for end users who require quick and easy methods of application.

## REFERENCES

1. R. López-López, R. Arteaga-Ramírez, M. A. VázquezPeña, I. López-Cruz and I. Sánchez-Cohen, "Índice de Estrés Hídrico Como un Indicador del Momento de Riego en Cultivos Agrícolas," Agricultura Técnica en México, Vol. 35, No. 1, 2009, pp. 92-106.

2. R. D. Jackson, S. B. Idso, R. J. Reginato and W. L. Ehrler, "Crop Temperature Reveals Stress," Crop Soils, Vol. 29, No. 8, 1977, pp. 10-13.

3. R. D. Jackson, "Canopy Temperature and Crop Water Stress," In D. Hillel, Eds., Advances in Irrigation, Academic Press, New York, 1982, pp. 43-85.

4. S. B. Idso, "Non-Water-Stressed Baselines: A Key to Measuring and Interpreting Plant Water Stress," Agricultural Meteorology, Vol. 27, No. 1-2, 1982, pp. 59-70.

5. K. L. Clawson and B. L. Blad, "Infrared Thermometry for Scheduling Irrigation of Corn," Agronomy Journal, Vol. 74, No. 3, 1982, pp. 311-316.

6. M. Usman, A. Ahmad, S. Ahmad, M. Arshad, T. Khaliq, A. Wajid, K. Hussain, W. Nasim, T. Mehmood Chattha, R. Trethowan and G. Hoogenboom, "Development and Application of Crop Water Stress Index for Scheduling Irrigation in Cotton (Gossypium hirsutum L.) under Semiarid Environment," Journal of Food Agriculture and Environment, Vol. 7, No. 3-4, 2009, pp. 386-391.

7. B. L. Blad, D. G. Gardner, N. J. Watts and N. J. Rosenberg, "Remote Sensing of Crop Moisture Status," Remote Sensing, Quart. 3, 1981, pp. 4-20.

8. M. S. Moran, T. R. Clarke, Y. Inoue and A. Vidal, "Estimating Crop Water Deficit Using the Relation between Surface-Air Temperature and Spectral Vegetation Index," Remote Sensing Environment, Vol. 49, No. 3, 1994, pp. 246-263.

9. J. C. Price, "Using Spatial Context in Satellite Data to Infer Regional Scale Evapotranspiration," IEEE Transactions on Geoscience and Remote Sensing, Vol. 28, No. 5, 1990, pp. 940-948.

10. T. N. Carlson, R. R. Gillies and T. J. Schmugge, "An Interpretation of Methodologies for Indirect Measurement of Soil Water Content," Agricultural and Forest Meteorology, Vol. 77, No. 3-4, 1995, pp. 191-205.

11. I. Sandholt, K. Rasmussen and J. Andersen, "A Simple Interpretation of the Surface Temperature/Vegetation Index Space for Assessment of Surface Moisture Status," Remote Sensing Environment, Vol. 49, No. 3, 2002, pp. 246-263.

12. L. Wang, G. Y. Qiu, X. Zhang and S. Chen, "Application of a New Method to Evaluate Crop Water Stress Index," Irrigation Science, Vol. 24, No. 1, 2005, pp. 49-54.

13. R. R. Gillies, T. N. Carlson, J. Cui, W. P. Kustas and K. S. Humes, "A Verification of the 'Triangle' Method for Obtaining Surface

Fluxes from Remote Measurements of the Normalized Difference Vegetation Index (NDVI) and Surface Radiant Temperature," International Journal of Remote Sensing, Vol. 18, No. 15, 1997, pp. 3145-3166.

14. L. Jiang and S. Islam, "A Methodology for Estimation of Surface Evapotranspiration over Large Areas Using Remote Sensing Observations," Geophysical Research Letters, Vol. 26, No. 17, 1999, pp. 2773-2776.

15. S. Stisen, I. Sandholt, A. Norgaard, R. Fensholt and K. H. Jensen, "Combining the Triangle Method with Thermal Inertia to Estimate Regional Evapotranspiration—Applied to MSG-SEVIRI Data in the Senegal River Basin," Remote Sensing Environment, Vol. 112, No. 3, 2008, pp. 1242-1255.

16. L. Jiang and S. Islam, "Estimation of Surface Evaporation Map over Southern Great Plains Using Remote Sensing Data," Water Resources Research, Vol. 37, No. 2, 2001, pp. 329-340.

17. C. H. B. Priestley and R. J. Taylor, "On the Assessment of Surface Heat Flux and Evaporation Using Large-Scale Parameters," Monthly Weather Review, Vol. 100, No. 2, 1972, pp. 81-92.

18. M. Galleguillos, F. Jacob, L. Prévot, A. French and P. Lagacherie, "Comparison of Two Temperature Differencing Methods to Estimate Daily Evapotranspiration over a Mediterranean Vineyard Watershed from ASTER Data," Remote Sensing Environment, Vol. 115, No. 6, 2011, pp. 1326-1340.

19. J. L. Monteith and M. Unsworth, "Principles of Environmental Physics," 2nd Edition, Butterworth-Heinemann, Burlington, 1990, 304p.

20. R. J. Granger, "An Examination of the Concept of Potential Evaporation," Journal of Hydrology, Vol. 111, No. 1-4, 1989, pp. 9-19.

21. W. Brutsaert and H. Stricker, "An Advection-Aridity Approach to Estimate Actual Regional Evapotranspiration," Water Resources Research, Vol. 15, No. 2, 1979, pp. 443-450.

22. V. Venturini, S. Islam and L. Rodríguez, "Estimation of Evaporative Fraction and Evapotranspiration from MODIS Products Using a Complementary Based Model," Remote Sensing Environment, Vol. 112, No. 1, 2008, pp. 132-141.

23. V. Venturini, L. Rodríguez and G. Bisht, "A Comparison among Different Modified Priestley and Taylor's Equations to Calculate Actual Evapotranspiration with MODIS Data," International Journal of Remote Sensing, Vol. 32, No. 5, 2010, pp. 1319-1338.

24. J. V. Straschnoy, C. M. Di Bella, F. R. Jaimes, P. A. Oricchio and C. M. Rebella, "Caracterización Espacial del Estrés Hídrico y de las Heladas en la Región Pampeana a Partir de Información Satelital y Complementaria," Revista de Investigaciones Agropecuarias, Vol. 35, No. 2, 2006, pp. 117-141.

25. W. J. Shuttleworth, "Insight from Large-Scale Observational Studies of Land/Atmosphere Interactions," Surveys in Geophysics, Vol. 12, No. 1-3, 1991, pp. 3-30.

26. J. M. Lewis, "The Story behind the Bowen Ratio," Bulletin of the American Meteorological Society, Vol. 76, No. 12, 1995, pp. 2433-2443.

27. C. O. Justice, J. R. G. Townshend, E. F. Vermote, E. Masuoka, R. E. Wolfe, N. Saleous, D. P. Roy and J. T. Morisette, "An Overview of MODIS Land Data Processing and Product Status," Remote Sensing Environment, Vol. 83, No. 2, 2002, pp. 3-15.

28. E. F. Vermote, N. Z. Saleous and C. O. Justice, "Atmospheric Correction of MODIS Data in the Visible to Middle Infrared: First Results," Remote Sensing Environment Vol. 83, No. 2, 2002, pp. 97-111.

29. Z. Wan and J. A. Dozier, "A Generalized Split-Window Algorithm for Retrieving Land-Surface Temperature from Space," IEEE Transactions on Geoscience and Remote Sensing, Vol. 34, No. 4, 1996, pp. 892-905.

30. V. Venturini, G. Bisht, S. Islam and L. Jiang, "Comparison of Evaporative Fractions Estimated from AVHRR and MODIS Sensors over South Florida," Remote Sensing Environment, Vol. 93, No. 1-2, 2004, pp. 77-86.

31. G. Kar and A. Kumar, "Surface Energy Fluxes and Crop Water Stress Index in Groundnut under Irrigated Ecosystem," Agricultural and Forest Meteorology, Vol. 146, No. 1-2, 2007, pp. 94-106.

32. P. D. Colaizzi, E. M. Barnes, T. R. Clarke, C. Y. Choi, M. Peter, P. M. Waller, J. Haberland and M. Kostrzews, "Water Stress Detection Under High Frequency Sprinkler Irrigation with Water Deficit Index," Journal of Irrigation and Drainage Engineering, Vol. 129, No. 1, 2003, pp. 36- 43.

33. N. R. Patel, A. N. Mehta and A. M. Shekh, "Canopy Temperature and Water Stress Quantificaiton in Rainfed Pigeonpea (Cajanus cajan (L.) Millsp.)," Agricultural and Forest Meteorology, Vol. 109, No. 3, 2001, pp. 223- 232.

34. R. Fensholt and I. Sandholt, "Derivation of a Shortwave Infrared Water Stress Index from MODIS Nearand Shortwave Infrared Data in a Semiarid Environment," Remote Sensing Environment, Vol. 87, No. 1, 2003, pp. 111-121.

35. W. Wang, D. Huang, X. G. Wang, Y. R. Liu and F. Zhou, "Estimation of Soil Moisture Using Trapezoidal Relationship between Remotely Sensed Land Surface Temperature and Vegetation Index," Hydrology and Earth System Sciences, Vol. 15, No. 5, 2011, pp. 1699-1712.

# Chapter 11

# Mangrove Restoration and Regeneration Monitoring in Gulf of Kachchh, Gujarat State, India, Using Remote Sensing and Geo-Informatics

**Ravi Upadhyay[1], Nischal Joshi[1], Atul Chandrakant Sampat[1], Arun Kumar Verma[1], Ajay Patel[2], Vijay Singh[2], Jaydipsinh Kathota[2], Manik H. Kalubarme[2*]**

[1]*Gujarat Ecological Commission (GEC), Forest and Environment Department, Government of Gujarat, Gandhinagar, India*

[2]*Bhaskarcharya Institute for Space Applications and Geo-Informatics (BISAG), Department of Science & Technology, Government of Gujarat, Gandhinagar, India*

## ABSTRACT

Indian coast harbors richly diverse and critical coastal habitats like coral reefs and mangroves. Mangroves form one of the most important ecosystems of coastal and marine areas. It safeguards the ecology of the coastal areas and provides livelihood opportunities to the fishermen and pastoral families living in these areas. In real sense, mangrove is the Kalpvriksh (divine tree which fulfills all the desires) for the coastal communities. The restoration and plantation of mangroves have received a lot of attentions worldwide. To assess the impact of mangrove plantation activities and to monitor the mangrove regeneration and restoration in various villages, a joint study under the Integrated Coastal Zone Management Project (ICZMP) was taken up by Gujarat Ecology Commission (GEC) and Bhaskaracharya Institute for Space Applications and Geo-Informatics (BISAG) in the Gulf of Kachchh, Gujarat State. The major objective of this study was to monitor the increase in mangrove cover in coastal areas of Gulf of Kachchh using the Indian Remote Sensing Satellite data of 2005, 2011 and 2014. The mangrove regeneration was monitored using multi-temporal Indian Remote Sensing Satellite (IRS) LISS-III and LISS-IV digital data covering Gulf of Kachchh region. The multi-temporal IRS LISS-III data covering Gulf of Kachchh of October-2005, November- 2011 and LISS-IV data of April-2014 was analyzed. The mangrove density and mangrove area in different talukas was estimated based on the analysis of IRS LISS-III digital data. The mangroves have been delineated based on the pink colour observed on satellite images and the area was estimated in the Geographic Information System (GIS) environment. The taluka or block-level mangrove areas were estimated and changes in the areas were monitored during the period of six years from 2005 to 2011. It was observed that the areas where mangrove regeneration activities were carried out with active participation of Community Based Organizations (CBOs), mangrove density as well as mangrove area have substantially increased in the Gulf of Kachchh region.

**Keywords:** Remote Sensing, Geo-Informatics, Mangrove Ecosystem, Change Monitoring, Indian Remote Sensing Satellite, Mangrove Regeneration, Community Based Organizations (CBOs)

## 1. INTRODUCTION

Since the 2004 tsunami, mangrove restoration has become a priority for national as well as State governments in their quest to provide security against sea storms and other natural hazards. The restoration and plantation of mangroves have received a lot of attention worldwide. The awareness regarding the ecological and economic importance of mangroves has grown through the years. As a result, numerous countries are engaged to restore this valuable resource, as reflected by an increasing number of publications and workshops dedicated to this subject (e.g. [1] -[3] ). Gujarat has the second largest mangrove cover (1046 sq. km) of the country (4639 sq. km) [4]. The mangrove cover of the state is distributed over four regions viz. Kori creeks (74.1%), Gulf of Kachchh (15.2%), Saurashtra (10.1%) and South Gujarat (0.6%), covering 11 coastal districts, unevenly distributed over these four regions [4]. There are a substantial number of plantation and restoration projects in Gujarat in which mangroves are planted on intertidal mudflats. Over 130 sq. km of inter-tidal mudflats has been afforested by the Gujarat Forest department from first five-year plan till year 1998-99 [5].

## 1.1. Importance of Mangrove Ecosystem

It is well known that coastal population succumbs to disasters of cyclones and Tsunamis, incurring heavy losses to their properties and live-stock. Mangrove plantation along the coast serves as a barrier to cyclones and Tsunamis and protects the people living in Coastal areas. Mangroves are considered as one of the fragile ecosystem in coastal areas. Mangroves are unique plants capable of surviving under extreme saline environment. Mangroves are evergreen trees and shrubs that grow in saline coastal waters in tropics and also sometimes in subtropics, mainly between 25°N and 25°S. The mangrove ecosystem is highly productive and plays critical role in economic and social development. Mangroves not only have environmental benefits like purifying water, they act as a buffer zone between land and sea and protect land from erosion and play an invaluable role as nature's shield against cyclones, ecological disasters and a protector of shorelines.

## 1.2. Characteristics of Gujarat Coastline

The State of Gujarat enjoys the distinction of having the longest coast line in the country. It is more than 1650 km and extends right from Jakhau in Kachchh District to Umargam in Valsad district. There are two prominent indentations in this stretch namely the Gulf of Kachchh and the Gulf of Khambhat. The coastal environment of India plays a critical role in the economy of the nation by virtue of its resources and ecosystems such as mangroves, coral reefs, salt marshes, sand dunes, estuaries etc. Gujarat coast has about 1.6 lakh sq. km on the continental shelf. The coast, due to its varied physiographic features, geomorphology, coastal processes and river dis- charge into the sea, provides a wide variety of coastal features. Based on the distinct variation in the land form categories, the Gujarat coast has been broadly classified into five regions:

1) The Rann of Kachchh, 2) Gulf of Kachchh, 3) The Saurashtra Coast, 4) Gulf of Khambhat, and 5) The South Gujarat Coast. The coast lines of the Gulf of Khambhat and Kachchh together form about 65% of the coast line of the state. The Rann of Kachchh is a saline desert flooded by the monsoon but remains dry for the rest of the year. There are two major creeks which connect the Rann: Surajbari creek in the Little Rann and Kori creek in the Great Rann. The Gulf of Kachchh has an area of 7350 sq. km stretching in the east-west direction and about 170 km long. The shore line of the Gulf of Kachchh has extensive mudflats and mangroves.

### 1.3. Role of Remote Sensing and Geo-Informatics

Remote sensing technique due to its synoptic, multi-temporal, coverage and multi-spectral ability in whole range from optical to microwave regions can effectively act as tool par excellence providing advance and reliable information on mangrove extent and status of its growth along the coastal areas. The reflectance pattern of vegetation in visible & NIR spectral region give information on condition of vegetation cover.

Gujarat Ecology Commission (GEC) along with Bhaskaracharya Institute for Space Applications and Geo- Informatics (BISAG) carried out mapping of mangroves in the coastal districts of Gujarat State using Indian Remote Sensing Satellite (IRS) LISS-III digital data of the 2006-07 period [6]. The maps have been interpreted at 1:10,000 scale and the mangroves have been classified into two density classes namely, dense and spares mangroves. Mangrove atlas for 13 coastal districts with 35 coastal talukas in Gujarat, have been prepared. The comparison of mangrove area estimates carried out by BISAG and Forest Survey of India (FSI) indicates a net increase in the area under mangrove cover. The assessment by BISAG, has recorded 996.3 sq. km under mangrove cover, which shows a rise of 88 sq. km. In addition to the existing Mangrove cover, the present assessment also gives the availability of potential area of 1153 sq. km, where mangrove regeneration program can be taken up.

Classification and mapping of the mangroves in Sunderban Biosphere Reserve (SBR) in the West Bengal province of India using IRS 1D LISS-III satellite data, was carried out using different classification approaches, viz., on-screen visual interpretation, supervised and unsupervised classifications [7]. The study showed that four mangrove classes, and eight non-mangrove classes could be delineated using all the three approaches. It was observed that among the three approaches, on-screen visual interpretation yielded higher classification accuracy (91.7%) compared to supervised (79.9%) and unsupervised classifications (71.1%). The results obtained through on-screen visual interpretation showed that all mangrove categories together cover 23.2% of the total geographical area of SBR, of which the mixed mangrove category covers maximum area (18.3%).

Mangroves plantation development is largely dependent not only the proper selection of site but also the selection of planting species and planting method. Kachchh district has a large inter-tidal mudflat which is the potential site for mangroves plantation but all these inter-tidal mudflats are not suitable. A study was carried out to provide a preliminary overview of the coastal segment which was suitable for mangroves plantation. The suitable sites for mangrove plantation are identified based on four variables: coastal land use and land cover, coastal geomorphology, mean tidal range and significant wave height [8] . The significant variables leading to identify suitable plantation sites were land use and land cover and coastal geomorphology. The study suggested that number of variables should be considered such as soil water, ground water and sea water salinity, soil pH etc. for more accurate site selection. The study found that the major problems for mangrove plantation are high temperature and evapo-transpiration, low rainfall/freshwater supply and high salinity. The study also demonstrates that the remote sensing derived variables and GIS techniques are highly useful for suitable site selection for mangroves plantation.

## 1.4. Integrated Coastal Zone Management Project (ICZM)

Gujarat Ecology Commission (GEC) is actively involved in the development and design of the Integrated Coastal Zone Management Project (ICZMP) with Forest & Environment Department (MoEF), which is a World Bank-funded project. GEC has been appointed as the State Project Management Unit (SPMU) of the ICZMP. Under the ICZMP, GEC is working with several Project Executing Agencies (PEAs) to achieve the specific objectives of the project. GEC is also working with Kachchh Forest Circle for mangrove regeneration and conservation. As the

SPMU, GEC shall also coordinate among the Project Executing Agencies (PEAs), ensure effective utilization of allocated funds and provide guidelines to the agencies for better implementation of their respective components thereby providing management services for successful operation of the ICZM program. GEC is the implementing agency for Component A-1 of ICZM i.e. Restoration, Plantation and Conservation of Mangrove along the coast of Gulf of Kachchh. The major responsibilities of GEC are as follows:

- Increase the sensitivity, understanding and acceptance of local communities, Government and coastal industries towards the need to protect, conserve and regenerate mangroves.
- Capacity building of coastal communities and staff of Forest Department for community-based mangrove regeneration.
- Increase the mangrove cover in coastal areas of Gulf of Kachchh through extensive plantation of mangroves in potential coastal and inland areas.
- Implement eco-development program for villages adjoining the Gulf of Kachchh to increase people's participation.

## 1.5. Restoration and Plantation of Mangrove along the Coast of Gulf of Kachchh Carried out by GEC

GEC has taken up an area of 75 ha in the year 2010-11 and 75 ha in 2011-12. So far 150 ha have been covered in village Tuna with active participation of Community Based Organisations (CBOs) Tuna Dariakantha Vistar Vikas Samiti. Village Tuna falls in Anjar block of Kachchh district. During 2012-13, an area of 100 ha has been planted. The main role of GEC is to build the capacity of CBOs in various technical aspects of the plantations. GEC facilitates the local CBO in preparing the micro plan and other technicalities. In addition to this, entry point activity has also been planned based on the need of the village keeping in view the environmental issues faced by the village. Kachchh Forest Circle has been roped in to carry out mangrove re-generation through Forest Development Agencies (FDAs) falling within the Kachchh district. FDA Kachchh has carried out mangrove plantation in villages like Lakki, Bhutau in Lakpath block and villages Jangi and Ambaliyara in Bhachau block. A table showing plantation activities at various villages through participatory approach is given in Table 1.

## 1.6. Need for Mangrove Restoration

It has become clear that mangrove restoration activities need to be monitored from the standpoint of their societal values, and how they can contribute to improving the health of coastal ecosystems and conserving biodiversity within the context of integrated coastal area management. Despite recent improvements in mangrove restoration, the state still has large areas throughout the coast that have a potential for mangrove regeneration. The mangrove Restoration Project through public-private partnership was launched in 2007 by Gujarat Ecology Commission [9]. The important issues which have been considered for mangrove restoration along the Gujarat coast are as follows:

Ecological and Livelihood Security: Mangroves form one of the most important ecosystems of coastal and marine areas. They safeguard the ecological security of the coastal belt and also provide livelihood opportunities to coastal communities, especially fishermen and pastoral families.

Increased Degradation: Due to arid/semi-arid climate in the state, there has been an ever-increasing dependency over this natural resource by local communities. Further, coastal erosion, pollution and anthropogenic pressure are also on the rise with large tidal-amplitude and reduction in the fresh-water inflow into the sea. This has adversely affected the survival of mangroves, leading to their degradation.

## 1.7. Objectives

To Increase Mangroves cover, GEC has undertaken Mangrove plantation and regeneration in several coastal villages in the Kachchh district. The major objectives of study on Mangrove Restoration and Regeneration Monitoring in Gulf of Kachchh in Gujarat State are as follows:

- To monitor Restoration, Plantation and Conservation of Mangrove along the coast of Gulf of Kachchh using multi-temporal Indian Remote Sensing Satellite (IRS) data.
- Monitoring of growth of mangrove on different regeneration sites in different villages using multi-date IRS satellite data of 2005 and 2011.

- Taluka or block-level area estimation under mangroves using GIS techniques and change monitoring of mangroves during the period of six years from 2005 to 2011.

**Table 1**. Plantation activities at various villages through participatory approach.

| District name | Taluka/Village name | Plantation achievement/Target (ha) | | Total area (ha) |
|---|---|---|---|---|
| | | 2010-11 | 2011-12 | |
| Kachchh | Lakhpat/Lakki | 500/500 | 750/750 | 1250 |
| | Lakhpat/Bhutau | 500/500 | 750/750 | 1250 |
| | Bhachau/Amaliyara | 0 | 300/300 | 300 |
| | Bhachau/Jangi | 0 | 200/300 | 200 |
| Total | | 1000/3000 | 2000/2000 | 3000 |

## 2. STUDY AREA

The Gulf of Kachchh (GOK), a large marine ecosystem in the State of Gujarat, has unique ecosystems like coral reefs and mangroves at its southern coast and mangroves at selected locations along the northern coast. The GOK has tidal amplitude ranging from 3 - 8 m. The developmental activities like construction of ports and human interventions like mining of corals, destruction of mangroves to expand the salt pan activities etc., have extensively damaged the corals and the mangroves in the region. Due to its proximity to the Arabian Gulf, the GOK has attracted establishment of facilities like ports, Single Point Mooring (SPM) stations for handling large amount of crude oil. The gulf is rich in marine wealth and biodiversity. The region was notified as a Marine National Park and Sanctuary (MNPS) in 1982. The map of the Gulf of Kachchh (GOK), showing study area is shown in Figure 1.

## 3. METHODOLOGY

### 3.1. Indian Remote Sensing Satellite Data Used

Indian Remote Sensing Satellite (IRS-P6 and IRS-RS2) digital data of for three different time periods, i.e. October-2005, November-2011 and Apri-2014 covering Gulf of Kachchh region were acquired for monitoring & mapping of mangrove regeneration activity in various villages. The details of IRS digital data acquired are given in Table 2.

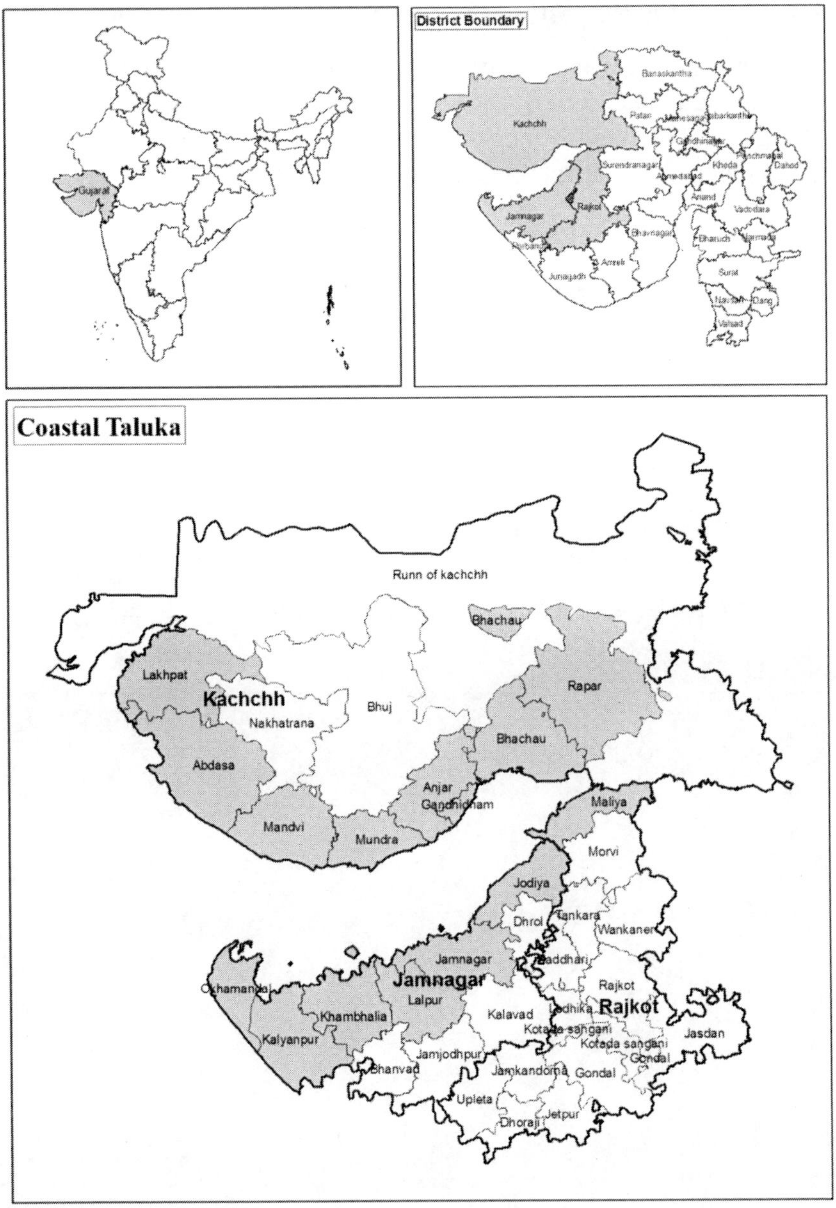


**Figure 1**. Location map of the Gulf of Kachchh (GOK), in Gujarat State (India).

## 3.2. Mangrove Plantation Carried out by GEC

The potential sites for mangrove plantation were identified based on the detailed field survey carried out in various villages of different blocks by GEC. One of the site selected in Tuna village of Anjar taluka (block) of Kachhch district is given in Figure 2.

**Table 2**. Satellite data used in mapping monitoring mangrove regeneration.

| Sr. No. | Sensor | Path/Row | Date of pass |
|---|---|---|---|
| 1. | LISS-III | 89/55 | 14-Oct-2005 |
| 2. | LISS-III | 89/56 | 14-Oct-2005 |
| 3. | LISS-III | 90/55 | 18-Oct-2005 |
| 4. | LISS-III | 90/56 | 18-Oct-2005 |
| 5. | LISS-III | 89/55 | 12-Nov-2011 |
| 6 | LISS-III | 89/56 | 12-Nov-2011 |
| 7. | LISS-III | 90/55 | 16-Nov-2011 |
| 8. | LISS-III | 90/56 | 16-Nov-2011 |
| 9. | LISS-IV | 91/56 | 18-Apr-2014 |
| 10. | LISS-IV | 90/56 | 13-Apr-2014 |

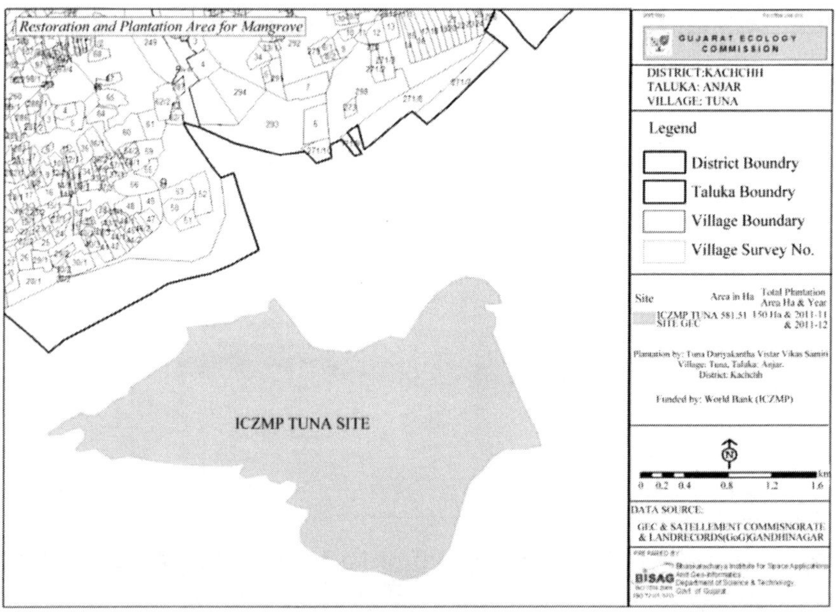


**Figure 2**. Mangrove plantation site Tuna, Taluka-Anjar (Gulf of Kachchh) identified under ICZM project.

**Figure 3**. Mangrove seedings raised in nursery for plantation by participatory approach in Gulf of Kachchh.

**Figure 4**. Mangrove plantation by participatory approach in Gulf of Kachchh.

Local communities are trained in technical aspects so that they are able to carry out various operations like raising of nursery, seed sowing, seedling collection and plantation in the field. After plantation, a constant monitoring is carried out to ensure better survival rate. In addition to monitoring by PEAs, a third part monitoring was also carried out. The field photographs of mangrove seedlings raised in nursery and plantation of mangrove seedlings by local communities by participatory approach are given in Figure 3 & Figure 4.

## 3.3. IRS LISS-III Data Analysis

IRS LISS-III and LISS-IV digital data covering study area were analyzed and it broadly consists of following steps:

1) Multi-date data preparation and geo-referencing, 2) administrative boundary superimposing, 3) generation of spatial information in GIS environment at the desired scale, 4) superimposing GPS locations of mangrove sites collected during GT data collection on the registered LISS-III digital data, 5) identification & delineation of mangrove into different density classes and 6) area estimation under mangroves and change detection in the mangrove vegetation and density.

### 3.3.1. Extraction of IRS LISS-III Data for Mangrove Mapping

IRS LISS-III and LISS-IV digital data of different dates was registered using Ground Control Points (GCPs) identified on image and Global Positioning System (GPS) measurements. The administrative boundaries like district, taluka and village were superimposed on the registered satellite data. From the registered data-set study areas in different talukas was extracted for monitoring mangrove regeneration, mangrove vegetation density as well as area estimation in different coastal taluka. One of the images covering part of Kachchh district along with taluka and village boundaries is given in Figure 5.

### 3.3.2. Identification of Potential Sites for Mangrove on IRS LISS-IV Data

GEC initiated the Community Based mangrove Restoration across ten locations situated in six coastal districts of the Gujarat State through village-level community based organizations. The suitable sites for mangrove plantations across various sites were identified by the GEC and their GPS measurements were recorded. These GPS measurements of various sites were transferred on the IRS LISS-IV images covering various sites for monitoring the mangrove regeneration and growth and density of mangrove using multi-date satellite data. The locations of potential sites identified on IRS LISS-IV satellite images covering part of Jodia taluka in Jamnagar district are given in Figure 6.

### 3.3.3. Mangrove Mapping Procedure

The multi-date satellite LISS-III digital data have been interpreted visually for mapping mangrove cover in the selected villages. Intensity of pink tone observed on LISS-III images was considered as basis for mangrove mapping along the coastal areas. The mangrove mapping was carried out using IRS LISS-III images of 2005 and 2011. The Quantum GIS software was used for mangrove delineation and computation of area. The area under mangrove in Jamnagar and Anjar talukas in Jamnagar and Kachchh districts was estimated for 2005 and 2011 as well as 2014.

**Figure 5**. IRS LISS-III image covering part of Kachchh district: 12-Nov-2011.

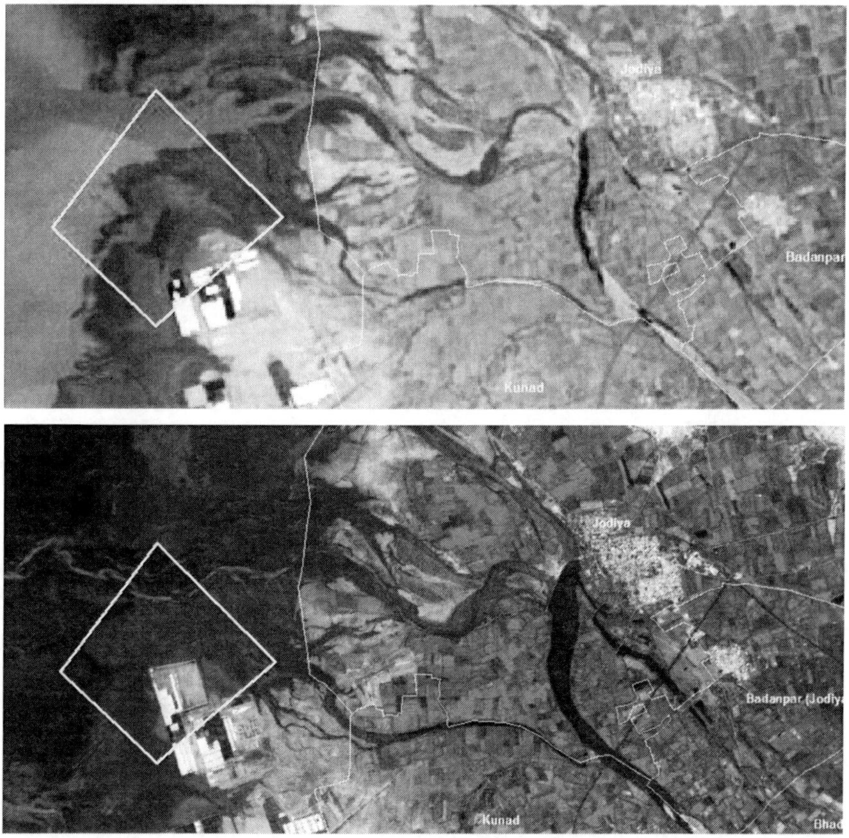


**Figure 6**. Mangrove regeneration site marked on IRS LISS-IV images near Jodia village.

## 4. RESULTS AND DISCUSSIONS

The multi-date Indian Remote Sensing Satellite (IRS-P6) LISS-III digital data of October-2005 and November-2011 covering coastal areas of Anjar and Jamnagar talukas were interpreted visually for mapping mangrove cover. In these two talukas extensive mangrove regeneration programme was carried out by the GEC. The areas where mangroves regeneration activity was carried out have been visually interpreted for mapping of mangrove cover based on the intensity of pink tone observed on LISS-III images. The mangroves have been classified into two density classes namely dense mangrove and sparse mangrove based on the intensity of pink tone observed on LISS-III images. On IRS

LISS-III False Colour Composites (FCC) mangroves occur in bright red colour if it is in continuous patch. When mangroves are scattered then instead of bright red colour, a light brick red colour may be seen.

### 4.1. Mangrove Mapping in Anjar Taluka

In Anjar taluka, coastal areas adjoining Tuna and Rampart villages were identified for mangrove regeneration under ICZM project by the GEC (Figure 2). The mangrove plantation by participatory approach was carried out by GEC on this selected site. Indian Remote Sensing satellite (IRS-P6) LISS-III data of October-2005 and November-2011 covering Anjar taluka was analysed for mapping and monitoring the mangrove plantation areas. The IRS-P6 LISS-III images of October-2005 and November-2011 covering coastal areas of Anjar taluka is given in Figure 7. The LISS-III image of November-2011 shows large patches of mangroves in the costal areas as compared to October-2005 image. This indicates there is increase in Mangrove area in the Anjar taluka during the period of last 5 to 6 years (2005 to 2011). The area under mangrove was estimated using GIS techniques. The comparison of two satellite images in Figure 7 indicates that the estimated area covering part of coastal areas of Anjar taluka has increased from 11.76 sq. km during October-2005 to 13.42 sq. km during November- 2011.

### 4.2. Mangrove Mapping in Jamnagar Taluka

In Jamnagar taluka, two areas were studied one covering coastal areas of Rozibet, Jabuda and Khijadiya villages and other covering Sarmet and Bed villages. IRS LISS-III images covering these villages were studied and areas under mangrove were estimated using GIS. The IRS-P6 LISS-III image of October-2005 and November-2011 covering coastal areas of Jamnagar taluka is given inFigure 8. The analysis of two satellite images in Figure 8 indicates that the estimated area covering part of coastal areas of Jamnagar taluka has increased from 78.76 sq. km during October-2005 to 80.66 sq. km during November-2011.

### 4.3. Mangrove Mapping in Maliya Taluka

In Maliya taluka of Rajkot district, coastal areas adjoining Navalakhi, Bodki and Varsamedi villages were identified for mangrove regeneration under ICZM project. The mangrove plantation by participatory approach was carried out by GEC on the selected site. Indian Remote Sensing satellite (IRS-P6) LISS-III data of October-2005 and November-2011 covering Maliya taluka was analysed for mapping and monitoring the mangrove plantation areas. The IRS-P6 LISS-III images of October-2005 and November-2011 covering coastal areas of Maliya taluka are given in Figure 9.

**Figure 7**. IRS LISS-III images of October-2005 & November-2011 coving Anjar taluka, Kachchh.

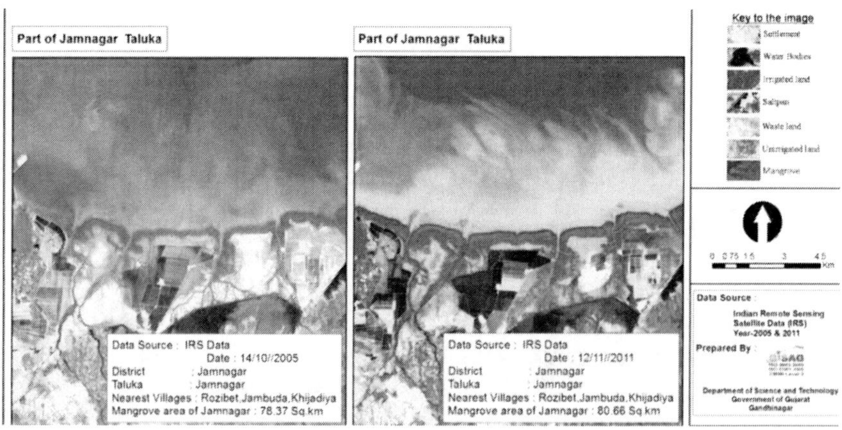


**Figure 8**. IRS LISS-III images of October-2005 & November-2011 coving Jamnagar taluka, Kachchh.

The LISS-III image of November-2011 shows large patches of dense mangroves areas as compared to October-2005 image. This indicates there is increase in Mangrove area in the Anjar taluka during the period of last 5 to 6 years (2005 to 2011) mainly due to mangrove regeneration activity carried out by the GEC. The estimated area under mangroves indicates that the estimated has increased from 8.82 sq. km during October-2005 to 22.95 sq. km during November-2011, in the coastal area of Maliya taluka in Rjakot District.

## 4.4. Change Monitoring of Area under Mangrove Plantation

The IRS LISS-III images clearly indicate that the mangrove area is increased substantially in the various taluka of Gulf of Kachchh over the period of 6 years from 2005 to 2011. The results indicated in the Figures 7-9 are tabulated in Table 3 to bring out the increase in mangrove area due to the efforts of mangrove regeneration programme carried out by GEC with active participation of Community Based Organizations (CBOs) in the Gulf of Kachchh region.

The results of present study on change monitoring reveals that mangrove area has considerably increased in study talukas mainly because of the mangrove regeneration activity carried out by the GEC with active partici- pation of Community Based Organizations (CBOs).

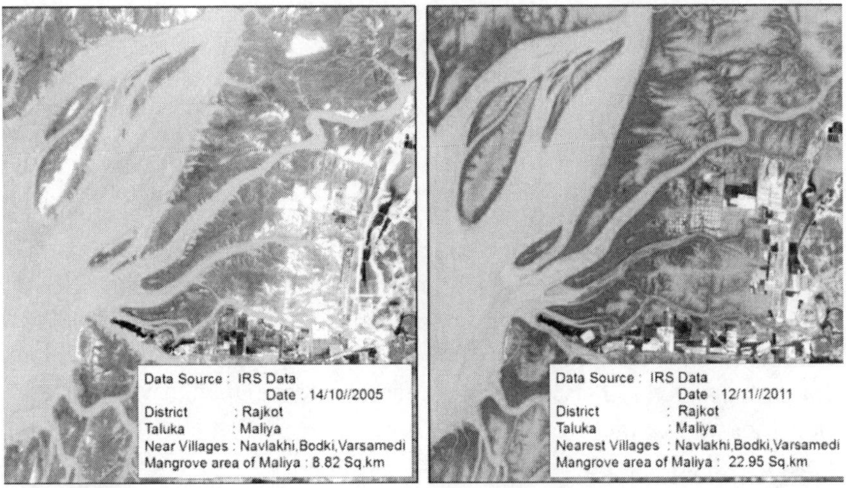


**Figure 9**. IRS LISS-III images of October-2005 & November-2011 coving Maliya taluka, Rajkot.

**Table 3**. Change monitoring of mangrove regeneration activity in Gulf of Kachchh region.

| District | Taluka | Mangrove density | Mangrove area (ha) | |
|---|---|---|---|---|
| | | | 2005 | 2011 |
| | | Dense | 486.7 | 636.9 |
| Kachchh | Anjar | Sparse | 689.3 | 705.2 |
| | | Total | 1176.0 | 1342.1 |
| | | Dense | 272.1 | 790.9 |
| Jamnagar | Maliya | Sparse | 609.4 | 1504.1 |
| | | Total | 881.5 | 2295.0 |
| | | Dense | 5033.6 | 2423.1 |
| Jamnagar | Jamnagar | Sparse | 2803.6 | 5643.0 |
| | | Total | 7837.2 | 8066.1 |

This increase in mangrove cover in most of the places is mainly due to the better understanding and acceptance of the need to protect, conserve and regenerate mangroves by the local communities. At the same time, capacity building of coastal communities for community-based mangrove regeneration as part of integrated coastal zone management has helped in substantial increase of the mangrove cover. The other reason is attributed to the fact that as a part of Coastal Zone clearances, most of the coastal industries were required to raise and regenerate mangrove plantations. Monitoring of mangrove regeneration activity was carried out independently by the private sector which enforced strict compliance of the mangrove regeneration guidelines by all the coastal industries which has also helped to increase the mangrove cover in the study areas.

# 5. CONCLUSIONS

Gujarat Ecology Commission (GEC) has taken up extensive mangrove restoration in various villages of different blocks in Gulf of Kachchhh with active participation of Community Based Organizations (CBOs) under the Integrated Coastal Zone Management Project (ICZMP) with Forest & Environment Department (MoEF), which is a World Bank-funded project. To assess the impact of these mangrove plantation activities and to monitor the mangrove regeneration and restoration in various villages, a joint study under the ICZM project was taken up by GEC and Bhaskarcharya Institute for Space Applications and Geo-

Informatics (BISAG) Monitoring mangrove areas in the Gulf of Kachchh, Gujarat State using Remote Sensing and Geo-Informatics.

The achievements and sustainability of mangrove restoration activity by GEC were monitored by analysing Indian Remote Sensing Satellite (IRS-P6) LISS-III and LISS-IV digital data covering Gulf of Kachchh for the period of 2005 and 2011. Remote sensing data was analyzed using ENVI Software and the mangrove area was delineated using the Quantum GIS software. The changes in mangrove cover over a period six years in the selected coastal areas of different talukas in Gulf of Kachchh were analysed. The results of present study on change monitoring revels that mangrove area has considerably increased in study talukas mainly because of the mangrove regeneration activity carried out by the GEC with active participation of Community Based Organizations (CBOs).

## ACKNOWLEDGEMENTS

The authors like to express their sincere thanks to Shri T. P. Singh, Director, Bhaskarcharya Institute for Space Applications and Geo-Informatics (BISAG), Department of Science & Technology, Government of Gujarat, Gandhinagar 382007, for his keen interest and encouragement in successfully executing this project.

The authors are very much thankful to the Forest Department, Government of Gujarat, Chief Conservator of Forest (CCF), Marine National Park (MNP), Jamnagar and Chief Conservator of Forest (CCF), Kachchh, for providing detailed information and logistic support to carry out this study.

Data on plantation sites for mangrove regeneration activities carried out with active participation of Community Based Organizations (CBOs), was provided by Shri Lomesh Brahmbhatt and Shri Ishwar Desai, as well as Project coordinators and community mobilisers, GEC team (Socio-Economic component, ICZM Project), Gujarat Ecology Commission (GEC). The authors sincerely thank them for their kind cooperation and active support to carry out this study.

## REFERENCES

1. Field, C.D. (1996) Restoration of Mangrove Ecosystems. International Society for Mangrove Ecosystems, Okinawa.

2. Field, C.D. (1999) Mangrove Rehabilitation: Choice and Necessity. Hydrobiologia, 413, 47-52.

3. Khemnark, C. (1995) Ecology and Management of Mangrove Restoration and Regeneration in East and Southeast Asia. Proceeding of the ECOTONE IV, Surat Thani, 18-22 January 1995.

4. India State of Forest Report (2009) Forest Survey of India.

5. Singh, H.S. (2007) Mangroves in Gujarat (Current Status and Strategy for Conservation). Gujarat Ecological Education & Research (GEER) Foundation.

6. Ajay, P., Vijay, S., Mehmood, K., Jaydipsinh, K., Kalubarme, M.H., Pandya, C.H., Joshi, N. and Brahmabhatt, L. (2014) Mapping and Monitoring of Mangroves in Coastal Districts of Gujarat State Using Remote Sensing and Geo- Informatics. Asian Journal of Geoinformatics, 14, 15-26.

7. Nandy, S. and Kushwaha, S. (2011) Study on the Utility of IRS-1D LISS-III Data and the Classification Techniques for Mapping of Sunderban Mangroves. Journal of Coastal Conservation, 15, 123-137.

8. Mahapatra, M., Ratheesh, R. and Rajawat, A.S. (2013) Potential Site Selection for Mangrove Plantation along the Kachchh District, Gujarat, India Using Remote Sensing and GIS Techniques. International Journal of Geology, Earth & Environmental Sciences, 3, 18-23.

9. Gujarat Ecology Commission (2012) Mangrove Restoration through Public-Private Partnership. 86.

# Chapter 12

# A Dynamic Remote Sensing Data-Driven Approach for Oil Spill Simulation in the Sea

**Jining Yan [1,2], Lizhe Wang [1,3], Lajiao Chen [1], Lingjun Zhao [1,2] and Bomin Huang [4]**

[1]*Institute of Remote Sensing and Digital Earth, Chinese Academy of Sciences, Beijing 100094, China*

[2]*University of Chinese Academy of Sciences, Beijing 100049, China*

[3]*School of Computer Science, China University of Geoscience, Wuhan 430074, China*

[4]*Space Science and Engineering Center, University of Wisconsin-Madison, Madison, WI 53706, USA*

## ABSTRACT

In view of the fact that oil spill remote sensing could only generate the oil slick information at a specific time and that traditional oil spill simulation models were not designed to deal with dynamic conditions, a dynamic data-driven application system (DDDAS) was introduced. The DDDAS entails both the ability to incorporate additional data into an executing application and, in reverse, the ability of applications to dynamically steer the measurement process. Based on the DDDAS, combing a remote sensor system that detects oil spills with a numerical simulation, an integrated data processing, analysis, forecasting and emergency response system was established. Once an oil spill accident occurs, the DDDAS-based oil spill model receives information about the oil slick extracted from the dynamic remote sensor data in the simulation. Through comparison, information fusion and feedback updates, continuous and more precise oil spill simulation results can be obtained. Then, the simulation results can provide help for disaster control and clean-up. The Penglai, Xingang and Suizhong oil spill results showed our simulation model could increase the prediction accuracy and reduce the error caused by empirical parameters in existing simulation systems. Therefore, the DDDAS-based detection and simulation system can effectively improve oil spill simulation and diffusion forecasting, as well as provide decision-making information and technical support for emergency responses to oil spills.

**Keywords:** DDDAS; remote sensing; oil spill; detection; simulation

# 1. INTRODUCTION

In recent years, frequent oil spills have affected the environment, economy and quality of life for coastal inhabitants [1]. For example, the Deepwater Horizon oil spill, which occurred on 20 April 2010, killed 11 people and presented an unprecedented threat to Gulf of Mexico marine resources [2]; the Dalian Xingang oil spill incident, which happened on 16 July 2010, spread an estimated 1500 metric tons of crude oil into the Yellow Sea and stretched over 430 $km^2$ of ocean, with 12 $km^2$ severely affected [3]; on 4 and 17 June 2011, two separate oil spill accidents occurred at platforms B and C of the Bohai Penglai 19-3 oilfield, and they caused 5500 $km^2$ of seawater pollution, roughly equivalent to 7% of the Bohai Sea area; and in the same year, on 12 July, the Liaoning Suizhong oil spill incident spread an estimated 0.1 to 0.15 cubic meters of crude oil into the Bohai Sea and stretched over 1 $km^2$ of ocean, according to the China State Oceanic Administration (SOA) [4].

Remote sensing technology plays an important role in monitoring marine pollution caused by oil spills. There are numerous broad uses of oil spill remote sensing [5,6], such as mapping of spills [7,8], surveillance and general slick detection [9–11], direction of oil spill countermeasures [12] and determination of slick trajectories [13]. Due to the long revisit time for satellites, it is difficult to have oil slick time series for long periods after the first observation. Therefore, many scholars have combined remote sensing detection with numerical models, so as to provide continuous tracking of the oil range and spill diffusion direction. For example, for the first time, the combination of satellite images of surface oil slicks with Lagrangian trajectory models has been implemented in the operational oil spill trajectory hindcast/forecast for the Deepwater Horizon oil spill [2,14,15]; Zodiatis *et al.*(2012) used the MEDSLIKoil spill model to forecast the drift and spreading of oil slicks detected from satellite images using MyOcean ( www.myocean.eu.org) forecasting data, through the implementation of MEDESS-4MS(www.medess4ms.eu) project services [16]; and Dominicis *et al.* (2013) used the MEDSLIK-II Lagrangian marine model to simulate oil slick transport and transformation processes for realistic oceanic cases and the model initialized using the slick position and slick shape provided by satellite systems, both SAR (synthetic aperture radar) and optical images [17].

However, the growth and evolution mechanisms of oil particles are very complicated, and the fate of the oil spilled into the ocean depends on many factors, including transport and dispersion by ocean circulation along with chemical transformations and biological consumption of the oil itself. A system for tracking the oil, both at the surface and at depth, was nonlinear, time-varying and multivariable, and it was difficult to analyze and forecast the complex movement of oil particles in an exact way [18]. As for the combined model, the location and size of the surface oil slick are inferred from satellite images, and they should be frequently re-initialized to reduce the forecast errors, which could be accumulated from the initial locations (conditions), as learned from the rapid response to the Deepwater Horizon oil spill in the Gulf of Mexico [2,14,15]. However, other related model parameters, such as data on wind forcing, sea surface temperature and sea currents, are from empirical values, model calculations or some meteorological data centers [17], and they may not be precise. Then, the inaccurate data could affect the model's predictive accuracy. Therefore, we should explore some new methods to minimize the influence of accumulated errors.

DDDAS is a new paradigm for application simulations and measurements. It entails both the ability to incorporate additional data into an executing application and, in reverse, the ability of applications to dynamically steer the measurement process. Through the synergistic feedback and control loop between application simulations and measurements, the paradigm offers the promise of improving modeling methods and augmenting the analysis and prediction capabilities of application simulations and the effectiveness of measurement systems [19]. We introduced the DDDAS into oil spill remote sensing detection and numerical simulation in order to minimize the influence caused by inaccurate parameters and to improve the predictive accuracy of oil spill simulations. In order to test our DDDAS-based oil spill simulation approach, we chose the Penglai 19-3B platform, Dalian Xingang and Liaoning Suizhong oil spill accidents as experimental cases. For conveniently describing the introduced methodology completely, we chose the Penglai 19-3B platform oil spill incident as the key experimental case, and the other two cases were selected as method validation cases.

# 2. OIL SPILL REMOTE SENSING DETECTION TECHNIQUES

## 2.1. Oil Spill Remote Sensing

Remote sensor data can provide information about the oil spill area, type and thickness in a large area, so as to guide maritime surveillance vessels and aircraft to enforce monitoring [20]. Among satellite sensors, active microwave sensors, like synthetic aperture radar (SAR), which has relatively wide coverage and day/night/all-weather imaging capabilities, have been widely used to provide valuable synoptic information about the position and size of oil spills and seeps [21–23]. Envisat (Environmental Satellite), launched on 1 March 2002, is an Earth-observing satellite still in orbit. Envisat Advanced Synthetic Aperture Radar (ASAR) operates in the C band in a wide variety of modes. The wavelength is 5.6 cm, and it has many unique properties, such as multipolarity, variable viewing angles and wide imaging. The wide swath (WS) mode, the only one specifically designed for detecting offshore oil spills by international radar satellites, provides better coverage range and radiation accuracy [24,25].

Due to the slick's damping of small-scale capillary waves and short gravity waves, areas covered by oil appear as dark features if the wind speed, wave height, SAR observation incidence angle and radar beam are suitable [26]. In addition, as the oil spill depth decreases, the gray values increase slightly [27]. Thus, we can use the gray difference between the oil spill and the seawater to extract the contour and area of oil slicks with threshold methods [28].

## 2.2. Oil Spill Detection

The Penglai 19-3 oil field is located in the south central Bohai Sea 11/05 contract area and the Tan-Lu fault zone, which lies in the northeast end of the Bonan uplift [29]. The exploitation range is 120°01′E to 120°01′E, 38°17′N to 38°27′N. The Bohai Sea is semi-closed, with an average water depth of 18 m [30] (Figure 1). The 2011 oil spills were the first large-scale subsea oil spills in Mainland China in recent years. According to ConocoPhillips China statistics, these accidents resulted in the release of approximately 700 barrels of oil and 2500 barrels of mineral oil-based drilling mud onto the seabed [33].

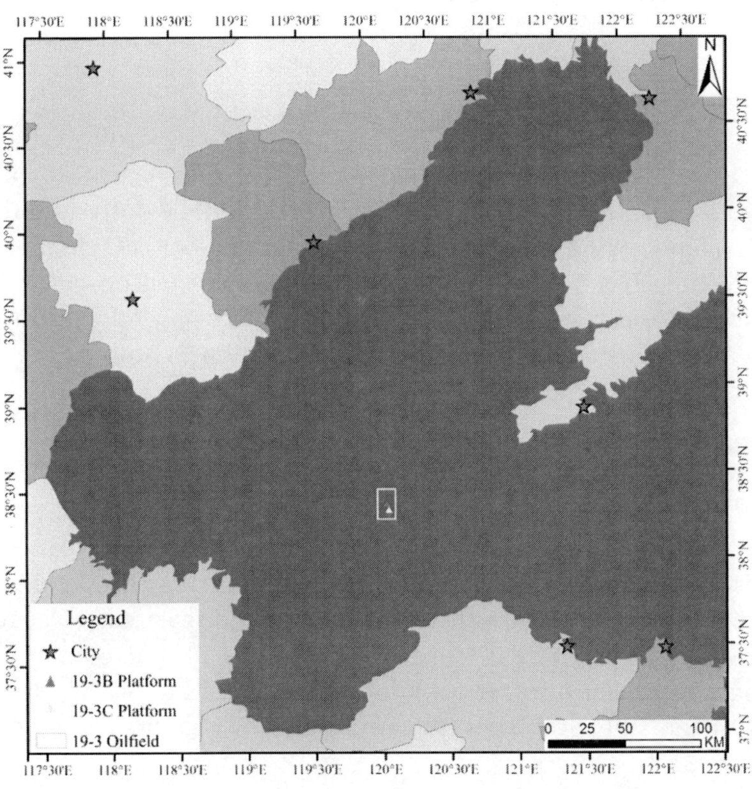


**Figure 1.** The location of Penglai 19-3 Oilfield and Penglai19-3B, C Platform.

In this study, we chose three ASAR WS mode images as the Bohai Penglai 19-3 monitoring data of the oil spill picked up by remote sensors, obtained on 11, 14 and 22 June 2011. After the geometric precision correction and an enhanced Lee filter were applied [31,32], a subset of the image area (800 × 800) was extracted at the Penglai 19-3B and C oil platforms (Figure 2).

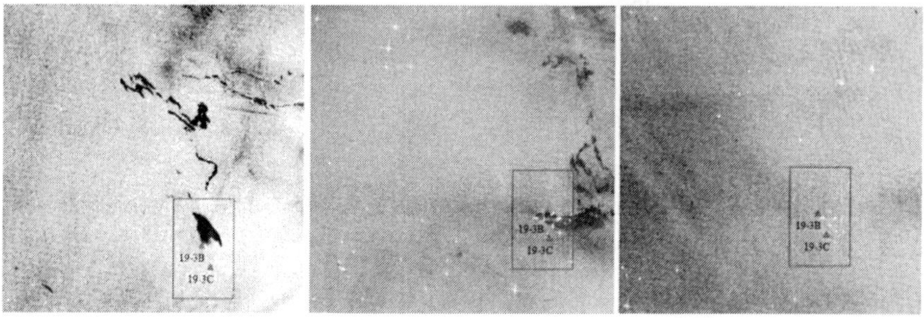


**Figure 2.** ASAR images of Penglai 19-3 Oilfield: (**a**) 11 June 2011; (**b**) 14 June 2011; (**c**) 22 June 2011.

As Figure 2 shows, numerous oil slicks were observed northeast of the 19-3B platform on 11 June (Figure 2a), and they drifted southeast on 14 June (Figure 2b); however, the integrated oil slicks disappeared by 22 June (Figure 2c). The floating oil was cleaned up by 22 June, which is consistent with the China State Oceanic Administration's 2011 Bohai Bay Oil Spill Accident Investigation Report by the Joint Investigation Team.

Given the grayscale and specific outlines of the oil slick, the single threshold method was used to extract information about the oil spill from the ASAR images [34,35]. Since there are generally two peaks in the histograms of ASAR images that contain oil spill information, we chose the minimum between the two peaks as the threshold values to distinguish the oil slicks from the background sea area. Because the oil was cleaned up by 22 June, we chose the 11 and 14 June ASAR images from which to extract information about the oil. By using the ENVI band statistical tools, the two image histograms can be drawn (Figure 3).

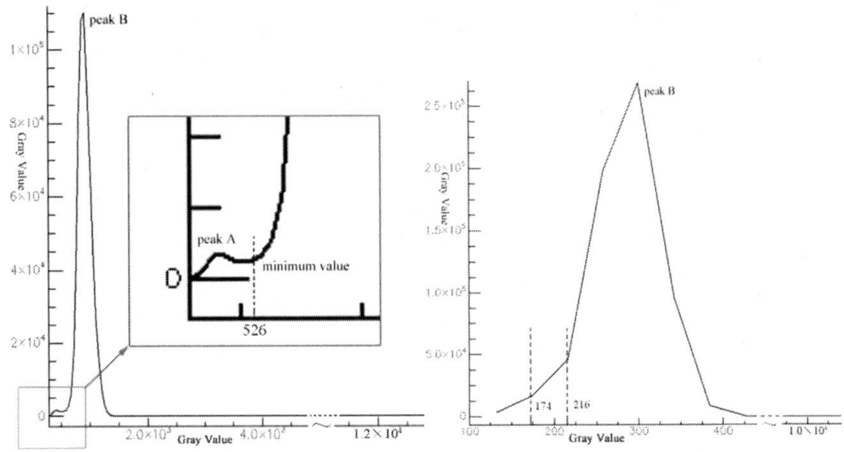


**Figure 3.** ASAR images' histograms: (**a**) 11 June 2011; (**b**) 14 June 2011.

**Figure 4.** Oil spill information extracted results: (**a**) 11 June 2011; (**b**) 14 June 2011.

As Figure 3a shows, there were two significant peaks, A and B, and we selected the minimum gray value 526 between the two peaks as the segmentation threshold to extract information about the oil spill. However, for the 14 June ASAR image histogram (Figure 3b), the smaller gray peak is not obvious, because the oil spill area was small compared with the entire image. There was only one peak in the 14 June ASAR image histogram. We cannot use the minimum gray value between the two peaks to determine the segmentation threshold. Nevertheless, due to the oil slick's "dark" features in the SAR images, the 14 June ASAR image histogram shows two distinct turning points at the 174 and 216 gray levels. We chose 174 and 216 as alternative segmentation thresholds. Through many experiments and comparisons, 216 is the segmentation threshold of the 14 June image for extracting information about the oil spill.

After the single threshold segmentation, the results for the 11 and 14 June ASAR images are as shown in Figure 4.

To verify the results, oil slick samples were selected from the original ASAR images. The overall classification accuracy was 92.1726%, and the kappa coefficient was 0.9425. As we know, the overall classification accuracy is calculated by summing the number of pixels classified correctly and dividing by the total number of pixels. The kappa coefficient (K) is a discrete multivariate technique used in accuracy assessment. $K > 0.80$ represents strong agreement and good accuracy; $0.40 < K < 0.80$ is in the middle, and $K < 0.40$ is poor. The extraction results could serve as evaluation standards of oil spill forecast models [36].

# 3. OIL SPILL SIMULATION TECHNIQUES

## 3.1. Oil Spill Simulation Basic Theory

To simulate the trajectory of an oil spill, a three-dimensional oil transport model, including a variety of processes, such as spreading, stranding, evaporation and emulsion, was developed based on the ECOM (the Estuarine and Coastal Ocean Model). ECOM, with a setup of bathymetry, initial conditions, open boundary and gridding, was developed from the marine hydrodynamic Princeton Ocean Model (POM) [37], which was used for shallow-water 3D hydrodynamic simulation. The initial conditions include the oil type, oil density, spill site, release depth, number of oil particles, tidal constituents, wind field data, and so on. More details on the ECOM can be found on HydroQual's website ( www.hydroqual.com) [30,38].

The ECOM adopts the Lagrangian discrete particle algorithm, and the oil slick was taken as being composed of numerous equal mass droplets. The small droplets went into the seawater at a certain rate in the oil spill incident point, then advected and diffused randomly. We define the movement velocity Vt→ of the oil droplets as follows:

$$\vec{V_t} = \vec{V} + \vec{V}' \tag{1}$$

where V⃗ represents the drift velocity of oil particles and V→′ simulates the turbulent fluctuations of velocity of oil particles in the water column. The drift velocity V⃗ is affected by the combined forces of wind, currents and waves. It was mentioned previously that the Chinese Bohai Sea is a C-shaped semi-enclosed sea. Compared with tidal currents and wind forcing, in the offshore area, wave-induced currents were not considered as the most dominant factor to drift the spilled oil on the sea surface [39,40]. In this paper, we emphasize tidal currents and wind as dominant driving forces; wave-induced currents were ignored during simulations [30]. The current speed can be calculated in the simulation model, and the wind field data may be from the Quickscat/NCEP hybrid data. The turbulent fluctuation of velocity V→′ is mainly affected by the turbulent diffusivity in the horizontal and the vertical directions, respectively. What is more, other related parameters, such as random angle, also affect the velocity V→′. However, these parameters will be assumed empirically to be constants in the proceeding model for simplicity.

At each time step $\Delta t$, the displacement $\Delta S$ of oil particles is calculated by making a time integration for Vt→. If $\Delta t$is too wide, the subinterval $\delta t_k$ can be used to calculate the displacement $\Delta S$ of oil particles to satisfy the precision demands. Then, the displacement $\Delta S$ of oil particles within the $\Delta t$ time range is as follows:

$$\Delta S = \sum^{k} V_{t,k} \delta t_k \tag{2}$$

where $V_{t,k}$ is the movement velocity of oil particles within the $\delta t_k$ time range, ∑kδtk=Δt. Additionally, $\delta t_k$should satisfy the following conditions:

$$\delta t_k \leq \left[\frac{u_k}{\Delta x} + \frac{v_k}{\Delta y}\right]^{-1} \tag{3}$$

where $u_k$ and $v_k$ are the X-velocity component and the Y-velocity component, respectively, of the oil particles' velocity $V_k$.

Within each time step, the oil particles will spread after advection and diffusion, and the spreading is also an important part of oil particles' migration at the early stage. Due to evaporation, emulsion, *etc.*, the quality of the oil particles gradually decreases. When the oil particles arrive at the shore, they are adsorbed on the shore or partially re-enter the water, based on the conditions of the coast.

After all calculation processes are completed, all processes, such as advection, diffusion, evaporation and emulsion, of the oil particles in a time step have been finished. In the next time step, only the temperature, wind and flow field conditions should be changed, and the entire calculation process is repeated.

### 3.2. Model Setup and Oil Spill Simulation

The study area in the Bohai Sea is 117.5° E–122.5° E, 37° N–41° N. Using rectilinear grids, the study area was cut into 25,920 (180 × 144) small grids in the horizontal direction, and the grid cell was 0.0277° × 0.0277° of longitude and latitude, respectively. Ten layers were designated vertically. The coastal line and bathymetry data were derived from the nautical chart made by the People's Liberation Army Navy Command Nautical Assurance. The temperature was 20 °C, and the salinity was 35. The wind field data is from the Quickscat/NCEP hybrid data. To meet ECOM requirements, the original 6-h interval was interpolated for a 3-h interval. The 3-h interval wind field data from 2 to 12 June is shown in Table 1. For the present simulations, amplitudes and phases for four major harmonic constants ($M_2$, $S_2$, $K_1$ and $O_1$) were interpolated along the open boundary to generate initial conditions driving the tide. The open boundary was defined along the rightmost vertical line at 122.5° E (Figure 1). The oil type was assumed as heavy oil with a density of 0.8 g/cm$^3$, and the density-driven flows were not considered. The instantaneous water level and current speed can be calculated in the model. According to modeling results, the flows in the Bohai Sea were almost completely tidal-induced loop or reciprocating currents, and the surface currents' speed averaged 0.5 m/s.

**Table 1.** Wind field data for the Estuarine and Coastal Ocean Model (ECOM) simulation (part).

| Time ($h$) | Wind Speed (m/s) | Wind Direction (°) |
|---|---|---|
| 54 | 5 | 112.5 |
| 57 | 1 | 247.5 |
| 60 | 1 | 247.5 |
| 63 | 3 | 315 |
| 66 | 2 | 112.5 |
| 69 | 1 | 90.0 |
| 72 | 3 | 90.0 |
| 75 | 4 | 112.5 |
| 78 | 7 | 180.0 |
| 81 | 4 | 157.5 |
| 84 | 5 | 180.0 |
| 87 | 4 | 202.5 |
| 90 | 4 | 157.5 |
| 93 | 5 | 157.5 |

Based on the scope of the Penglai 19-3 oil field and the time of the accident, we chose an oil slick near the 19-3B platform for the simulation demonstration. We assumed there was an oil spill point, which was located at (93, 52) of the horizontal rectangular grid (Figure 5). For the stability of currents developed in the model, the simulation start time was assigned to be 0:00 a.m. on 2 June, two days ahead of the spill time.

**Figure 5.** Oil film for ECOM simulation: (**a**) 11 June 2011; (**b**) 14 June 2011.

**Table 2.** Initial parameters of the ECOM simulation.

| Initial Parameters | Initial Values |
|---|---|
| Oil spilling point X | 93 |
| Oil spilling point Y | 52 |
| Step size | 30 |
| Oil particles number | 100 |
| Start step | 961 |
| End step | 7860 |

The oil particles' release time interval was assigned to be 961 to 7860, and each interval was 180 s. We also supposed that 100 oil particles were released every 30 steps. Because the first ASAR image is on 11 June after the platform B accident, the first simulation time span was from 2 to 12 June. Using the Lagrangian algorithm, surface oil particles drifted by both wind and tidal currents. The initial parameters are shown in Table 2.

To better visualize the ECOM simulation results, the resulting image for 0:00 11 June 2011 was overlaid with the ASAR image, which was obtained at 2:11:56 on 11 June 2011 (Figure 6). It is worth noting that operational model forecasts three days after the model initialization are generally not reliable at all and, thus, should not be used, as learned from the rapid response to the Deepwater Horizon oil spill in the Gulf of Mexico [14].

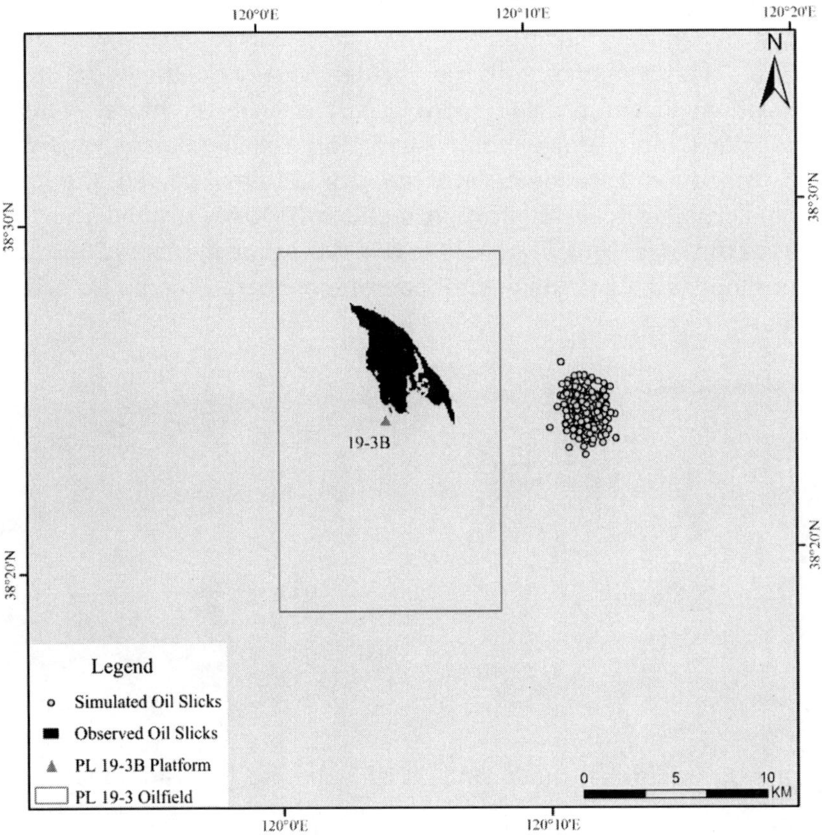


**Figure 6.** 11 June 2011 overlaid map of simulation results and RS extraction results.

As can be seen in Figure 6, the simulated oil particles are far from the remote sensor extraction results, and the simulation area is also smaller. Therefore, we introduced the DDDAS theory to optimize the ECOM simulation.

# 4. DDDAS-BASED OIL SPILL SIMULATION

## 4.1. DDDAS Basic Theory

Traditionally, theory, experiment/test and simulation are separate and conducted in series (Figure 7a). Since large, complex and uncertain systems are generally nonlinear, time-varying and uncertain, it is difficult to establish exact calculations and simulation models [41]. If simulations are combined with experiments or tests, which dynamically incorporate new data, either archival or from online measurements of the actual systems, the application simulations might provide more accurate analysis, more accurate predictions, more precise control and more reliable outcomes (Figure 7b). Based on this idea, Dr. Frederica Darema coined the term dynamic data-driven application systems (DDDAS) around the time she led the organizing effort of the NSF Workshop in March 2000 [19], which opened up a new method for conducting complex system research.

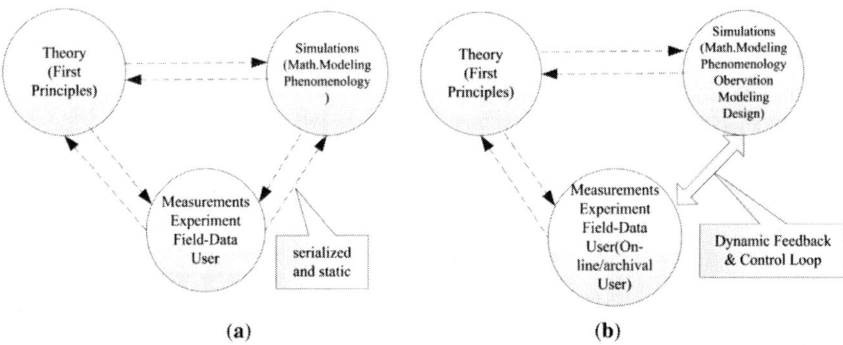


**Figure 7.** Traditional mode and DDDAS theory: (**a**) traditional mode (**b**) DDDAS.

DDDAS is a new paradigm for application simulations and measurements. The simulations receive and respond to online physical data and measurements and/or control the measurements. The DDDAS counterbalances incompleteness in the model and enhances application models by providing additional information to the model, since runtime additional data are used to selectively enhance or refine the original model. The simulations and real systems constitute a symbiotic, operational and dynamic feedback control system [42–44].

### 4.2. DDDAS-Based Oil Spill Simulation

The DDDAS-based ECOM can receive oil slick information extracted from data collected by dynamic remote sensors in the simulation process. Through comparison validation, information fusion and updates, optimized ECOM predictions are obtained; thus, the simulation's initial parameters are updated. Additionally, this idea also coincides with what has been learned from the rapid response to the Deepwater Horizon oil spill in the Gulf of Mexico: operational oil spill trajectory forecasting should be frequently re-initialized to reduce the forecast errors [2,14]. We then used the updated parameters to calculate more accurate simulation results. The flow chart of the oil spill simulation based on the DDDAS is shown in Figure 8.

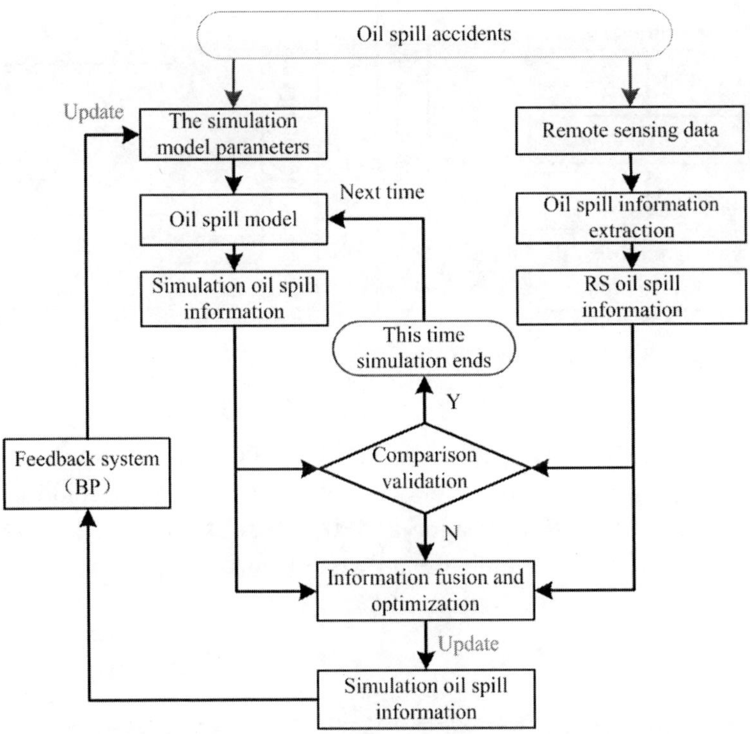


**Figure 8.** DDDAS-based oil spill simulation.

Due to the non-linear characteristics of the ECOM, it is difficult to reverse the thinking about the initial conditions and wind field data of the ECOM using the optimized prediction results. In terms of the nonlinear characteristics of artificial neural networks, we chose the BP neural network to back step the initial conditions of the oil spill to optimize the simulation. There are two stages in the BP neural network model: the training process and the forecasting process. At the training stage, ECOM simulation parameters (initial parameters and wind field data included) and simulation results are the input and output parameters of the BP model, respectively; at the forecasting stage, the updated ECOM simulation results are imported into the trained BP neural network, and the output is the updated oil spill initial parameters and wind field data [45,46] (Figure 9).

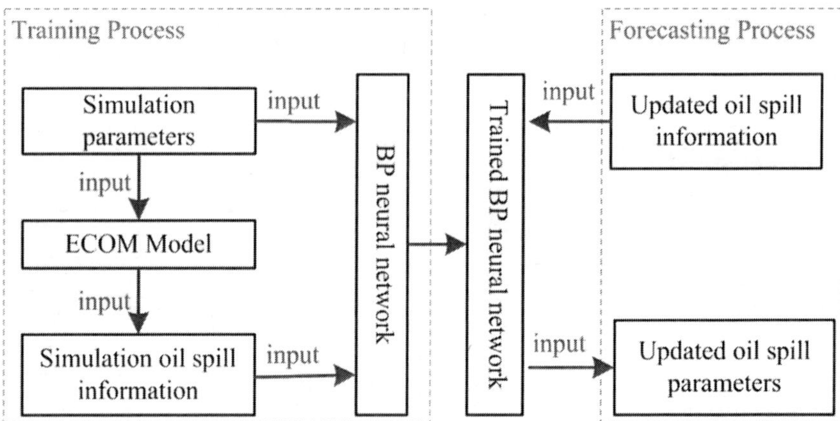


**Figure 9.** BP feedback system.

After four loop iterations, the updated wind field data and simulation initial parameters are as shown in Tables 3 and 4. The overlaid map of the ECOM simulation results and the results extracted from the remote sensors on 11 June are shown in Figure 10.

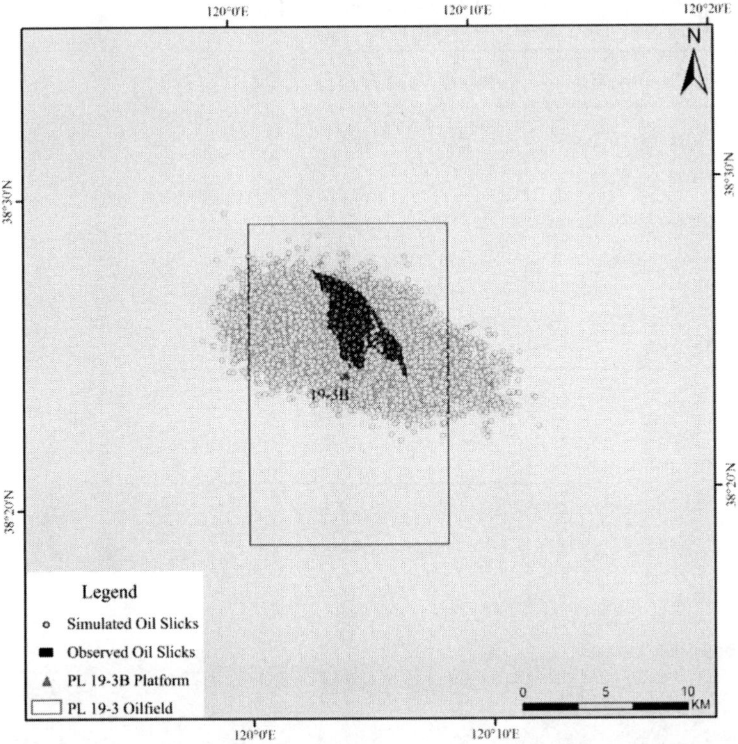


**Figure 10.** 11 June 2011 overlaid map of final simulation results and RS extraction results.

**Table 3.** Updated wind field data for ECOM simulation (part).

| Time ($h$) | Updated Wind Speed (m/s) | Updated Wind Direction (°) |
|---|---|---|
| 54 | 4 | 220.0 |
| 57 | 3 | 347.5 |
| 60 | 5 | 312.5 |
| 63 | 4 | 210.0 |
| 66 | 6 | 120.0 |
| 69 | 3 | 90.0 |
| 72 | 5 | 140.0 |
| 75 | 7 | 242.5 |
| 78 | 9 | 337.5 |
| 81 | 6 | 210.5 |
| 84 | 8 | 270.0 |
| 87 | 6 | 270.5 |
| 90 | 7 | 320.5 |
| 93 | 8 | 315.0 |

**Table 4.** Updated parameters for ECOM simulation.

| Updated Parameters | Updated Values |
|---|---|
| Oil spilling point X | 91 |
| Oil spilling point Y | 54 |
| Step size | 30 |
| Oil particles Number | 100 |
| Start step | 961 |
| End step | 4320 |

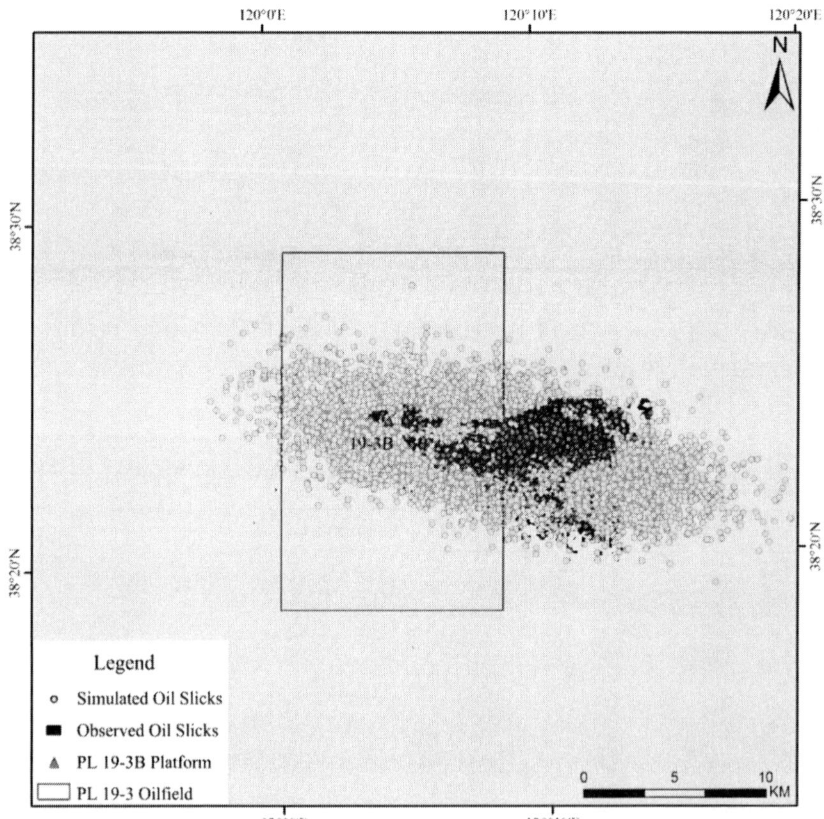


**Figure 11.** 14 June 2011 overlaid map of final simulation results and RS extraction results.

As can be seen in Figure 10, the oil film extracted from the ASAR image was completely covered by the simulated oil particles, which met our simulation requirements. Then, we entered the next step (14 June) oil spill simulation. After the wind field data from 11 to 13 June was entered, the ECOM simulation was run, and the simulation results for 14 June were generated. Through the BP neural network feedback, the wind field data were updated, and optimized simulated oil particles were obtained. The overlaid map of the ECOM simulation results and the results extracted from the remote sensors on 14 June are shown in Figure 11.

As can be seen in Figure 11, the oil film extracted from the ASAR image was completely covered by the simulated oil particles, and the center points almost overlapped. Therefore, the next oil spill simulation started. As the floating oil was cleaned up by 22 June, taking human intervention into account, the oil simulation dead time was defined by 0:00 a.m. on 20 June. Therefore, we could get the trajectory of the oil particles of 18 consecutive days, from 2 June to 20 June.

# 5. VALIDATIONS

## 5.1. Penglai Results Validation

Verification of oil spill forecasting is both a crucial issue and a difficult task to perform. In the case of the Penglai 19-3 oil spill, owing to the lack of *in situ* data, the only way to validate the DDDAS-based ECOM simulation is the comparison between actual oil locations inferred from satellite imagery and the model forecast positions from the latest forecast cycle similar to the work in [14]. In addition, due to the long revisit time for satellites and other reasons, we compared the simulation results with satellite images only by 11 and 14 June 2011. To show vividly the dynamic changing process of oil particles, we also partially listed the simulated trajectory of spilled oil (Figure 12).

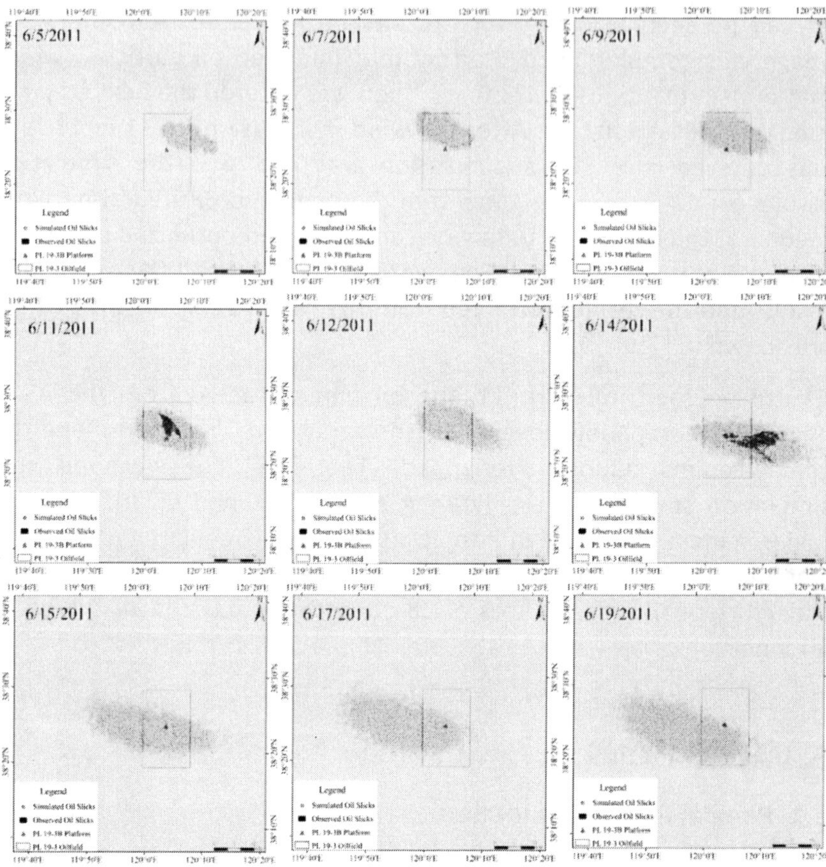


**Figure 12.** Trajectories of the Penglai 19-3B simulated oil spills.

As can be seen in Figure 12, a number of simulated oil particles distributed on the northeast of the 19-3B platform on 5 June 2011, which is consistent with the China State Oceanic Administration's 2011 Bohai Bay Oil Spill Accident Investigation Report: there were amounts of oil film floating on the surface on the northeast of the 19-3B platform on 4 June 2011. Due to the effect of the northwest wind force, particles as simulated mainly tended to transport southeasterly as time progressed. However, affected by the reciprocating flow, as mentioned above, oil particles moved a small distance to the west or north sometimes, as seen in Figure 12 (from 7 to 14 June). In addition, this trajectory of the simulated oil spill could be validated by oil locations inferred from satellite imageries on 11 and 14 June 2011. With the wind speed decreasing and direction changing, as well as the reciprocating flow influence, some simulated oil particles drift towards the west or north, and the overall distribution also became relatively scattered (from 15 to 19 June). This is consistent with the publication by Xu *et al.* [4].

## 5.2. Other Oil Spill Validation Cases

In addition, in order to further verify the feasibility and practicality of our introduced methodology, we also provided two other oil spill validation cases. The first case was the Dalian Xingang oil spill incident, which happened on 16 July 2010. As the report of the China State Oceanic Administration (SOA) [33], the Dalian Xingang oil spill incident spread an estimated 1500 metric tons of crude oil into the Yellow Sea and stretched over 430 km$^2$ of ocean, with 12 km$^2$ severely affected. Until 26 July, the floating oil was basically cleaned up [3]. Referring to the Penglai 19-3B model setup, we could get the simulated trajectory of spilled oil from 14 to 26 July 2010 (Figure 13), using our introduced methodology. Furthermore, we chose two high-resolution SAR images as the Dalian Xingang monitoring data of the oil spill picked up by remote sensors, obtained on 18 and 21 July 2010. The overlaid map of the ECOM simulation results and the results extracted from the remote sensors on 18 and 21 July are also shown in Figure 13.

As can be seen in Figure 13, a number of simulated oil particles mainly tended to transport southwesterly and northwesterly along the coast as time went on. Additionally, this may be mainly affected by the south and southeast wind force. On 18 and 21 July 2010, the oil films extracted from the SAR images were almost covered by the simulated oil particles; on 19 July 2010, the covered area of the simulated oil film was about 400 km$^2$, calculated using the ArcGIS statistical tool, and this was consistent with the monitoring results of Chinese patrols at 13:30 on 19 July 2010. In a word, using our introduced methodology, the ECOM simulation results of the Dalian Xingang oil spill incident met our simulation requirements.

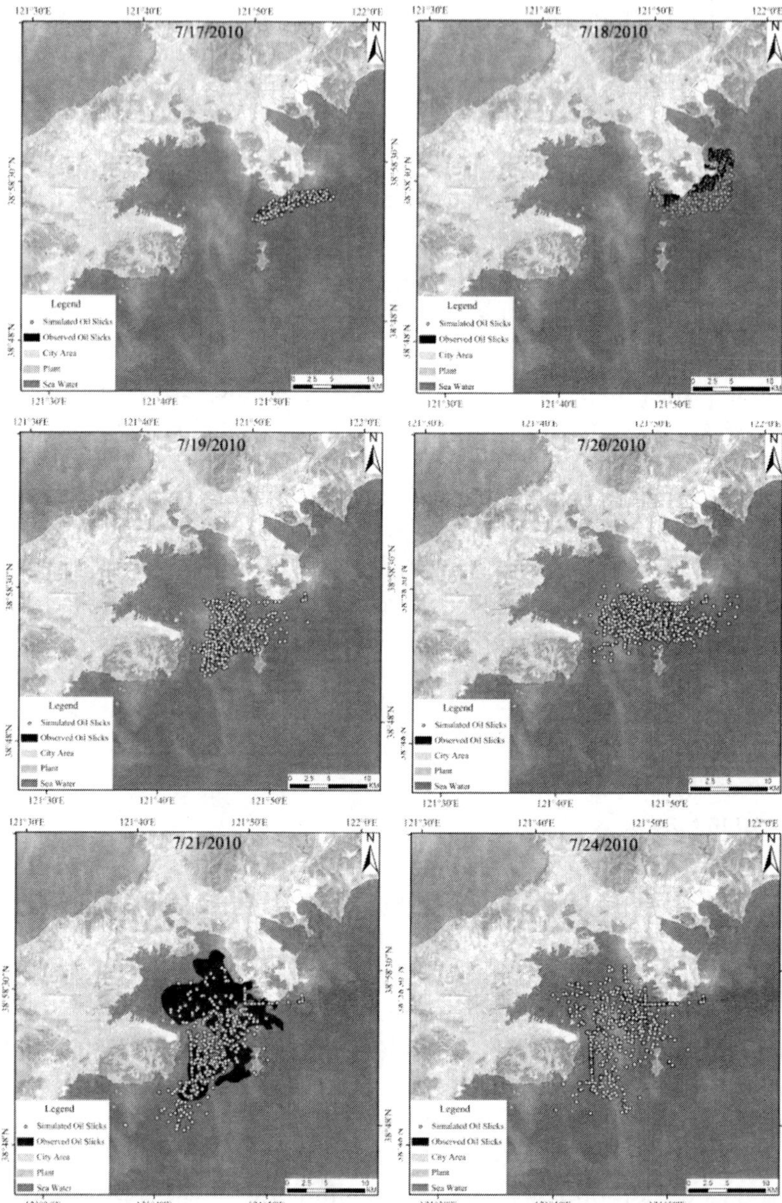


**Figure 13.** Trajectories of the Dalian Xingang simulated oil spills.

The second case was the Liaoning Suizhong oil spill incident, which happened on 12 July 2011. It was caused by the control faults of the Suizhong 36-1 oilfield center platform and spread an estimated 0.1 to

0.15 cubic meters of crude oil into the Bohai Sea. According to the 5:00 p.m. report of the Liaoning Suizhong City Office of Emergency Management on 12 July 2011, the crude oil drifted about five kilometers north away from the Suizhong 36-1 oilfield and stretched over 1 $km^2$ of ocean. Until the midday of 13 July, the floating oil was basically cleaned up [33]. Hence, our simulation time span was from 0:00 a.m. on 10 July to 0:00 a.m. on 15 July, and the simulation was based on an assumption that no cleanup was carried out. The simulated trajectory of spilled oil on 12, 13 and 14 July were listed in (Figure 14). Due to the fact that the Liaoning Suizhong oil spill incident was small and it only persisted for one and a half days, we could not find the appropriate remote sensing data to monitor the oil slick's movement, and the simulation result validation mainly depended on official reports.

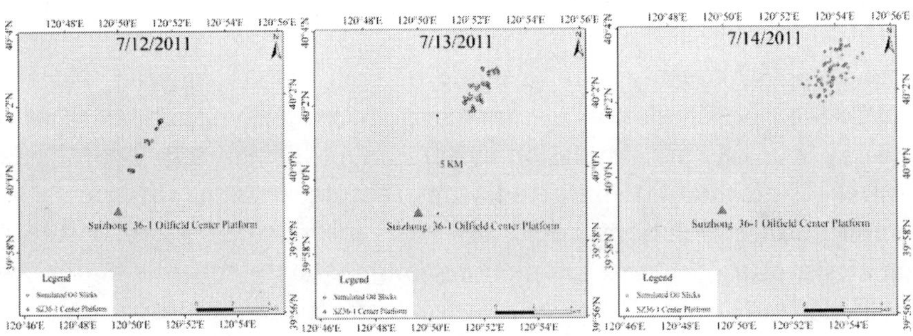


**Figure 14.** Trajectories of the Liaoning Suizhong simulated oil spills.

As can be seen in Figure 14, the simulated oil particles mainly tended to transport northeasterly as time went on, and the simulated oil film was about five kilometers north away from the Suizhong 36-1 oilfield, which is consistent with the report of the Liaoning Suizhong City Office of Emergency Management.

As a consequence, the two oil spill validation cases have shown the feasibility and practicality of our introduced methodology. However, a lack of sufficient field data at the time of the oil spill event hampered the efforts to validate the oil spill modeling. In the future, it would be good to deploy some surface drifters and validate the modeled trajectory with observations, similar to those in Liu and Weisberg (2011) [47] and Liu *et al.*(2014) [48].

# 6. CONCLUSIONS

Based on the defect analysis of the oil spill remote sensing and the current oil spill simulation models, the DDDAS was introduced to simulate a marine oil spill. The DDDAS can be viewed as a methodology to counterbalance incompleteness in the model and the capability to enhance the simulation models by imparting additional information into the model, as at runtime, additional data are used to selectively enhance or refine the original model. Combining oil spill data detected by remote sensors with the numerical simulation model, we established an integrated data processing, analysis, forecasting and emergency response system. Taking the oil film extracted from SAR images as the oil spill simulation validation criteria, the initial parameters of the simulation model were updated. Therefore, the prediction accuracy of the simulation model increased, and the error caused by empirical parameters reduced in existing simulation systems.

The Penglai, Xingang and Suizhong oil spill results showed that the DDDAS-based simulation system had good predictive capacity. That is to say, our oil spill simulation system could dynamically incorporate oil slick information extracted from remote sensing data, and the application simulations could offer the promise of more accurate analysis, more accurate predictions, more precise controls and more reliable outcomes. This system can effectively improve oil spill simulation and diffusion forecasting and improve decision making during marine oil spill emergencies.

For the ECOM model, we emphasize tidal currents and wind as dominant driving forces; wave-induced currents were ignored during simulations. However, some literature suggests that adding wave effects in ocean model systems is likely to increase the predictability of surface drifter trajectories [49]. Therefore, adding wave effects in simulations and obtaining more accurate simulation results require further study. Furthermore, due to the scarcity of *in situ* data, a sensitivity analysis or a skill score could not be performed. As a consequence, some future work, such as more extensive model validations [47,48] and multiple model ensembles [2,14], should be done.

## ACKNOWLEDGMENTS

Dr. Lizhe Wang's work is supported by the National High Technology Research and Development Program of China ("863" Program) (No. 2013AA12A301). The authors would also like to acknowledge the editors and anonymous reviewers for their valuable comments on the manuscript.

## AUTHOR CONTRIBUTIONS

Jining Yan and Lizhe Wang developed the oil spill detection and simulation methodology, its relative software, as well as the main part of the present manuscript. Lajiao Chen, Lingjun Zhao and Bormin Huang participated in the methodology development on the basis of their previous research in oil spill detection and simulation.

## REFERENCES

1. Li, Y.; Wang, L.; Chen, L.; Ma, Y.; Zhu, X.; Chu, B. Application of DDDAS in Marine Oil Spill Management: A New Framework Combining Multiple Source Remote Sensing Monitoring and Simulation as a Symbiotic Feedback Control System. Proceedings of the 2013 33rd IEEE International Geoscience and Remote Sensing Symposium (IGARSS), Melbourne, Australia, 21–26 July 2013; pp. 4526–4529.
2. Liu, Y.; Weisberg, R.H.; Hu, C.; Zheng, L. Tracking the Deepwater Horizon oil spill: A modeling perspective. *Eos Trans. Am. Geophys. Union* **2011**, *92*, 45–46.
3. Lv, Y.; Zhang, W.; Gao, Y.; Ning, S.; Yang, B. Preliminary study on responses of marine nematode community to crude oil contamination in intertidal zone of Bathing Beach, Dalian. *Mar. Pollut. Bull* **2011**,*62*, 2700–2706.
4. Xu, Q.; Li, X.; Wei, Y.; Tang, Z.; Cheng, Y.; Pichel, W.G. Satellite observations and modeling of oil spill trajectories in the Bohai Sea. *Mar. Pollut. Bull* **2013**, *71*, 107–116.
5. Leifer, I.; Lehr, W.J.; Simecek-Beatty, D.; Bradley, E.; Clark, R.; Dennison, P.; Hu, Y.; Matheson, S.; Jones, C.E.; Holt, B.; *et al.* State of the art satellite and airborne marine oil spill remote

sensing: Application to the BP Deepwater Horizon oil spill. *Remote Sens. Environ* **2012**, *124*, 185–209.

6. Fingas, M.; Brown, C. Review of oil spill remote sensing. *Mar. Pollut. Bull* **2014**, *83*, 9–23.
7. Plaza, J.; Pérez, R.; Plaza, A.; Martínez, P.; Valencia, D. Mapping oil spills on sea water using spectral mixture analysis of hyperspectral image data. *Proc. SPIE* **2005**, *5995*.
8. Ivanov, A.Y.; Zatyagalova, V.V. A GIS approach to mapping oil spills in a marine environment. *Int. J. Remote Sens* **2008**, *29*, 6297–6313.
9. Jha, M.N.; Levy, J.; Gao, Y. Advances in remote sensing for oil spill disaster management: state-of-the-art sensors technology for oil spill surveillance. *Sensors* **2008**, *8*, 236–255.
10. Pisano, A.; Bignami, F.; Santoleri, R. Oil Spill Detection in Glint-Contaminated Near-Infrared MODIS Imagery. *Remote Sens* **2015**, *7*, 1112–1134.
11. Ramsey, E., III; Rangoonwala, A.; Suzuoki, Y.; Jones, C.E. Oil detection in a coastal marsh with polarimetric synthetic aperture radar (SAR). *Remote Sens* **2011**, *3*, 2630–2662.
12. Ventikos, N.P.; Vergetis, E.; Psaraftis, H.N.; Triantafyllou, G. A high-level synthesis of oil spill response equipment and countermeasures. *J. Hazard. Mater* **2004**, *107*, 51–58.
13. Azevedo, A.; Oliveira, A.; Fortunato, A.; Bertin, X. Application of an Eulerian-Lagrangian oil spill modeling system to the Prestige accident: trajectory analysis. *J. Coast. Res* **2009**, *1*, 777–781.
14. Liu, Y.; Weisberg, R.H.; Hu, C.; Zheng, L. Trajectory forecast as a rapid response to the Deepwater Horizon oil spill. In *Monitoring and Modeling the Deepwater Horizon Oil Spill: A Record-Breaking Enterprise*; Elsevier: Amsterdam, The Netherlands, 2011; pp. 153–165.
15. Liu, Y., MacFadyen, A., Ji, Z.G., Weisberg, R.H., Eds.; Monitoring and Modeling the Deepwater Horizon Oil Spill: A Record Breaking Enterprise. In *Geophysical Monograph Series*; American Geophysical Union: Washington, DC, USA, 2011; Volume 195, pp. 1–271.

16. Zodiatis, G.; Lardner, R.; Solovyov, D.; Panayidou, X.; de Dominicis, M. Predictions for oil slicks detected from satellite images using MyOcean forecasting data. *Ocean Sci* **2012**, *8*, 1105–1115.

17. De Dominicis, M.; Pinardi, N.; Zodiatis, G.; Archetti, R. MEDSLIK-II, a Lagrangian marine surface oil spill model for short-term forecasting—Part 2: Numerical simulations and validations. *Geosci. Model Dev* **2013**, *6*, 1871–1888.

18. Elhakeem, A.; Elshorbagy, W.; Chebbi, R. Oil spill simulation and validation in the Arabian (Persian) Gulf with special reference to the UAE coast. *Water Air Soil Pollut* **2007**, *184*, 243–254.

19. Darema, F. Dynamic data driven applications systems: A new paradigm for application simulations and measurements. Proceeding the 4th International Conference on Computational Science-ICCS 2004, Kraków, Poland, 6–9 June 2004; pp. 662–669.

20. Hu, C.; Müller-Karger, F.E.; Taylor, C.J.; Myhre, D.; Murch, B.; Odriozola, A.L.; Godoy, G. MODIS detects oil spills in Lake Maracaibo, Venezuela. *Eos Trans. Am. Geophys. Union* **2003**, *84*, 313–319.

21. Migliaccio, M.; Nunziata, F.; Brown, C.E.; Holt, B.; Li, X.; Pichel, W.; Shimada, M. Polarimetric synthetic aperture radar utilized to track oil spills. *Eos Trans. Am. Geophys. Union* **2012**, *93*, 161–162.

22. Zhang, Y.; Li, Y.; Lin, H. Oil-Spill pollution remote sensing by Synthetic Aperture Radar. In *Advanced Geoscience Remote Sensing*; InTech: Rijeka, Croatia, 2014; pp. 27–50.

23. Amoon, M.; Bozorgi, A.; Rezai-rad, G.A. New method for ship detection in synthetic aperture radar imagery based on the human visual attention system. *J. Appl. Remote Sens* **2013**, *7*.

24. Brekke, C.; Solberg, A.H. Classifiers and confidence estimation for oil spill detection in ENVISAT ASAR images. *IEEE Geosci. Remote Sens. Lett* **2008**, *5*, 65–69.

25. Marghany, M.; Cracknell, A.P.; Hashim, M. Comparison between radarsat-1 SAR different data modes for oil spill

detection by a fractal box counting algorithm. *Int. J. Digit. Earth* **2009**, *2*, 237–256.

26. Li, X.; Li, C.; Yang, Z.; Pichel, W. SAR imaging of ocean surface oil seep trajectories induced by near inertial oscillation. *Remote Sens. Environ* **2013**, *130*, 182–187.

27. Brekke, C.; Solberg, A.H. Oil spill detection by satellite remote sensing. *Remote Sens. Environ* **2005**, *95*, 1–13.

28. Shirvany, R.; Chabert, M.; Tourneret, J.Y. Ship and oil-spill detection using the degree of polarization in linear and hybrid/compact dual-pol SAR. *IEEE J. Sel. Top. Appl. Earth Obs. Remote Sens* **2012**, *5*, 885–892.

29. Lu, Y.; Tian, Q.; Wang, X.; Zheng, G.; Li, X. Determining oil slick thickness using hyperspectral remote sensing in the Bohai Sea of China. *Int. J. Digit. Earth* **2013**, *6*, 76–93.

30. Liu, X.; Guo, J.; Guo, M.; Hu, X.; Tang, C.; Wang, C.; Xing, Q. Modelling of oil spill trajectory for 2011 Penglai 19-3 coastal drilling field, China. *Appl. Math. Model* **2014.**

31. Liu, P.; Huang, F.; Li, G.; Liu, Z. Remote-Sensing image denoising using partial differential equations and auxiliary images as priors. *IEEE. Geosci. Remote Sens. Lett* **2012**, *3*, 358–362.

32. Liu, P.; Eom, K.B. Restoration of multispectral images by total variation with auxiliary image. *Opt. Lasers Eng* **2013**, *7*, 873–882.

33. The China State Oceanic Administration, *2011 Bohai Bay Oil Spill Accident Investigation Report by the Joint Investigation Team*; The China State Oceanic Administration: Beijing, China, 2011; in Chinese.

34. Ganta, R.R.; Zaheeruddin, S.; Baddiri, N.; Rao, R. Segmentation of oil spill images with illumination-reflectance based adaptive level set model. *IEEE J. Sel. Top. Appl. Earth Obs. Remote Sens* **2012**, *5*, 1394–1402.

35. Moctezuma, M.; Parmiggiani, F. Adaptive stochastic minimization for measuring marine oil spill extent in synthetic aperture radar images. *J. Appl. Remote Sens* **2014**, *8*, 083553.

36. Chaudhuri, D.; Samal, A.; Agrawal, A.; Mishra, A.; Gohri, V.; Agarwal, R.; *et al.* A statistical approach for automatic detection of ocean disturbance features from SAR images. *IEEE J. Sel. Top. Appl. Earth Obs. Remote Sens* **2012**, *5*, 1231–1242.

37. Blumberg, A.F.; Mellor, G.L. A description of a three-dimensional coastal ocean circulation model. In *Three-Dimensional Coastal Ocean Models*; American Geophysical Union: Washington, DC, USA, 1987; pp. 1–16.

38. Guo, J.; Liu, X.; Xie, Q. Characteristics of the Bohai Sea oil spill and its

39. Tkalich, P.; Chan, E.S. Vertical mixing of oil droplets by breaking waves. *Mar. Pollut. Bull* **2002**, *44*, 1219–1229.

40. Mu, L.; Wu, S.; Song, J.; Li, H.; Liu, S.; Li, Y.; Gao, J. Numerical model research on Emergency Warning & Predicting system of ocean oil spill: I. Research on predicting of ocean dynamical factors. *Mar. Sci. Bull***2011**, *5*, 502–508, in Chinese.

41. Denham, M.; Cortés, A.; Margalef, T.; Luque, E. Applying a dynamic data driven genetic algorithm to improve forest fire spread prediction. Proseedings of the Computational Science—ICCS 2008, Kraków, Poland, June 23–25 2008; pp. 36–45.

42. Madey, G.R.; Barabási, A.; Chawla, N.V.; Gonzalez, M.; Hachen, D.; Lantz, B.; Pawling, A.; Schoenharl, T.; Szabó, G.; Wang, P.; *et al.* Enhanced situational awareness: Application of DDDAS concepts to emergency and disaster management. Proseedings of the Computational Science—ICCS 2007, Beijing, China, 27–30 May 2007.

43. Darema, F. Dynamic data driven applications systems: New capabilities for application simulations and measurements. Proseedings of the Computational Science—ICCS 2005, Atlanta, GA, USA, 22–25 May 2005; pp. 610–615.

44. Wang, L.; Chen, D.; Liu, W.; Ma, Y.; Wu, Y.; Deng, Z. DDDAS-Based Parallel Simulation of Threat Management for Urban Water Distribution Systems. *Comput. Sci. Eng* **2014**, *16*, 8–17.

45. Garcia-Pineda, O.; MacDonald, I.R.; Li, X.; Jackson, C.R.; Pichel, W.G. Oil spill mapping and measurement in the Gulf of Mexico with Textural Classifier Neural Network Algorithm (TCNNA). *IEEE J. Sel. Top. Appl. Earth Obs. Remote Sens* **2013**, *6*, 2517–2525.

46. Topouzelis, K.; Karathanassi, V.; Pavlakis, P.; Rokos, D. Dark formation detection using neural networks.*Int. J. Remote Sens* **2008**, *29*, 4705–4720.

47. Liu, Y.; Weisberg, R.H. Evaluation of trajectory modeling in different dynamic regions using normalized cumulative Lagrangian separation. *J. Geophys. Res.: Oceans* **2011**, *116*, C09013.

48. Liu, Y.; Weisberg, R.H.; Vignudelli, S.; Mitchum, G.T. Evaluation of altimetry-derived surface current products using Lagrangian drifter trajectories in the eastern Gulf of Mexico. *J. Geophys. Res.: Oceans* **2014**,*119*, 2827–2842.

49. Röhrs, J.; Christensen, K.H.; Hole, L.R.; Broström, G.; Drivdal, M.; Sundby, S. Observation-based evaluation of surface wave effects on currents and trajectory forecasts. *Ocean Dyn* **2012**, *62*, 1519–1533.

# Index